PRINCIPLES OF QUANTUM MECHANICS

PRINCIPLES OF QUANTUM MECHANICS

THE NON-RELATIVISTIC THEORY WITH SOME ILLUSTRATIVE APPLICATIONS

W. V. HOUSTON and G. C. PHILLIPS

William Marsh Rice University,
Houston, Texas

1973

NORTH-HOLLAND PUBLISHING COMPANY–AMSTERDAM

Library of Congress Catalog Card Number 69–18379
North-Holland I.S.B.N. 07204 0215 8
American Elsevier I.S.B.N. 0444 10483 6

PUBLISHERS:

NORTH-HOLLAND PUBLISHING COMPANY - AMSTERDAM
NORTH-HOLLAND PUBLISHING COMPANY, LTD. - LONDON

SOLE DISTRIBUTORS FOR THE U.S.A. AND CANADA:

AMERICAN ELSEVIER PUBLISHING COMPANY, INC.
52 VANDERBILT AVENUE
NEW YORK, N.Y. 10017

PRINTED IN THE NETHERLANDS

PREFACE

This book is intended as a text book, not an introductory text, but a book designed to give one who studies it carefully an overall view of the ways in which the various aspects of quantum mechanics fit in to a unified formulation.

It is expected that a student will have had a good introduction to descriptive quantum mechanics before starting this book. It would also be hoped that after finishing it an experimental physicist would be prepared to read and understand the theoretical interpretations of experimental results and that the theoretical student would be prepared to move on to the more elaborate procedures for dealing with complicated problems.

The pedagogical objective of the book has led to several features.

After a brief review of the experimental results which made quantum mechanics necessary, and a brief review of Hamiltonian mechanics, the quantum mechanics is introduced through wave mechanics. Wave mechanics provides a visualization of many phenomena, particularly those involving only one particle, even though this visualization is somewhat misleading, as is clear from treatments of two or more identical particles. Usually a wave mechanical treatment is less of a shock to a student who thinks in physical terms than is the more abstract analysis of non-commuting operators.

The text is interspersed with problems which are both illustrations and amplifications of the material presented. Many of them contain significant portions of the detailed theory, and their inclusion is based on the belief that a real understanding of quantum mechanics is attained only through development of some skill in its use in situations at least slightly different from those in terms of which the theory has been presented.

Although there is no specific chapter devoted to the interpretation, or the philosophical significance, of the theory, the whole is written from a specific point of view. This is, we believe, what is ordinarily referred to as the

"Copenhagen Interpretation"; although it would be manifestly incorrect to attribute to the Copenhagen group any of our misconceptions as to their thinking. It is emphasized throughout that the initial state of a system should be looked upon as an eigenstate of a set of commuting operators; and that its development with time leads to a state which is still an eigenstate of a set of commuting operators, but which can also be interpreted in terms of a statistical distribution over the eigenvalues of other operators.

Starting with Chapter 8 the whole subject is presented again in terms of non-commuting operators operating on vectors in an abstract space and using the notation introduced by Dirac. Again only minimal attention is paid to the details of approximation and perturbation procedures, but emphasis is laid on the equivalence of the method with that of the wave mechanics. In the later applications the two notations are used interchangeably. Chapters 11 and 12 treat the quantization of the electromagnetic field using the so-called "creation" and "destruction" operators already introduced in Chapter 8 for dealing with a simple oscillator.

Finally three major fields of application are illustrated. Atomic spectroscopy is of importance in its own right because of its great role in the development of quantum mechanics, and because of its current application to nuclear spectroscopy. The chapters on collision theory are of obvious importance in the extensive present day studies of nuclear reactions and the collisions of particles at high energy. The last chapter on electrons in solids gives a brief introduction to a field whose small beginnings some thirty years ago have given rise to a tremendous amount of recent important work. It also provides some application of the preceding chapter on statistical quantum mechanics, in which an attempt is made to distinguish clearly between the statistical interpretation of quantum mechanics and the statistical treatments of problems in which the initial conditions are not as sharply defined as possible.

Special thanks are due to students and others too numerous to mention with whom this matter has been gone over and who have raised many helpful questions and made many helpful suggestions. Our thanks also go to Mrs. Mary Comerford who has accurately and carefully typed the manuscript.

W. V. HOUSTON
G. C. PHILLIPS

Before the manuscript was finished Professor Houston passed away. We acknowledge the assistance of Dr. du Chattel of the Solid State Physics Group of the University of Amsterdam, who revised the manuscript of Chapter 18.

CONTENTS

CHAPTER 1. EXPERIMENTAL NECESSITY FOR QUANTUM MECHANICS 1
1. Experiments Analyzed by Means of the Concept of Elementary Particles Described by Ordinary Mechanics. 1
1.1. Experiments to Determine e/m 1
1.2. Experiments to Determine e 2
1.3. Experiments on the Scattering of Alpha Particles 3
1.4. Observation of Particle Tracks. 4
2. Experiments Analyzed by Means of the Electromagnetic Theory of Light . . 5
3. Experiments Interpreted in Terms of Light Quanta 7
3.1. Black-body Radiation 7
3.2. The Photoelectric Effect 12
3.3. The Compton Effect. 13
3.4. Light Quanta in an Interference Pattern 15
4. Experiments on the Wave Properties of Particles 15
5. Discrete Values of Energy and Magnetic Moment in Atoms 16
5.1. Line Spectra 17
5.2. Experiments of Franck and Hertz 18
5.3. Stern-Gerlach Experiment 18
6. Purpose of Quantum Mechanics 19

CHAPTER 2. ANALYSIS OF CLASSICAL MECHANICS 21
1. Outline of Classical Mechanics 21
1.1. Hamilton's Principle. 22
1.2. The Hamiltonian Equations of Motion. 23
1.3. Canonical Transformations 24
1.4. The Poisson Brackets 26
1.5. Infinitesimal Canonical Transformations 27
1.6. The Hamilton-Jacobi Equation 28
2. Analysis of the Applicability of Classical Mechanics 30
3. Analysis of Representative Methods of Observation 31
3.1. The Collimation of a Beam of Particles 31
3.2. Gamma-ray Microscope 32
3.3. General Considerations 35
4. The Principle of Indetermination 36

CHAPTER 3. FORMULATION OF WAVE MECHANICS 38
1. Momentum in Wave Mechanics 38
2. State of a System Represented by a Wave function 41

3. The First Postulate of Wave Mechanics . . . 44
3.1. Normalization of Wave Functions . . . 45
3.2. Functions of the Momenta . . . 46
3.3. Expectation Values . . . 47
3.4. Principle of Indetermination . . . 50
4. Dynamical Quantities other than Coordinates and Momenta . . . 52
5. The Equation of Motion . . . 54
5.1. Necessary Initial Conditions . . . 54
5.2. Form of the Hamiltonian Operator . . . 55
5.3. Computation of Time Derivatives and the Conservation Laws . . . 56
5.4. Hamiltonian Operator in the Presence of an Electromagnetic Field . 58
5.5. Gauge Invariance . . . 60
6. Criteria for the Validity of the Theory . . . 61

CHAPTER 4. CHARACTERISTIC STATES OR EIGENSTATES . . . 63
1. Definition of an Eigenstate . . . 63
2. Energy Eigenfunctions . . . 66
3. The One-Dimensional Harmonic Oscillator . . . 68
4. Degenerate States and the Two-Dimensional Oscillator . . . 70
5. Measurements in Quantum Mechanics . . . 73

CHAPTER 5. THE CLASSICAL APPROXIMATION . . . 75
1. Ehrenfest's Theorem . . . 75
2. Motion of a Free Particle . . . 76
3. Expansion in Powers of $\hbar$ . . . 78
4. Application to the Harmonic Oscillator . . . 80

CHAPTER 6. MOTION OF A PARTICLE IN A CENTRAL FIELD . . . 83
1. The Problem of Two Bodies . . . 83
2. The Central-Field Problem . . . 85
3. Angular Momentum in a Central Field . . . 86
4. Parity in a Central Field . . . 88
5. The Energy in a Coulomb Field . . . 89
5.1. Negative Energies . . . 89
5.2. Positive Energies . . . 91

CHAPTER 7. METHODS OF APPROXIMATION . . . 93
1. The Use of Discontinuous Potentials . . . 93
2. W.K.B. Approximation in One Dimension . . . 97
3. Perturbation Method for Eigenfunctions . . . 100
4. General Expansion in a Series of Eigenfunctions . . . 104
5. The Variation Method . . . 106
6. Development of Wave Function with Time . . . 108

CHAPTER 8. A GENERAL FORMULATION OF QUANTUM MECHANICS . . . 115
1. Vectors in Abstract Space . . . 115
2. Linear Operators . . . 119
3. Eigenvectors and Eigenvalues . . . 122
4. Relationship between the Mathematics and the Physical Situation . . . 124
5. Relationship with the Wave Function Notation . . . 126
6. The Schroedinger Representation . . . 129
7. The Momentum Representation . . . 131

8. Unitary Transformations 132
9. The Equation of Motion 133
10. The Harmonic Oscillator 136

CHAPTER 9. ANGULAR MOMENTUM AND SPIN 140
1. Eigenvalues of Angular Momentum 140
2. Matrix Representation of Angular Momentum Operators 143
3. Electron Spin 146
4. Electron in a Rotating Magnetic Field 149
5. An Electron with Spin in a Central Field 155
6. Vector Addition of Angular Momenta 159

CHAPTER 10. SYSTEMS OF IDENTICAL PARTICLES 162
1. Systems of Two Identical Particles 162
2. The Pauli Exclusion Principle 164
3. Separation of Spin and Orbital Motion 165
4. Systems of Three Electrons 167
5. Orbital Functions of Three Electrons 169
6. Spin Functions of Three Electrons 171
7. Combined Spin and Orbital Functions 172

CHAPTER 11. QUANTIZATION OF ELECTROMAGNETIC RADIATION IN EMPTY SPACE . 176
1. Classical Mechanics of a Vibrating String 176
2. Quantum Mechanics of a Vibrating String 179
3. The Electromagnetic Field in Empty Space 187

CHAPTER 12. INTERACTION OF RADIATION AND MATTER 193
1. Vibrating String Coupled to a Vibrating Particle 193
2. The Hamiltonian Treatment of a System Composed of Charged Particles and the Radiation Field 203
3. Quantum Mechanical Treatment of the Combined System 205
3.1. The Interaction Operator 206
3.2. Radiation from an Atom 207
3.3. The Dipole Transitions 208
3.4. Interpretation of the Wave Function of the State Vector 213
3.5. Emission of Quadrupole Radiation 214
3.6. Scattering by a Free Electron 216

CHAPTER 13. HYDROGEN-LIKE SPECTRA 220
1. The Spectrum of a Single Electron without Spin 221
1.1. General Scheme of Levels 221
1.2. The Zeeman Effect 224
1.3. Selection Rules and Intensities 225
2. The Spectrum of a Single Electron with Spin 227
2.1. General Scheme of Levels 227
2.2. The Anomalous Zeeman Effect 229
2.3. Selection Rules and Intensities 233

CHAPTER 14. TWO ELECTRON SPECTRA 236
1. The Hamiltonian Function 236
2. The Electrostatic Interaction 238
2.1. Operators that Commute with the Electrostatic Interaction 238

2.2. Eigenfunctions of the Operators that Commute with $1/r_{1,2}$ . . . 239
2.3. Eigenvalues from the Operator Relationships 241
2.4. Spin Variables and the Pauli Exclusion Principle 243
2.5. Case of Equivalent Electrons 243
2.6. Application of the Electrostatic Interaction 244
2.7. An Illustration of the Electrostatic Interaction 245
3. Spin-Orbit Interaction in *LS* Coupling 248
3.1. Operators Commuting with H_2 249
3.2. Application of H_2 as a Perturbation 250
3.3. Configuration *sl*. 251
3.4. General Two-Electron Configuration 253
4. Spin-Orbit Interaction in Arbitrary Coupling 255
5. The Parity Operator 259
6. Summary of Operators 259
7. Selection Rules 260

CHAPTER 15. QUANTUM MECHANICAL SCATTERING THEORY 263
1. Separation of the Center of Mass 264
2. Transformation to Laboratory Coordinates 265
3. General Time Dependent Operator Treatment of Scattering; Time Propagators and Lippmann-Schwinger Equation 267
3.1. The Initial State 267
3.2. The Scattering 268
4. The Born Series. 273
5. Feynman Diagrams for the Born Series 274
6. Momentum and Coordinate Bases for Matrices and Dyadic Operators . . 279
7. Time Development of the Scattering States of Momentum 284
8. The Time Propagator at Large Times 285
9. Unitarity of the Scattering State 286
10. Scattering Cross Section 288
11. Schroedinger Representation of Scattering 289
12. First Born Approximation for Local Central Potentials 290
13. The Optical Theorem 290
14. Unitarity, the R- and S-Matrices, and Phase Shifts 291
15. Relationship of Phase Shift to Causality, Time Delay, and Resonances . . 293
16. Partial Wave Representations. 295
17. Partial Wave Phase Shifts, Unitarity, and Dispersion Relationships . . 298
18. Deduction of Phase Shifts from Cross Sections 300
19. The Relationship between the Time Dependent Treatment and the Wave-function Treatment 301

CHAPTER 16. SOME SPECIAL CASES OF SCATTERING 302
1. Use of Spherical Coordinates 302
2. A Central Scattering Potential 304
3. Scattering of Identical Particles 306

CHAPTER 17. QUANTUM STATISTICAL MECHANICS 308
1. The Phase Space in Classical Mechanics 309
2. Pure States and Mixed States in Quantum Mechanics 313
3. The Density Operator or Density Matrix 315
4. The Canonical Density Operator 319
5. Use of the Sum over States 323

6. Systems of Harmonic Oscillators . . . 324
7. Systems of Identical Particles without Interactions . . . 326
7.1. The Fermi Statistics . . . 326
7.2. The Einstein-Bose Statistics . . . 329

CHAPTER 18. ELECTRONS IN SOLIDS . . . 331
1. General Form of the Wave Functions . . . 332
1.1. One-Dimensional Case . . . 333
1.2. Three-Dimensional Case . . . 337
2. General Properties of the Energy Spectrum . . . 344
3. Nearly Free Electrons and Tight Binding Approximations . . . 345
3.1. Nearly Free Electron Approximation . . . 346
3.2. Tight Binding Approximation . . . 349
4. The Fermi Energy; Conductors, Insulators and Semiconductors . . . 354

MATHEMATICAL APPENDIX . . . 358
A.1. The Hermite Polynomials . . . 358
A.2. The Associated Legendre Polynomials . . . 359
A.3. Surface Spherical Harmonics . . . 360
A.4. The Laguerre Polynomials . . . 362
A.5. Bessel Functions . . . 363
A.6. Spherical Bessel Functions . . . 365
A.7. Evaluation of Integrals Containing the Fermi Function . . . 366
A.8. The Clebsch-Gordon Coefficients . . . 369

REFERENCES . . . 370

SUBJECT INDEX . . . 371

CHAPTER 1

EXPERIMENTAL NECESSITY FOR QUANTUM MECHANICS

Quantum mechanics often seems very abstract in its formulation, but it is not merely a system of mathematics. It is needed to describe the results of experimental work. It has been devised to describe the results of observations and experiments made in the study of atomic physics where it is found that a straightforward application of ordinary means of description leads to inconsistencies and contradictions. For this reason it is desirable to begin the study of quantum mechanics with a compilation and analysis of some of the principal types of experiments that have led to the development of the theory, and it is helpful at all times to keep in mind the physical implications of the various features of the mathematical formulation.

1. Experiments analyzed by means of the concept of elementary particles described by ordinary mechanics

The first class of experiments includes those dealing with electrons, protons, alpha particles, or atomic nuclei, whose natural interpretation is that such things are small, electrically charged, material particles, and whose behavior can be described by ordinary mechanics, including relativity.

1.1. *Experiments to determine e/m*

A "particle" of atomic dimensions is characterized almost uniquely by its specific charge. The measurement, by J. J. Thomson (for example, THOMPSON [1897]), of the specific charge of electrons gave rise to the custom of referring to them as particles.

An ordinary method of measuring the specific charge of such particles is to set up an arrangement of electric and magnetic fields to accelerate the particle in various ways. If the electrons are emitted from a localized source, such as a hot filament, two electrodes can be arranged so that an electric field between them accelerates the particles through a known potential difference. If the particles then pass into a magnetic field so that they move in a circle and the radius of the circle can be measured, the ratio of the charge to the mass can be inferred. Such inference assumes a picture of a material particle with a diameter smaller than the uncertainty in the specification of its path. It assumes that Newton's laws of motion can be applied with the forces given by ordinary electromagnetic theory. The validity of these assumptions must then be judged from the results of their use in various situations. The facts are that when this kind of analysis is applied to experiments on electrons from different kinds of sources, and with many different arrangements of electric and magnetic fields, the results agree in giving a characteristic value of $e/m = 1.759 \times 10^{11}$ coulomb per kilogram. The same analysis has been applied to experiments with protons and alpha particles, and to the study of atomic masses with a mass spectrograph. All such experiments give highly precise, consistent results and tend to justify the assumptions made.

1.2. *Experiments to determine e*

MILLIKAN [1917] developed the methods for measuring the electric charge of a small drop of oil until he could show, with a high degree of precision, that all the charges are integral multiples of an elementary charge, $e = -1.6 \times 10^{-19}$ coulomb. In his experiments the oil drop was allowed to fall between the two horizontal plates of a condenser, and its velocity of fall, v_1, was measured while observing it in a microscope. The velocity was constant and, according to the law of motion in a viscous medium, was proportional to the weight of the particle. A potential difference was then applied to the plates of the condenser in such a direction as to cause the drop to rise, and this velocity, v_2, was measured. These two velocities are related by the equation

$$\frac{v_1}{v_2} = mg/(Eq - mg)$$

where m is the mass of the drop, g is the acceleration of gravity, q is the charge on the particle and E is the electric field strength. Solved for q, this equation gives

$$q = \frac{mg}{E} \frac{v_1 + v_2}{v_1}.$$

The actual evaluation of q from this equation required a knowledge of the mass for the oil drop under observation, which could be determined from the viscosity of air and the velocity v_1. But with the above equation alone, without a knowledge of m, it was possible to conclude from the observations that all of the values of q for a given drop were integral multiples of a unit charge. The charge on a drop could be changed by exposing it to X-rays, and the values of v_1 and v_2 were obtained for different charges on the same drop. In this way the discrete nature of the electrical charges was demonstrated with high precision. Later measurements by this and other methods have given $e = -1.601 \times 10^{-19}$ coulomb.

In the oil drop experiment, the laws of mechanics are applied to the oil drops and not to the electrons. Nevertheless the results are convincing evidence that electrical charge exists only in discrete and indivisible units. They might be called "atoms" or "quanta" of charge. When this conclusion is combined with the fact of a unique value of e/m for electrons, the picture of an electron as a "particle" with a charge e and a mass m is almost irresistible.

1.3. *Experiments on the scattering of alpha particles*

In a series of experiments suggested by RUTHERFORD[1911] and carried out by GEIGER and MARSDEN [1913], the distribution in angle of the alpha particles scattered from a thin foil was shown to be that expected on the basis of Newtonian mechanics, with the assumption that both the alpha particle and the scattering center are of small extent.

If a particle of charge Ae is shot at a fixed point charge of magnitude Ze so as to pass, if undeflected, at a distance Δ from the fixed charge, it can be shown that the angle ϑ through which the particle is deflected will be given by

$$\cot \tfrac{1}{2}\vartheta = \frac{8\pi\varepsilon_0 \Delta m v^2}{2AZe^2},$$

where m is the mass and v the velocity of the incident particle. This is calculated by ordinary mechanics for a particle moving under the influence of a fixed center, which attracts or repels it with a force inversely proportional to the square of the distance. The same equation holds for both attraction and repulsion. In the case of attraction the particle passes around the attracting center, while in the case of repulsion it is deflected away from it. The determination of the number scattered at the angle ϑ, when the particles are incident on a thin film, is simply the determination of the number of particles so directed as to pass at a distance Δ from a nucleus. If the scattering nuclei are assumed to be uniformly distributed and so far apart that an incident particle passes close to not more than one nucleus, the distribution of scattered particles can be computed. The fraction of the total number of incident particles that is scattered, per unit solid angle, in a direction making an angle ϑ with the incident beam is

$$f = \left[\frac{AZe^2}{8\pi\varepsilon_0 mv^2}\right]^2 \frac{Nt}{\sin^4 \frac{1}{2}\vartheta},$$

where N is the number of nuclei per unit volume and t is the thickness of the film.

The experiments of Geiger and Marsden, as well as later experiments, showed the above equation to be an accurate description of the scattering probability. In addition they observed scattering at angles so large that the alpha particles must have approached very close to the scattering centers.

In 1912 the interest in such scattering experiments was to determine the nature of the scattering atom. On the basis of the results, Rutherford proposed his atomic model with a very heavy nucleus at the center surrounded by moving electrons, as opposed to the model of J. J. Thompson in which the negative electrons were embedded in a jelly of positive charge. The results justified the conclusion that the nucleus acts as a heavy, charged, particle of matter with a diameter not greater than 10^{-12} cm.

For light nuclei, and very high energy incident particles, deviations from the above distribution are observed. They are interpreted in terms of the nuclear structure and are an important means of investigating such structure. However, the deviations appear only when the incident particle has sufficient energy to approach very close to the scattering nucleus. Although the first alpha particle scattering experiments were done to study the size of an atomic nucleus, the interpretation also assumes the alpha particle to be a small piece of electrically charged matter. The first experiments were done with a scintillation screen on which the impact of each individual alpha particle was observed with a microscope. Such direct counting of the scintillations, one at a time, made it unthinkable that any radically different interpretation could be given to the observations.

1.4. *Observation of particle tracks*

The invention in 1911 by WILSON [1912] of his "cloud chamber" made possible the observation of the paths of individual particles, and strengthened the conviction that the laws of ordinary mechanics could be used to describe such things as alpha particles and electrons. In particular, the conservation of energy and momentum could be verified for collisions between them, and the deflection of charged particles in electric and magnetic fields could be directly observed. The behavior was always found to be that expected of small massive particles.

PROBLEM 1. Work out a schematic design for an apparatus to measure e/m. Estimate the errors introduced by inaccuracies in describing the path, and by the velocity with which particles are emitted from the source.

PROBLEM 2. What sizes of oil drops can be used in the experiment to measure e? What factors determine the best size?

PROBLEM 3. Work out suitable dimensions for an apparatus in which the scattering of alpha particles from an aluminum foil 0.001 cm thick can be observed. If scintillations can be counted at rates between 0.1 and 2 per second in a microscope, what intensity source is necessary?

PROBLEM 4. Show that if a proton collides with a stationary proton in a cloud chamber, the deflected track and the recoil track will be at right angles to each other.

2. Experiments analyzed by means of the electromagnetic theory of light

The Maxwell-Lorentz equations of the electromagnetic field, in a region in which there is neither charge nor current, are

$$\text{(I)}\ \operatorname{curl} \boldsymbol{E} = -\frac{\partial \boldsymbol{B}}{\partial t} \qquad \text{(II)}\ \operatorname{curl} \boldsymbol{B} = \mu_0 \varepsilon_0 \frac{\partial \boldsymbol{E}}{\partial t}$$

$$\text{(III)}\ \operatorname{div} \boldsymbol{E} = 0 \qquad \text{(IV)}\ \operatorname{div} \boldsymbol{B} = 0 \qquad (1.1)$$

$$\text{(V)}\ \boldsymbol{F} = q(\boldsymbol{E} + \boldsymbol{v} \times \boldsymbol{B}).$$

Equation (V) for the force on a charge q can be regarded as the defining equation for the fields $\boldsymbol{E}$ and $\boldsymbol{B}$. The fields are presumed to be measured by observing the force on a test charge, q, so small that it does not disturb the field. μ_0 and ε_0 are the constants introduced in the mks system of units. $\mu_0 \varepsilon_0 = 1/c^2$, where c is the velocity of light in free space. From the above equations it follows that

$$\nabla^2 \boldsymbol{E} = \mu_0 \varepsilon_0 \frac{\partial^2 \boldsymbol{E}}{\partial t^2} \quad \text{and} \quad \nabla^2 \boldsymbol{B} = \mu_0 \varepsilon_0 \frac{\partial^2 \boldsymbol{B}}{\partial t^2}. \qquad (1.2)$$

These equations admit solutions that represent waves of electric and magnetic intensity, and by the use of the expression for the energy density, $\frac{1}{2}(\boldsymbol{E} \cdot \boldsymbol{D} + \boldsymbol{H} \cdot \boldsymbol{B})$, and by using the Poynting vector $\boldsymbol{E} \times \boldsymbol{H}$ for the density of energy flow, it is possible to interpret all of the various experiments on diffraction, interference, and polarization of light. It has been possible to produce an unbroken sequence of frequencies of electromagnetic waves, from the very low frequencies due to ordinary alternating currents to the very high frequencies of X-rays and gamma rays, and all of these appear to be described, in their propagation, by means of the field equations.

In a diffraction experiment, the diffracting surfaces of the apparatus need to be designed with reference to the wavelength of the waves to be diffracted. For visible light narrow slits can be used, or the rulings on a diffraction grating can be of the order of 10^{-4} cm apart. For X-rays and γ rays it is necessary to use the periodic structure of crystals. But in all cases the observation is of a diffraction pattern, with varying intensity, related to the diffracting apparatus. One of the simplest arrangements is the diffraction of light through a narrow slit. The light passing through the slit appears on a photographic plate, or visibly on a screen, as a series of bands parallel to the slit. The central band is the most intense and is located on the line from the source through the slit to the screen. The side bands have a diminishing intensity and a separation proportional to the wavelength of the light. Figure 1.1 indicates in a rough way, the intensity distribution.

In another arrangement the light from a line source, or from a single slit, passes through two slits and then on to the screen. The resulting pattern is indicated in figure 1.2. The important thing is that this pattern depends on both slits being open. If one slit is closed by a shutter, the pattern reverts to

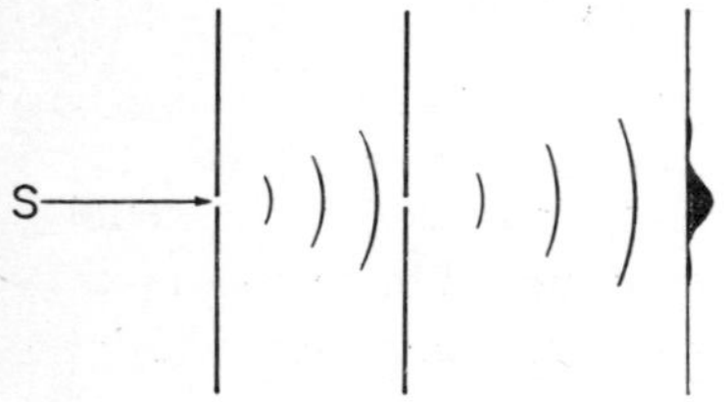

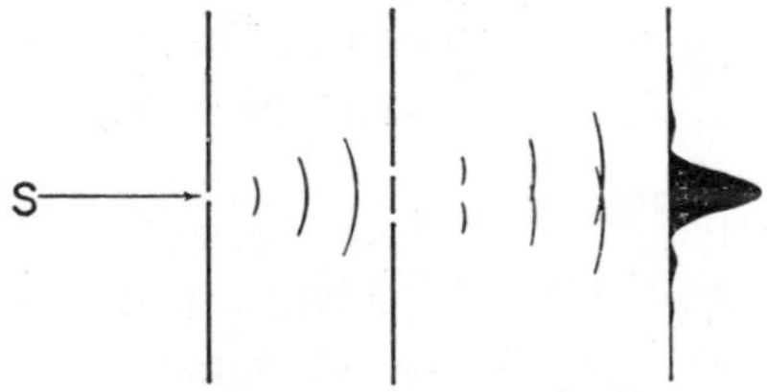

Fig. 1.1. A beam of light starting from a source S and collimated by passing through a pair of slits spreads out and strikes the screen on the right with an intensity distribution somewhat as indicated. The scale of the spreading is proportional to the wavelength. The fact on which the wave mechanics is based is that the same diffraction pattern appears if the source S emits electrons, protons, or even molecules. In both of these cases the intensity represented at the right is built up as a succession of individual spots, and only after a large number of spots have appeared does the whole pattern smear together as an intensity distribution.

Fig. 1.2. A schematized diffraction experiment for either light or material particles. It differs from the one shown in fig. 1.1 in having two slits through which the wave or particles can pass and showing a large maximum of intensity at the center point where would be a relatively low intensity if only one slit were open.

the form in figure 1.1. It cannot be said, when both slits are open, that the light passes through one slit or the other. It passes through both, and the interference of the two parts of the wave gives the observed pattern. This shows a fundamental distinction between "particles" and "waves". The interference experiments lead to the conclusion that the light is a wave phenomenon, not a stream of particles.

PROBLEM 5. Compute the intensity pattern due to a wave passing through two slits. Take the width of the slits to be much less than a wavelength but let their separation be several wavelengths. Consider only the intensity near the center of the pattern.

3. Experiments interpreted in terms of light quanta

3.1. *Black-body radiation*

The existence of a radiation pressure makes it possible to apply thermodynamic arguments to radiation in equilibrium with matter. If radiation is completely enclosed by matter at a given temperature, it will come to an equilibrium characteristic of the temperature but independent of the nature of the material. Such radiation is called "black-body" radiation, and the temperature is known as the temperature of the radiation. The distribution in frequency is such as would be radiated by a completely black or completely absorbing body at this temperature. Any enclosed cavity in a piece of matter contains black-body radiation at the temperature of the body. The spectral distribution can be determined experimentally by allowing a small amount of light to escape from a small hole in the cavity wall in to a spectrometer.

The fact that the spectral distribution of energy is independent of the material can be established by thermodynamic reasoning, but the distribution itself must be calculated in some other way. It was at first expected that the spectrum could be calculated on the basis of the electromagnetic theory of light. The electromagnetic field inside an enclosed volume can be described by an infinite number of normal coordinates. The statistics of such a system is relatively simple, and it is easy to express the average energy in such a cavity as a function of the temperature.

An electromagnetic field that satisfies equation (1.1) can be expressed in terms of the electromagnetic vector potential A by means of the equations

$$\boldsymbol{E} = -\frac{\partial \boldsymbol{A}}{\partial t}, \qquad \boldsymbol{B} = \operatorname{curl} \boldsymbol{A}, \qquad \operatorname{div} \boldsymbol{A} = 0. \tag{1.3}$$

The last condition is necessary to specify $\boldsymbol{A}$ uniquely when $\boldsymbol{E}$ and $\boldsymbol{B}$ are given, for without it there are many different vector fields $\boldsymbol{A}$ that give the same $\boldsymbol{E}$ and $\boldsymbol{B}$. With $\boldsymbol{E}$ and $\boldsymbol{B}$ expressed in this way equations (I), (III) and (IV) are satisfied identically, while equation (II) requires that

$$\nabla^2 \boldsymbol{A} = \mu_0 \varepsilon_0 \frac{\partial^2 \boldsymbol{A}}{\partial t^2}. \tag{1.4}$$

According to equation (1.4), if $\boldsymbol{A}$ and $\partial \boldsymbol{A}/\partial t$ are given as functions of position at $t = t_0$, the electromagnetic field is completely determined as a function of the time. Consequently, in the case under consideration, when no charges or currents are present, a single vector potential $\boldsymbol{A}$ is sufficient to describe the whole electromagnetic field.

To study the equilibrium situation, consider for simplicity the case in which the radiation is enclosed in a cubical box of edge L. The exact form of the boundary conditions for a material box is very complicated, but all of the essential elements are retained if it is required merely that the fields on one side shall be the same as those on the opposite side. Then each component of the vector potential can be expressed as a triple Fourier series, and the potential itself is

$$\boldsymbol{A} = \sum_{\boldsymbol{K}} [\boldsymbol{a}_{\boldsymbol{K},1} \cos(\boldsymbol{K} \cdot \boldsymbol{r}) + \boldsymbol{a}_{\boldsymbol{K},2} \sin(\boldsymbol{K} \cdot \boldsymbol{r})]. \tag{1.5}$$

In this expression $\boldsymbol{r}$ is the vector from the origin to the point at which $\boldsymbol{A}$ is the vector potential, and $\boldsymbol{K}$ is a vector such that

$$\boldsymbol{K} = \frac{2\pi}{L}(k_1 \hat{\boldsymbol{\imath}} + k_2 \hat{\boldsymbol{\jmath}} + k_3 \hat{\boldsymbol{k}}) \tag{1.5a}$$

where k_1, k_2 and k_3 are positive or negative integers. The form of (1.5) is such that $\boldsymbol{K}$ and $-\boldsymbol{K}$ are not regarded as different, and the different vectors $\boldsymbol{K}$ fill a solid angle of 2π only. The sum is over all vectors that lie in this solid angle and satisfy equation (1.5a). Equation (1.5a) guarantees the periodicity of the vector potential, and hence of the fields, with the period L. Any arbitrary vector function of the coordinates can be expressed in the form of equation (1.5), if it is periodic with the period L along the three coordinate axes. To obtain a vector field that represents an electromagnetic vector potential, it is necessary to impose the restrictions that div $\boldsymbol{A} = 0$ and that equation (1.4) is satisfied. Thus

$$\operatorname{div} \boldsymbol{A} = \sum_{\boldsymbol{K}} [-\boldsymbol{K} \cdot \boldsymbol{a}_{\boldsymbol{K},1} \sin(\boldsymbol{K} \cdot \boldsymbol{r}) + \boldsymbol{K} \cdot \boldsymbol{a}_{\boldsymbol{K},2} \cos(\boldsymbol{K} \cdot \boldsymbol{r})] = 0. \tag{1.6}$$

Since the sine and cosine functions for different values of $\boldsymbol{K}$ are orthogonal to each other, equation (1.6) requires that each $\boldsymbol{a}_{\boldsymbol{K},\mu}$ ($\mu = 1$ and 2) be perpendicular to the $\boldsymbol{K}$ with which it is associated. Each of the $\boldsymbol{a}_{\boldsymbol{K},\mu}$ vectors can then be decomposed into component vectors, perpendicular to each other and to $\boldsymbol{K}$, and specified by $\lambda = 1$ and $\lambda = 2$. The three vectors $\boldsymbol{a}_{\boldsymbol{K},1,\mu}$, $\boldsymbol{a}_{\boldsymbol{K},2,\mu}$, $\boldsymbol{K}$ are related as are vectors along the axes of a right-hand coordinate system. With this notation, any periodic vector field, whose divergence is everywhere zero, may be written

$$\boldsymbol{A} = \sum_{\boldsymbol{K},\lambda} [\boldsymbol{a}_{\boldsymbol{K},\lambda,1} \cos(\boldsymbol{K}\cdot\boldsymbol{r}) + \boldsymbol{a}_{\boldsymbol{K},\lambda,2} \sin(\boldsymbol{K}\cdot\boldsymbol{r})]. \tag{1.7}$$

To have a vector suitable for a vector potential, it is also necessary to have a time variation such that equation (1.4) is satisfied. All of the time dependence in equation (1.7) is in the vectors $\boldsymbol{a}_{\boldsymbol{K},\lambda,\mu}$. These depend only on the time and are not functions of position. Substitution of (1.7) in (1.4) leads to

$$\mu_0\varepsilon_0 \sum_{\boldsymbol{K},\lambda} \left[\frac{\mathrm{d}^2\boldsymbol{a}_{\boldsymbol{K},\lambda,1}}{\mathrm{d}t^2} \cos(\boldsymbol{K}\cdot\boldsymbol{r}) + \frac{\mathrm{d}^2\boldsymbol{a}_{\boldsymbol{K},\lambda,2}}{\mathrm{d}t^2} \sin(\boldsymbol{K}\cdot\boldsymbol{r})\right] =$$

$$= -\sum_{\boldsymbol{K},\lambda} K^2 [\boldsymbol{a}_{\boldsymbol{K},\lambda,1} \cos(\boldsymbol{K}\cdot\boldsymbol{r}) + \boldsymbol{a}_{\boldsymbol{K},\lambda,2} \sin(\boldsymbol{K}\cdot\boldsymbol{r})]. \tag{1.8}$$

Again because of the orthogonality of the trigonometric functions this requires

$$\mu_0\varepsilon_0 \frac{\mathrm{d}^2\boldsymbol{a}_{\boldsymbol{K},\lambda,\mu}}{\mathrm{d}t^2} + K^2\boldsymbol{a}_{\boldsymbol{K},\lambda,\mu} = 0, \tag{1.9}$$

where K^2 is the square of the length of the vector $\boldsymbol{K}$. Equation (1.9) shows that these vectors are simple harmonic functions of the time and that the whole electromagnetic field can be described by giving the values of the $\boldsymbol{a}_{\boldsymbol{K},\lambda,\mu}$ and their derivatives at some initial time. Although the expansion as a Fourier series assumes a periodicity in the field, this is no essential restriction, since the period can be made as large as is desired.

With the vector potential written in terms of the $\boldsymbol{a}_{\boldsymbol{K},\lambda,\mu}$ it is possible to transform the expression for the energy in the field from an integral over the volume to a sum over the various $\boldsymbol{a}_{\boldsymbol{K},\lambda,\mu}$ and their time derivatives. The result is

$$W = \frac{1}{2}\int \left[\varepsilon_0 E^2 + \frac{B^2}{\mu_0}\right] \mathrm{d}v = \tfrac{1}{4}\varepsilon_0 V \sum_{\boldsymbol{K},\lambda,\mu} \left[\dot{\boldsymbol{a}}^2_{\boldsymbol{K},\lambda,\mu} + \frac{K^2}{\mu_0\varepsilon_0} \boldsymbol{a}^2_{\boldsymbol{K},\lambda,\mu}\right]. \tag{1.10}$$

This last expression is obtained by inserting the series expressions for $\boldsymbol{E}$ and $\boldsymbol{B}$ and carrying out the integrals. Because of the orthogonality of the trigonometric functions, most terms in the integral vanish, and the remaining ones integrate to $\frac{1}{2}V$. Since this expression contains only the squares of the coordinates and their time derivatives, the $\boldsymbol{a}_{\boldsymbol{K},\lambda,\mu}$ are normal coordinates. Equation (1.10) is an exact and general expression for the energy in the field. It contains no assumptions as to the nature of the field except the one of periodicity. If there were only the field present, the distribution of energy among the different normal coordinates would remain constant and there would be no approach to an equilibrium distribution. Since the $\boldsymbol{a}_{\boldsymbol{K},\lambda,\mu}$ satisfy equation (1.9), each term in the sum (1.10) is a constant. However, black-body radiation is radiation in equilibrium with matter, and the presence of the matter provides a means by which energy can be transferred from one normal coordinate to another and by which a state of equilibrium can be attained. The state of equilibrium can be described in statistical terms without reference to the process by which it is reached. According to the equipartition theorem of classical statistical mechanics, the average energy associated with each normal vibration, when equilibrium has been attained, is kT. Thus with each set of subscripts, $\boldsymbol{K}$, λ and μ, must be associated the energy kT, where k is the molecular gas constant.

Equation (1.9) shows that each normal coordinate vibrates as a simple harmonic oscillator of frequency ν where $\nu = K/2\pi\ (\mu_0\varepsilon_0)^{\frac{1}{2}} = Kc/2\pi$. In order to determine the energy of the electromagnetic field as a function of frequency, it is necessary to find how many normal vibrations have frequencies lying between ν and $\nu+\mathrm{d}\nu$. This is equivalent to finding the number of vectors $\boldsymbol{K}$ whose lengths are between $2\pi\nu/c$ and $(2\pi/c)(\nu+\mathrm{d}\nu)$. The components of the $\boldsymbol{K}$'s are restricted by equation (1.5a), so that if one plots k_1, k_2, k_3 along the three axes of a system of cartesian coordinates, every point with integral coordinates will be the end of a vector from the origin associated with a permitted $\boldsymbol{K}$. The length of such a vector is

$$(k_1^2+k_2^2+k_3^2)^{\frac{1}{2}} = L\nu/c. \tag{1.11}$$

The number of such vectors in a solid angle of 2π and in a range of lengths corresponding to $\mathrm{d}\nu$ is $2\pi(L/c)^3\nu^2\,\mathrm{d}\nu$. For each vector $\boldsymbol{K}$ there are two values of λ and two values of μ, so that

$$\mathrm{d}N = \frac{8\pi V}{c^3}\nu^2\,\mathrm{d}\nu \tag{1.12}$$

gives the number of normal coordinates whose frequencies lie between ν and $\nu+\mathrm{d}\nu$. Then according to the classical statistics the energy distribution in black-body radiation is given by

$$E_\nu \, d\nu = \frac{8\pi V}{c^3} kT\nu^2 \, d\nu . \tag{1.13}$$

From equation (1.13) it follows that

1. There is an infinite total energy in the cavity. This cannot be immediately disproved experimentally because the total energy is not a directly observable quantity; but it is an unpleasant result.

2. The radiation in the cavity has an infinite specific heat. This is definitely not true, since the temperature can be raised with a finite amount of energy and the necessary energy can be measured.

3. The specific heat is independent of the temperature, and the total energy is proportional to the temperature. This again is definitely not true since it can be shown thermodynamically, as well as observed experimentally, that the total energy in the cavity is proportional to the fourth power of the temperature T.

4. The energy is concentrated in the high frequencies, and the energy per unit frequency range is proportional to the square of the frequency. This is again definitely not true, since the experiments show that the energy as a function of the frequency has a definite maximum at a frequency which is a function of the temperature.

The obvious contradiction to the observed spectral distribution of black-body radiation obtained by the above straightforward application of the electromagnetic theory constituted a paradox which was eventually resolved by the introduction of Planck's constant h.

PLANCK [1901] assumed, quite arbitrarily to obtain the correct result, that each normal coordinate can vibrate with only such amplitudes as will give it an energy $nh\nu$, where n is an integer and ν is the frequency of the normal vibration in question. The constant h has the dimension required to make $h\nu$ an energy. Such an assumption has no basis in electromagnetic theory. It must be regarded as an additional restriction on the nature of the electromagnetic field, a restriction quite foreign to the well established ideas of electricity and magnetism.

With this novel assumption, the usual statistical treatment gives for the average energy of a normal coordinate

$$\langle\varepsilon\rangle = \frac{\sum_{n=0}^{\infty} nh\nu \exp(-nh\nu/kT)}{\sum_{n=0}^{\infty} \exp(-nh\nu/kT)} = \frac{-h\nu \frac{d}{dx} \sum_{n=0}^{\infty} \exp(-nx)}{\sum_{n=0}^{\infty} \exp(-nx)} = \frac{h\nu}{\exp(h\nu/kT)-1} . \tag{1.14}$$

From equations (1.12) and (1.14) it follows that the energy-distribution law is

$$E_\nu = (8\pi V/c^3)h\nu^3/[\exp(h\nu/kT)-1]. \tag{1.15}$$

This expression is known as Planck's law of black-body radiation, and it agrees closely with the observed distribution. Through this work Planck's constant h became a part of the theory of physics, and the idea of the existence of radiant energy in pieces, or quanta, of magnitude $h\nu$ was the beginning of the quantum theory. According to recent determinations $h = 6.623 \times 10^{-34}$ joule sec.

Considerable interest attaches to this treatment of black-body radiation because through it Planck's constant was introduced and the quantum theory was started. The significance of the first work is sometimes overlooked because of the striking novelty of some of the more recent experimental discoveries and theoretical developments. But the treatment of black-body radiation still remains essentially as Planck gave it. The more recent theoretical developments have served to incorporate into a general theory the assumption of discrete energy values.

3.2. *The photoelectric effect*

The discovery of the photoelectric effect and its subsequent detailed study by many different experimenters gave evidence of still further discrepancies between the facts of observation and the implications of the electromagnetic theory of light. The experiments on the photoelectric effect showed that the energy with which an electron is ejected from a metal surface is independent of the intensity of the light and depends on the frequency only. This is certainly not to be expected on the ordinary electromagnetic theory, since the energy density and the Poynting vector both depend upon the field strengths.

Shortly after Planck's derivation of the law of black-body radiation, EINSTEIN [1904] proposed to treat a beam of light as a stream of corpuscles, or light quanta, each with the energy $h\nu$. He proposed to take seriously the conclusion from the observations that the energy in electromagnetic radiation occurs in bundles. Perhaps, then, these bundles have a location, a mass, and a momentum also. In the photoelectric process he proposed that a light quantum be considered as transferring all of its energy to a single electron in a process thought of as a collision. The electron will lose a certain amount, b, of this energy in escaping from the metal, so that the equation describing the process is

$$\tfrac{1}{2}mv^2 = h\nu - b \tag{1.16}$$

where v is the velocity with which the electron escapes. This is Einstein's photoelectric law, which was verified by a series of careful measurements (MILLIKAN [1916]).

Einstein's photoelectric equation (1.16) is based on the conservation of energy only, although the model of corpuscular radiation would seem to require the conservation of momentum also. An apparent difficulty arises here, because the momentum ascribed to a corpuscle, or photon, is $h\nu/c$. It is impossible to conserve both energy and momentum in an interaction between a single photon and a single electron if the energy of the photon is entirely transferred to the electron. This is because the energy of a photon, $h\nu$, is always just c times its momentum, while the ratio of the energy of an electron to its momentum depends on the velocity. Consequently the photoelectric process can be understood only with the assumption that the atoms of the metal take part in the interaction and take up the necessary momentum along with a negligible amount of energy. Such an assumption was at first rendered dubious when Millikan showed that the connection between the quantity b and the contact potential of the metal was such as to indicate that the electrons ejected photoelectrically were the free conduction electrons. Later theoretical work reconciled these facts by showing that it is just the interaction of the conduction electrons with the surface of the metal, and hence with the forces responsible for the work function, that suffices to conserve the momentum. Hence the idea of "particles" of radiant energy is necessary and also satisfactory for the understanding of the photoelectric effect.

3.3. *The Compton effect*

In 1923 A. H. COMPTON [1923] showed that the scattering of X-rays by matter could be described by the same picture as used by Einstein for the photoelectric effect. In this process the conservation of both energy and momentum must be considered, and because of the high energies involved, the interactions with the atoms of the substance can be neglected except in very precise work. The process is treated as a collision between corpuscles of radiation and free electrons.

If a beam of light or X-rays is composed of corpuscles of energy $h\nu$ and momentum $h\nu/c$, these quantities will be changed in collisions with electrons. The scattered photons will have energies and momenta, and hence frequencies and wavelengths, which depend on the direction of scattering. Each electron also will recoil in a direction and with a velocity connected by the conservation laws with the direction of scattering of the photon. Let the electron be initially at rest, let the frequency of the incident photon be ν, that of the scattered photon ν', and let the velocity of recoil of the electron be v. Then the conservation of energy requires that

$$h\nu = h\nu' + mc^2 \left\{ \frac{1}{(1-v^2/c^2)^{\frac{1}{2}}} - 1 \right\}. \tag{1.17}$$

If ϑ is the angle through which the photon is deflected, and if φ is the angle which the path of the recoil electron makes with the direction of the incident photon, the conservation of momentum requires that

$$\begin{aligned} \frac{h\nu}{c} &= \frac{h\nu'}{c}\cos\vartheta + \frac{mv\cos\varphi}{(1-v^2/c^2)^{\frac{1}{2}}} \\ 0 &= \frac{h\nu'}{c}\sin\vartheta - \frac{mv\sin\varphi}{(1-v^2/c^2)^{\frac{1}{2}}}. \end{aligned} \tag{1.18}$$

From these relations it follows that

$$\frac{c(\nu-\nu')}{\nu\nu'} = \lambda' - \lambda = \frac{h}{mc}(1 - \cos\vartheta). \tag{1.19}$$

This equation gives a relationship between the change of wavelength $(\lambda' - \lambda)$ and the angle of scattering ϑ, which has been verified by many experiments. The angle at which the electron recoils can also be evaluated in terms of the angle ϑ and the frequency of the incident radiation.

Although equation (1.19) follows directly from the picture of corpuscular radiation, it is less easy to show that the experimental results could not be explained in some other way. The classical electromagnetic theory also leads to a result similar to that in equation (1.19). According to the classical treatment, the electron begins to move forward under the radiation pressure of the incident wave. The scattered wave is then changed in wavelength because of the Doppler effect. The scattered wave spreads out in all directions, and the electron recoils in the direction of the incident beam, but the relationship between the change in wavelength and the direction of the radiation is much the same as that given in equation (1.19).

To justify completely the photon picture, it was very important to establish the relationship between the direction of scattering of the radiation and the direction of recoil of the electron. This was done in 1925 by COMPTON and SIMON [1925], who actually observed by means of a Wilson cloud chamber, the tracks of the recoil electrons and the action of the scattered X-rays in producing secondary electrons.

The above three well-established but surprising experimental results, the energy distribution in black-body radiation, the photoelectric effect, and the Compton effect, produced during the first quarter of the twentieth century a conviction that in spite of the evidence of the interference experiments, many of the properties of light and X-rays are corpuscular properties.

3.4. *Light quanta in an interference pattern*

When electrons are ejected from a metal surface by a beam of light the energy of each electron depends on the frequency only, as described previously. However, the number of electrons ejected each second is proportional to the intensity of the light. If the light is very weak only one electron, on the average, may be ejected each second. It then becomes possible to count the electrons, and by inference, to count the light quanta as they strike the metal surface. With visible light the efficiency of such a counting device is low and only a small fraction of the light energy is accounted for by the ejected electrons. For shorter waves, however, such as X-rays or gamma rays, the efficiency becomes higher and light quanta can be counted.

A very impressive experiment is that in which the intensity of an interference pattern is measured by counting the light quanta incident on the screen in much the same way that alpha particles are counted.

In order to understand the distribution of light quanta over the screen one must use a wave diffraction picture in which the energy is distributed over an advancing surface. To measure the intensity, however, the number of "impacts" per second on a small area of screen is counted. This kind of experiment presents the dilemma of atomic physics in a striking form.

A complete description of the phenomena of electromagnetic radiation requires both a "particle" and a "wave" basis. The appropriate description depends on the particular experimental conditions. This dual description is unified by asserting that radiation of wavelength λ has quanta associated with it of momentum p, where $\lambda = h/p$.

PROBLEM 6. Show from equation (1.15) that the total energy of the radiation in a cavity is proportional to T^4.

PROBLEM 7. Show from equation (1.15) that ν_m/T is independent of the temperature, where ν_m is the frequency at which E_ν is a maximum.

PROBLEM 8. Work out the direction of motion of the recoil electron in the Compton effect.

4. Experiments on the wave properties of particles

In spite of the impressive evidence that electrons, protons and alpha particles have the properties of small pieces of solid matter, it is possible to perform with them the same diffraction experiments as with a beam of light. If a beam of electrons is incident on a crystal, the beam is diffracted in just the manner of a beam of X-rays. A narrow pencil of electrons passing through a thin film of gold to impinge on a photographic plate, produces on it, not a single spot, but a series of concentric rings just as do X-rays under similar circumstances. From the diameter of the rings, and a knowledge of the atomic spacing in the film, it is possible to compute an effective wavelength for the electron beam. The effective wavelength is equal to h/mv (Davisson and Germer [1927], G. P. Thompson [1928]), where mv is the momentum of one of the electrons. Similar diffraction can be observed with gases, and the electron diffraction is now a standard method of studying the geometrical arrangement of atoms in molecules and solids. Perhaps even more striking is the fact that similar experiments can be performed with a beam of molecules, Estermann and Stern [1930].

In just the same way as the photons can be counted in an optical or X-ray diffraction pattern, the electrons can be counted when they are impinging on a screen in such a way as to form a diffraction pattern. It is a relatively simple matter to count the electrons with good efficiency and such an experiment brings out the paradox of waves and particles very clearly.

Suppose the arrangement suggested in fig. 1.2 represents a source of electrons so arranged that the electrons have access to the two slits as indicated. The electrons will impinge on the screen in such a way that the number per unit time at any point is represented by the curve of intensity shown. Now suppose one of the slits is covered by a shutter. The pattern will be different, as in fig. 1.1. In some places the closing of one slit will lead to an increase in the number of electrons striking the screen; in other places, to a decrease.

There is certainly no way in which such experiments can be explained or described on the classical picture of mass points. The difficulty here is not merely one of quantitative results but is a contradiction of the very concepts of the classical theory. The contradictory nature of the situation is exemplified by the use in one equation, $\lambda = h/mv$, of quantities pertaining to waves and to corpuscles of matter. The problem of quantum mechanics is to provide a set of concepts and rules of computation that correlates these apparently contradictory phenomena into a unified whole.

5. Discrete values of energy and magnetic moment in atoms

The assumption by Planck that an "oscillator" describing the electromagnetic field can have only such amplitudes of vibration as correspond to a discrete set of energy levels was the first introduction to the idea of discreteness in quantities which, in the classical theory, could take on any value. Later experiments have extended this idea in a number of ways.

5.1. *Line spectra*

The striking feature of the radiation emitted by free atoms and molecules is that the spectrum consists of well defined wavelengths and frequencies of vibration. During the first part of the twentieth century, there was steady improvement in spectroscopic apparatus and a steady increase in the resolving power with which such lines could be examined. In spite of this, the spectral lines emitted by stationary and isolated atoms were shown to be very sharp indeed, and to have a "natural width" beyond the resolving power of all but the best spectrographs. These spectral lines were, at first, supposed to correspond to natural frequencies of vibration of the various electrons in an atom. However, the very large number of the observed lines, and the lack of any harmonic relations between their frequencies made such a picture unsatisfactory.

About 1908, W. Ritz formulated his combination principle, according to which the frequency of a spectral line can be expressed as the difference between two spectral "terms". After Planck formulated his law for black-body radiation, it was natural to identify these spectral terms with an energy divided by h. Each line was then to be designated by indices representing the two terms and

$$\nu_{m,n} = \frac{W_m - W_n}{h}. \tag{1.20}$$

From this comes the idea that an atom can exist only in certain states of well defined energy, and the emission of radiation is associated with a transition between two states in which the energy difference is emitted.

Bohr's explanation of the hydrogen spectrum on this basis in 1913 was the beginning of an extensive theory of atomic structure and spectra. It proved possible to describe the hydrogen spectrum quantitatively, and other spectra qualitatively, in terms of discrete energy levels of the atoms. For atoms more complex than hydrogen, only the qualitative features of the energy-level scheme could be calculated, but the idea was sufficient for the analysis and classification of all atomic spectra.

The calculation of the discrete energy levels was made by treating classically the motion of the electrons around the heavy nucleus and then selecting, from the possible orbits, a few that were characterized by the so-called quantum rules. These rules had no rational basis, but their success in correlating the facts of spectroscopy made them of great importance. The general theory of quantum mechanics provides all of these results in a unified form and enables further spectral details to be worked out.

5.2. *Experiments of Franck and Hertz*

The experiments of FRANCK and HERTZ [1914] showed that when electrons collide with atoms they transfer to the atoms only certain discrete amounts of internal energy, which depend on the nature of the atoms. If a beam of very slow electrons is shot into a monatomic gas, only elastic collisions can take place. As the velocity is increased, inelastic collisions begin at the definite minimum energy of excitation that such an atom can absorb. Such collisions always represent a transfer of this same amount of energy. Only when the electron energy is increased to another definite level can more energy be absorbed by the atom.

These experiments are consistent with the idea that the atoms can exist in only certain definite energy states, and the energy of these states agrees with those deduced from spectroscopy. This idea of definite energy states is quite foreign to classical mechanics. It was sometimes described as indicating a peculiar stability of certain kinds of motion, but no description in terms of ordinary mechanics could be given for such stability.

5.3. *Stern-Gerlach experiment*

In 1921 STERN and GERLACH [1921] sent a beam of atoms through ad inhomogeneous magnetic field in such a way that an atom would be deflecten by the field to an extent proportional to the component of its magnetic moment parallel to the field gradient. According to the classical conception of such a situation, the magnetic moments of the atoms would be oriented completely at random, and the beam should be drawn into a broad band whose maximum of intensity would be at the center. The experiment showed, however, a series of discrete lines corresponding to discrete values of the component of the magnetic moment parallel to the field. Here again is a situation similar to that of the atomic energy levels. Only certain values of a dynamical quantity seem to be permitted, while an infinity of others are for some reason excluded.

Other experiments, such as the measurement of the magnetic susceptibilities of solutions of paramagnetic salts, have confirmed the fact that the angular momentum and the magnetic moment of an atomic system can have only a discrete set of values.

PROBLEM 9. What is the wavelength of an electron whose velocity is 10^8 cm per sec? What potential difference is necessary to produce an electron wavelength of 1 Å?

PROBLEM 10. What energy states are necessary to account for the first few doublets of the principal series of sodium? Express the energies in electron volts.

6. Purpose of quantum mechanics

The above experiments, and many others, when described in classical terms, give rise to contradictions. They lead to results at variance with the basic concepts of ordinary mechanics. It is therefore not surprising that a satisfactory theory requires a fundamental modification of the ideas and methods of mechanics. This modification must be a generalization, for it must in some way include all of the very satisfactory and useful applications of ordinary mechanics and electrodynamics.

Quantum mechanics is a theory, of which the classical mechanics is a limiting case, that suffices to correlate in a unified scheme the variety of experimental results described in this chapter, as well as many others. The remainder of this book will be devoted to the presentation and application of a set of concepts and principles in terms of which all of these experiments can be described. No attempt will be made to derive the principles from the observations. The principles are obtained by induction and assumption, not by deduction. No claim is made that the quantum mechanics to be presented is the only theory that could be devised. In fact, several different, but equivalent, forms of the theory will be presented. Each form has its advantages, and each can be used where its application makes calculation simplest.

Chapters 3 through 7 will be devoted to the wave properties of electrons and other objects generally thought of as particles. It will be of particular interest to note the method of resolution of the contradictions mentioned above. The first class of experiments is covered by the proof in Chapter 5 that classical mechanics is an approximation to quantum mechanics in the appropriate limit. The experiments in question are such that the approximation is good, and it will be of importance to make clear the extent of the region in which such an approximation can be used. It will appear that the conditions are very similar to those for the applicability of geometrical optics to the propagation of light.

Chapters 8 and 9 show that the "wave mechanics", while providing a possible pictorial representation for systems of one or possibly two particles, is not at all necessary as a formulation of the results that follow from it. The more abstract formulations, the notation due to Dirac, the non-commutative algebra of operators, the use of matrices as representations of physical quantities and their operators, all emphasize the symbolic nature of modern quantum mechanics.

Chapters 9 and 10 deal with angular momentum, electron spin and the Pauli exclusion principle. These are peculiarly suited to application of operator methods.

Chapters 11 and 12 take up the question of the "particle" nature of electromagnetic radiation, and show that Planck's basic assumption follows from the application to radiation of the general methods of quantum mechanics. The Compton effect and the photoelectric effect follow from similar considerations.

The idea of discrete values of energy and angular momentum are treated in Chapters 4 and 10, and shown again to fit in to the general scheme.

Although the quantum mechanics is adequate for the treatment of a wide range of phenomena that are beyond the range of classical mechanics, there are still limits to its validity and cases in which it is not completely satisfactory. Some of these will be pointed out from time to time as they arise. Nevertheless, the emphasis will be on the successes of the theory, since it has made possible for the first time an adequate electron theory of matter and has made possible an extensive, if not entirely adequate, understanding of atomic nuclei.

CHAPTER 2

ANALYSIS OF CLASSICAL MECHANICS

A careful study of experiments such as those discussed in the previous chapter makes it possible to describe conditions under which classical mechanics is applicable and conditions under which the more general quantum mechanics must be used. The precise formulation of these conditions requires the introduction of Planck's h, and in fact the classical mechanics can be understood as a limiting case of quantum mechanics, a case in which h is negligible.

1. Outline of classical mechanics[1]

Classical mechanics is based on the system of concepts and principles included in Newton's laws of motion and their relativistic generalization. To apply the mechanics one must know the forces involved. In cases of natural phenomena inaccessible to experimental control, such as the motions of planets, the forces are inferred from the observed motions. In the various experiments with atomic particles the same attempt has been made to apply classical mechanics to determine the forces from observed motions. However, the observations necessary to discover the forces are less direct than in the case of planetary motion; and the results of the experiments soon raised serious doubts as to whether the classical laws are applicable to all such situations.

In classical mechanics the state of a system of mass particles is specified, at a time t, by the values of all the coordinates and all the velocities or momenta. This is, of course, an idealization. A practical measurement of

[1] This brief outline of classical mechanics is inserted here only to establish a terminology and a notation to be used later, and to emphasize those portions to which quantum mechanics is closely analogous. For a more extensive discussion of these and other points, see some standard text on analytical dynamics.

coordinates and momenta is always subject to experimental error. However, until the discovery of quantum phenomena, dependent on the fact that $h \neq 0$, there seemed to be no reason in principle why the accuracy of such determinations should not be increased without limit. When the state of the system, as given by the values of the coordinates and the momenta, is taken as an appropriate idealization, the changes in this state are given by the equations of motion. The changes of the coordinates and momenta with time follow directly from a knowledge of the forces.

The equations of motion are differential equations of the second order, whose integrals give the coordinates and the velocities as functions of the time and the initial state. Because the equations of motion are of the second order, the specification of the state involves only the coordinates and their first derivatives with respect to the time. No higher derivatives can be arbitrarily specified. But to use the laws of motion, there must be a real (experimental or physical) significance to the idea of the state of the system at t_0, and then later at t, and in fact at all times in between.

1.1. *Hamilton's principle*

A convenient form in which the equations of motion can be written is that of Hamilton's principle. According to it,

$$\delta \int_{t_1}^{t_2} L(q_i, \dot{q}_i, t)\, \mathrm{d}t = 0, \tag{2.1}$$

where L is the Lagrangian function, and the integration is between fixed limits. q_i represents all of the independent coordinates and $\dot{q}_i$ all the corresponding rates of change. This variation principle is equivalent to Newton's equations and is frequently a more convenient form than the equations themselves in cartesian coordinates. It is a form independent of the coordinates used and from which the differential equations can be readily obtained by differentiation. The Lagrangian equations corresponding to equation (2.1) are

$$\frac{\mathrm{d}}{\mathrm{d}t}\frac{\partial L}{\partial \dot{q}_i} - \frac{\partial L}{\partial q_i} = 0. \tag{2.1a}$$

The knowledge of the physical system and of the forces is included in the form of the Lagrangian function L. For conservative systems in which the velocities are always very small compared with the velocity of light, $L = T - V$, where T is the kinetic energy and V is the potential energy. For cases in which the velocities become comparable with the velocity of light another form for the Lagrangian is necessary. In equation (2.1a) it is assumed that all constraints have been removed and that the Lagrangian function is expressed in terms of a set of independent coordinates q_i and their derivatives $\dot{q}_i$.

1.2. *The Hamiltonian equations of motion*

It is frequently convenient to transform the set of second-order equations into a set of twice as many first-order equations, known as Hamilton's canonical equations. To do this, let $p_i = \partial L/\partial \dot{q}_i$. Then according to equation (2.1a), $\dot{p}_i = \partial L/\partial q_i$. Let[1]

$$H = \sum_{i=1}^{n} p_i \dot{q}_i - L(q_i, \dot{q}_i, t) = H(p, q, t). \tag{2.2}$$

The last equality signifies that the Hamiltonian function, H, is to be written in terms of the p's and the q's and that the quantities $\dot{q}_i$ are to be eliminated by means of the definitions of the p_i. By taking the differential of H and of its expression in terms of L, it can be shown that

$$\dot{q}_i = \frac{\partial H}{\partial p_i} \quad \text{and} \quad \dot{p}_i = \frac{\partial L}{\partial q_i} = -\frac{\partial H}{\partial q_i}. \tag{2.3}$$

The usefulness of these canonical equations of motion is due in part to the possibility of making transformations to new coordinates and momenta which are such functions of the old coordinates and momenta that the equations are simplified. In using the Lagrangian equations, transformations can be made to simplify the solution, but these transformations can involve the coordinates only. The inclusion of the momenta in the transformations, and the consequent disappearance of any mathematical distinction between the coordinates and momenta, greatly increase the range of the transformations available. If, for instance, it is possible to find such a transformation that one or more coordinates are absent from the Hamiltonian function, the conjugate momenta are constants of the motion according to equation (2.3).

As a matter of fact it is rarely possible to find a neat transformation for any but the simplest cases. The form of the Hamiltonian equations, however, lends itself better to general arguments than does the form of the second order Lagrangian equations.

Equations (2.3) permit one to plot in the "phase space", whose coordinates are the p's and the q's, the trajectory of a point representing the state of the whole system. It is evident from the form of equations (2.3) that a

[1] The omission of the subscripts indicates that L and H are functions of all of the variables q_i and $\dot{q}_i$, or of q_i and p_i.

trajectory in the phase space will never intersect itself; i.e., the derivatives of the trajectory are all single-valued at each point.

1.3. *Canonical transformations*

A transformation from the coordinates and momenta (q_i, p_i) to new coordinates and momenta (Q_i, P_i) is canonical if the new variables satisfy the Hamiltonian canonical equations. Since these equations follow from Hamilton's principle, a necessary and sufficient condition for a canonical transformation is

$$\delta \int \left[\sum_{i=1}^{n} p_i \dot{q}_i - H(p, q) \right] \mathrm{d}t = \delta \int \left[\sum_{i=1}^{n} P_i \dot{Q}_i - H'(P, Q) \right] \mathrm{d}t = 0. \quad (2.4)$$

This condition does not require the integrands to be equal, but it does require their difference to be a derivative with respect to the time, of a function of the quantities p_i, q_i, P_i, Q_i, which are taken as fixed at the limits of integration. Only $2n$ of these quantities are independent, so that the function may be expressed in numerous ways as a function of $2n$ independent quantities in addition to the time.

Take now the special case in which this function, designated by V (not the potential energy), depends upon q_i, P_i and t. Then

$$\sum_{i=1}^{n} p_i \dot{q}_i - H(p, q) = \sum_{i=1}^{n} P_i \dot{Q}_i - H'(P, Q) + \frac{\mathrm{d}}{\mathrm{d}t} V(q, P, t), \quad (2.5)$$

where H' is that function of the new variables P_i and Q_i which will appear in the new equations of motion. Further, let

$$S(q, P, t) = \sum_{i=1}^{n} P_i Q_i + V(q, P, t). \quad (2.6)$$

This is possible because P_i and Q_i are functions of q_i and p_i and hence of q_i and P_i. Then it follows from (2.5) that

$$\sum_{i=1}^{n} p_i \dot{q}_i - H(p, q) = -\sum_{i=1}^{n} Q_i \dot{P}_i - H'(P, Q, t) + \frac{\mathrm{d}}{\mathrm{d}t} S(q, P, t). \quad (2.7)$$

By equating the coefficients of $\dot{q}_i$, $\dot{P}_i$, and the remaining terms, there follow the equations

$$p_i = \partial S / \partial q_i \quad (2.8a)$$

$$Q_i = \partial S / \partial P_i \quad (2.8b)$$

$$H' = H + \partial S / \partial t. \quad (2.8c)$$

If a transformation is formed from any function S by means of these equations, the new variables will satisfy the Hamiltonian equations of motion. S is frequently called the generating function of the transformation. Equations (2.8a) and (2.8b) will give the Q_i and the P_i in terms of the q_i and the p_i, and the equation (2.8c) will give the new Hamiltonian function.

It is clear from this method of obtaining the transformation that the property of a transformation being canonical does not depend on the particular problem under consideration; it does not depend on the Hamiltonian function. A transformation which is canonical for one problem is canonical for all problems.

As an example of a canonical transformation consider a particle with three cartesian coordinates and the corresponding momenta. Then consider a transformation to $(Q_1, Q_2, Q_3, P_1, P_2, P_3)$ given by the generating function

$$S = (x \cos\lambda + y \sin\lambda)P_1 + (-x \sin\lambda + y\cos\lambda)P_2 - \tfrac{1}{3}m\sqrt{8g}(P_3 - z)^{\frac{3}{2}}.$$

From equations (2.8a) and (2.8b) it follows that

$$\begin{aligned} p_x &= P_1 \cos\lambda - P_2 \sin\lambda & Q_1 &= x\cos\lambda + y\sin\lambda \\ p_y &= P_1 \sin\lambda + P_2\cos\lambda & Q_2 &= -x\sin\lambda + y\cos\lambda \\ p_z &= (2m^2g)^{\frac{1}{2}}(P_3 - z)^{\frac{1}{2}} = -Q_3 & Q_3 &= -(2m^2g)^{\frac{1}{2}}(P_3-z)^{\frac{1}{2}} = -p_z. \end{aligned}$$

These can be solved to give also

$$\begin{aligned} x &= Q_1\cos\lambda - Q_2\sin\lambda & P_1 &= p_x\cos\lambda + p_y\sin\lambda \\ y &= Q_1\sin\lambda + Q_2\cos\lambda & P_2 &= -p_x\sin\lambda + p_y\cos\lambda \\ z &= P_3 - Q_3^2/2m^2g & P_3 &= z + p_z^2/2m^2g. \end{aligned}$$

To illustrate the use of this transformation for a particle in a gravitational field, take

$$H = \frac{1}{2m}(p_x^2 + p_y^2 + p_z^2) + mgz.$$

Since the generating function S is independent of the time, the transformation of the Hamiltonian involves only the change of variables:

$$H = \frac{1}{2m}(P_1^2 + P_2^2) + mgP_3.$$

Since this H is independent of the Q's, each of the P's is a constant, and

$$\dot{Q}_1 = P_1/m, \quad \dot{Q}_2 = P_2/m, \quad \dot{Q}_3 = mg.$$

1.4. *The Poisson brackets*

If F is a function of the coordinates and momenta of a dynamical system and of the time, its rate of change is

$$\frac{dF}{dt} = \frac{\partial F}{\partial t} + \sum_{i=1}^{n} \left(\frac{\partial F}{\partial q_i} \dot{q}_i + \frac{\partial F}{\partial p_i} \dot{p}_i \right)$$
$$= \frac{\partial F}{\partial t} + \sum_{i=1}^{n} \left(\frac{\partial F}{\partial q_i} \frac{\partial H}{\partial p_i} - \frac{\partial F}{\partial p_i} \frac{\partial H}{\partial q_i} \right) = \frac{\partial F}{\partial t} + [F, H]. \qquad (2.9)$$

The expression $[F, H]$ is called the Poisson bracket of F and H. It is invariant under a canonical transformation so that equation (2.9) is in a form independent of the coordinates used.

Poisson brackets involving pairs of functions not including the Hamiltonian are also useful. The Poisson brackets can be used to express the conditions for a canonical transformation. If Q_i and P_i are the coordinates and momenta derived by a transformation from q_i and p_i, the condition that the transformation be canonical can be written

$$[Q_i, Q_j] = 0; \qquad [P_i, P_j] = 0; \qquad [Q_i, P_j] = \delta_{ij}. \qquad (2.10)$$

It will be shown in later chapters that in quantum mechanics the role of the Poisson brackets is taken by the commutators of the corresponding operators.

PROBLEM 1. Work out the expressions for a canonical transformation when V in equation (2.5) is a function of (p, Q).

PROBLEM 2. Show that a transformation from cartesian to spherical polar coordinates is a canonical transformation. Write the generating function for it.

PROBLEM 3. Show that if $[Q_i, P_j] = \delta_{ij}$, $[A, B]$ is invariant to the transformation from (q_i, p_i) to (Q_i, P_i), where A and B are arbitrary functions of (q_i, p_i) and hence also of (Q_i, P_i).

PROBLEM 4. Show that if the conditions (2.10) are satisfied, the Hamiltonian equations are satisfied in the new coordinates and momenta.

PROBLEM 5. Find the transformation corresponding to the generating function $S = -\frac{1}{2} Q^2 \tan p$.

PROBLEM 6. Show that $[F, G] = -[G, F]$ where F and G are arbitrary dynamical quantities and that $[F, C] = 0$ when C is a constant.

PROBLEM 7. Show that

$$\frac{\partial F}{\partial p_i} = -[F, q_i] \quad \text{and} \quad \frac{\partial F}{\partial q_i} = [F, p_i].$$

1.5. *Infinitesimal canonical transformations*

A transformation of coordinates can be interpreted in two ways. On the one hand it may be regarded as a change in the coordinates by which a given function is described. A function $F(p_i, q_i)$ has the same value at a certain point in the phase space no matter whether the point is located by the co-ordinates (p_i, q_i) or by (P_i, Q_i). On the other hand, the transformation may be regarded as a transformation from one point to another in the phase space with a consequent possible change in the value of the function F.

The situation is most easily visualized in the case of a two dimensional vector. A rotation of the coordinates through the angle θ requires that a vector with the components (x, y) now have the components

$$\begin{aligned} x' &= x \cos \theta + y \sin \theta \\ y' &= -x \sin \theta + y \cos \theta \end{aligned}$$

even though it is still the same vector. On the other hand the transformation may be regarded as a rotation of the vector itself through the angle $(-\theta)$ so that the new components now describe a new vector in the old coordinate system.

In the same way, a canonical transformation may be regarded as moving each point (p_i, q_i) to a new point in the phase space (P_i, Q_i). If the movement is infinitesimal it can be described by an infinitesimal transformation whose generating function can have the form

$$S = \sum_i q_i P_i + \varepsilon G(q, P). \tag{2.11}$$

ε is an infinitesimal multiplying the arbitrary function $G(q, P)$, so that the whole generating function $S(q, P)$ differs only infinitesimally from the function $\Sigma_i q_i P_i$, which generates the identity transformation.

From equations (2.8a) and (2.8b)

$$p_i = P_i + \varepsilon \frac{\partial G}{\partial q_i} \quad \text{and} \quad Q_i = q_i + \varepsilon \frac{\partial G}{\partial P_i}. \tag{2.12}$$

Since the change from p_i to P_i is infinitesimal, the derivative may be taken with respect to p_i instead of P_i so the transformation equations are

$$\begin{aligned} Q_i &= q_i + \varepsilon \frac{\partial G}{\partial p_i}, \\ P_i &= p_i - \varepsilon \frac{\partial G}{\partial q_i}. \end{aligned} \tag{2.13}$$

When dealing with this kind of a transformation the function $G(Q,P) \to G(q,p)$ is often referred to as the generating function.

If $\varepsilon = \mathrm{d}t$ and $Q_i - q_i = \mathrm{d}q_i$, $P_i - p_i = \mathrm{d}p_i$ the transformation equations become

$$\frac{Q_i - q_i}{\varepsilon} = \frac{\mathrm{d}q_i}{\mathrm{d}t} = \frac{\partial H}{\partial p_i}$$
$$\frac{P_i - p_i}{\varepsilon} = \frac{\mathrm{d}p_i}{\mathrm{d}t} = -\frac{\partial H}{\partial q_i} \tag{2.14}$$

when G is taken as H. Thus the Hamiltonian function is the generator of a canonical transformation that describes the change of the coordinates and momenta with the time. Repeated application of this transformation gives the trajectory of the representative point in the phase space. This fact is of no great help in finding an explicit solution of the equations of motion, but it does give a point of view toward the canonical equations which has its analog in quantum mechanics.

1.6. *The Hamilton-Jacobi equation*

The repeated application of the infinitesimal transformation generated by the Hamiltonian function gives a transformation from the initial values of the coordinates and momenta to the values at some later time t. This "integrated" transformation is then a solution of the problem. Such a transformation has an inverse which is a transformation, a function of the time, from the coordinates and momenta at time t to the initial values. If the transformed function $H'(P, Q, t_0)$ is independent of t, it may be taken to be zero. Then, according to equation (2.8c),

$$\frac{\partial S}{\partial t} + H(p, q, t) = 0. \tag{2.15}$$

If equation (2.8a) is used to replace the p_i there results the Hamilton-Jacobi partial differential equation for $S(q, t, q_0)$:

$$H' = \frac{\partial S}{\partial t} + H\left(\frac{\partial S}{\partial q_i}, q_i, t\right) = 0. \tag{2.16}$$

A solution of equation (2.16) will contain $(n+1)$ arbitrary constants, of which one will be simply an additive constant. This can be ignored since only derivatives of S are important. Hence a solution will be

$$S(q_i, \ldots q_n, \alpha_i, \ldots \alpha_n, t).$$

The constants α_i, or any desired linearly independent combination of them, may be taken as the momenta P_i.

In case the Hamiltonian function does not contain the time, the partial differential equation for S shows that $\partial S/\partial t$ must be a constant. Call it $-W$. Then

$$H\left(\frac{\partial S}{\partial q_i}, q_i\right) = W, \tag{2.17}$$

where W is the total energy.

Equations (2.16) and (2.17) are both known as the Hamilton-Jacobi partial differential equation. The solution of this one equation gives the complete solution of the mechanical problem, but no general prescription can be given for dealing with such a partial differential equation.

If the Hamiltonian function is separable and does not contain the time explicitly, the function S will be a function of the time plus functions of each of the separated systems of variables.

Let

$$S = -Wt + \sum S_i(q_i, \alpha_i). \tag{2.18}$$

Equation (2.16) gives

$$-W + \sum_i H_i\left(\frac{\mathrm{d}S_i}{\mathrm{d}q_i}, q_i\right) = 0. \tag{2.19}$$

Each member of this sum must be a constant, hence

$$S_0(t) = -Wt + \beta_0, \tag{2.20a}$$

$$H_i\left(\frac{\mathrm{d}S_i}{\mathrm{d}q_i}, q_i\right) = \beta_i, \tag{2.20b}$$

$$-W + \sum \beta_i = 0. \tag{2.20c}$$

Equation (2.20b) is of the first order so it can be reduced to an integration. The separable case can, in principle, be solved.

PROBLEM 8. Treat the problem of simple harmonic motion by means of the Hamilton-Jacobi equation.

PROBLEM 9. Use the Hamilton-Jacobi equation to treat the problem of a pendulum with finite amplitude.

2. Analysis of the applicability of classical mechanics

A careful consideration of the concepts of classical mechanics shows several conditions that a system must satisfy if it is to be describable by such means.

1. Classical mechanics ostensibly applies to isolated systems, or to systems influenced by known forces. Application of the theory to an isolated system implies an observer, entirely apart from the system itself, who describes the motion under the acting forces but who does not in principle take any part in it. If the application is to a system subject to external forces, it is implied that the observer is unconcerned with the source of the external forces but still knows them as functions of the time. In particular the reaction of the system upon the source of these forces is not considered.

Classical mechanics is not expected to give correct results when outside influences are acting in an unknown manner on the system under consideration. On this account, a system cannot be described by classical laws and at the same time be observed, unless the observations can be made in such a way as not to influence the motion at all, or unless the influence of the method of observation can be precisely known.

2. The use of classical mechanics also implies the possibility of fixing, or discovering, the state of the system at an initial time t_0. The state of the system is given by the position and the velocity of each particle at this time, and the equations of motion give the way in which both the position and the velocity change as the time goes on.

Further consideration shows that these two requirements are not independent. To find the initial conditions of a motion, it is necessary to make an observation on the moving system. Classically it is assumed possible to do this without interfering with the motion at all, so that the details of making the observation and of finding the state of the system, initially and at other times, are not considered as an essential part of the mechanics.

One might think it not at all necessary to observe a system during its motion in order to have the classical methods be of use. It should be sufficient to use the classical theory to connect the state of the system at some initial time, i.e., the coordinates and velocities at this time, with the state at some later time when an observation could be made and the later disturbed motion ignored. If this could be done, the theory might still be useful. But there is yet a difficulty in the fact that the determination of the initial state requires the measurement both of coordinates and of their rates of change. The measurement of such a rate of change involves the measurement of at least two positions and the time interval between them. If the measurement of the second position is accompanied by such an interaction between the measuring instrument and the moving system that an exchange of an unknown amount of momentum takes place, it is clear that the utility of the first position measurement is destroyed. Hence there arises the question as to whether it is ever possible to satisfy the conditions under which classical mechanics can be used.

A study of possible methods of observation, in the light of experimental knowledge of the interactions involved, indicates that there are no known methods of observation that are admissible. Although such conclusions must be based on the study of specific methods, BOHR [1935] has emphasized the generality of the result. He has pointed out the incompatibility, by definition, of the concepts of position and momentum, and has indicated that the precise specification of the initial state as required by classical mechanics is impossible. Classical mechanics must be regarded as only an approximation.

The emphasis on observations as the disturbing influences is based on the supposition that all other influences can be removed. This may be questionable, but even in the most favorable case the means of observation represent an irreducibel minimum of disturbance.

3. Analysis of representative methods of observation

Under the heading of observations are included all methods for ascertaining the values of any dynamical quantities that may be of interest. One ordinarily thinks of an observation as giving the position or the momentum of some specific particle, but the term may also be applied to methods for isolating those particles which have the desired properties and discarding the remainder. The first one to be described is of this latter type.

3.1. *The collimation of a beam of particles*

Consider the apparatus indicated in fig. 2.1. S is a source of particles, such as electrons, of which some pass through the slits A and B and strike the screen C. The slits A and B and the screen C are all solid bodies, whose molecular structure is ignored for the purposes of this analysis and which are all rigidly fastened to the base. This whole rigid body constitutes the coordinate system with reference to which all positions must be measured. According to classical mechanics, such a system should so fix the vertical motion of the particles passing through both slits that they will all strike the screen along a line P in the plane of the slits.

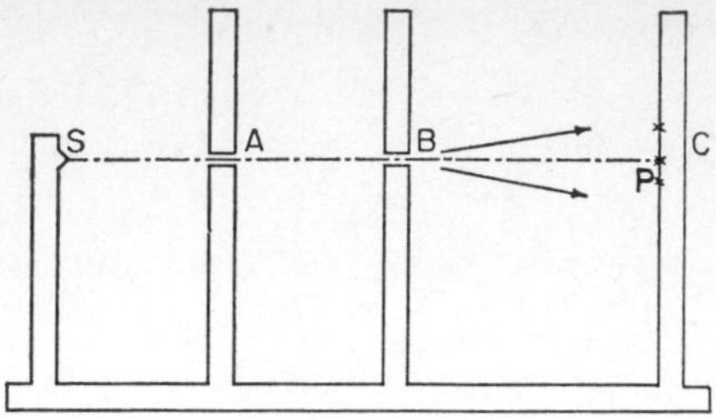

Fig. 2.1. Another sketch of the diffraction apparatus for particles or waves collimated by two slits. It is to be emphasized that in either case the arrival of the waves of or the particles is signalled by the appearance of "spots", photographic or otherwise counted, on the screen.

But it is known from experiment that the particles will not all strike along the line at P. Some will strike above it and some below it. The number of particles striking each point can be computed as proportional to the intensity of a wave that has been diffracted by the slits and whose wavelength is h/mv. Although this is a method for computing the distribution of the particles over the screen, it is certainly no explanation, in terms of classical mechanics, of the behavior of the particles. The only such interpretation is that the particles in passing through the slit B have so interacted with the slit as to transfer to it or to obtain from it various amounts of momentum in the vertical direction. This transfer is uncontrollable and unknown in amount for any one particle. It is not the same for all particles, and it cannot be measured for one particle by measuring the recoil of the slit. For the slit is part of the coordinate system with reference to which all positions are to be measured. If it is allowed to recoil, that recoil must be measured on some other coordinate system with which the same difficulty will arise.

This is an example of the impossibility of so fixing the initial state of a particle that its behavior can be predicted by classical mechanics. The slits A and B could be regarded as the apparatus for the establishment of the initial conditions, a position in the vertical direction along with a zero vertical velocity. During the passage from B to C the particle would be unobserved, and it could be expected to arrive at P. But it does not always do so. Thus, speaking in classical terms, it must be concluded that this apparatus is not adequate to fix the vertical velocity and position of the particles.

3.2. *Gamma-ray microscope*

A customary method of measuring the position of a particle is to observe it with a microscope. One ordinarily thinks of this as interfering less with the motion than any method in which there might be material contact between

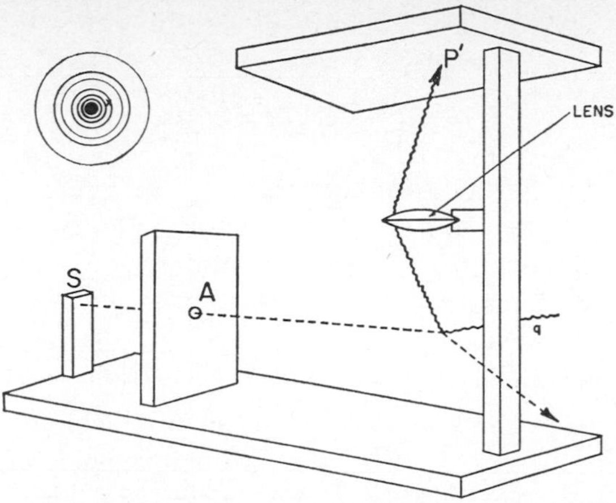

Fig. 2.2. A schematic arrangement of the gamma-ray microscope devised by Heisenberg. The location of the single spot on the photographic plate at P′ could be anywhere in the diffraction pattern schematized in the upper left corner.

the particle and the measuring apparatus. Consider the arrangement illustrated in fig. 2.2. Again a stream of particles comes from the source S and passes through the hole at A. If we are interested in motion in the horizontal plane and can make another measurement of this coordinate without disturbing the motion or by disturbing it by a known amount, then the classical mechanics should be adequate to describe the future motion of the particle in this direction.

If a beam of light is sent in along the path q, some of it may be scattered from the particle through the lens to the photographic plate. It is known experimentally that the light striking the plate will strike at definite points, and the process is described by saying that the light travels in the form of quanta, each one of which makes a spot on the plate. If the light is very weak, and if it is shut off after the first quantum hits the plate, the disturbance of the particle by the light will be a minimum. The problem is then to conclude from the location of the spot P′ the location of the particle at the time the light was scattered.

If many light quanta are scattered by the particle and pass through such a lens, the ordinary theory of diffraction gives the distribution of intensity on the plate. There will be a dense central spot surrounded by a series of rings. If but a single quantum goes through the lens, it may strike the plate anywhere within the pattern. The radius of the first diffraction ring may be taken as an arbitrary measure of the accuracy with which this one spot will locate the center of the pattern. The theory of diffraction gives this approximately as

$$\Delta x \approx \frac{\lambda}{\sin \varepsilon} \tag{2.21}$$

where λ is the wavelength of the light and ε is the angular aperture of the lens. From this expression it follows that the position can be measured as accurately as is desired if only light of small enough wavelength is used. Hence the name "gamma-ray microscope" has been applied to this conceptual method of measuring a position.

It still remains to find out to what extent this kind of measurement of position influences the motion and to what extent it involves an exchange of momentum. The interaction can be described by the theory of the Compton effect. The light is treated as a stream of particles with energy hc/λ and momentum h/λ, and the classical laws of conservation of energy and momentum are applied to the scattering process. If the momentum of the light quantum both before and after the impact is known, the difference will be the momentum transfer and the motion of the particle can be classically predicted. To consider the most favorable case, assume that the momentum before the impact is precisely known. To know the momentum after the impact, it is essential to know the direction in which the quantum has been scattered. It is clear that the quantum has gone through the lens, but through what part of the lens is entirely unknown. Hence there is an uncertainty as to the final momentum of the light quantum, and this uncertainty is proportional to the sine of the aperture of the lens and inversely proportional to the wavelength. For qualitative considerations it can be assumed that the wavelength is unchanged by the scattering. Then the uncertainty in the knowledge of the final momentum of the light quantum, and correspondingly, the uncertainty in the amount of momentum transferred to the particle, is approximately

$$\Delta p \approx \frac{h \sin \varepsilon}{\lambda}. \tag{2.22}$$

As the wavelength is made smaller, the undefined effect of the observation becomes larger, although conversely the disturbance can be made very small by using light of a long wavelength. The condition for a precise observation of position is just such as to exclude the condition for a small amount of unknown interference with the motion.

If light of an intermediate wavelength is used, the position can be roughly measured with only a limited uncertainty in the amount of the change in momentum. The two uncertainties from equations (2.21) and (2.22) are then connected by the relationship

$$\Delta x \, \Delta p \approx h \, . \tag{2.23}$$

If one of these quantities is precisely measured, all knowledge of the other is lost.

3.3. *General considerations*

The above examples illustrate two cases in which the means of observation so interfere with the system as to destroy the possibility of unambiguously defining the quantities required by classical mechanics. These illustrations show the way in which position and momentum are of such a nature that the determination of one automatically excludes the determination of the other. In special cases this might be regarded as a peculiarity of the means of observation, but since it is a general result, it must be regarded as a property of the definitions of coordinate and momentum. They must be regarded as quantities whose definitions are mutually incompatible.

Although only two illustrations are given above, Bohr has emphasized that the result is very general. The very concept of a coordinate involves a coordinate system, and this is in fact a large rigid body. To measure a coordinate requires an interaction of the mechanical system with the coordinate system in some way. This interaction will necessarily be accompanied by a possible transfer of momentum, and the amount of this transfer cannot be made infinitely small because of the finiteness of h. Before the advent of the quantum theory and the experimental study of the interactions, it was tacitly assumed that the uncertainties introduced in this way could be neglected; but the experimental discoveries leading to the quantum theory have shown that such is not the case. Furthermore this transfer of momentum cannot be controlled or known. To know the momentum transferred to the coordinate system it would be necessary to let it recoil and measure the momentum of the recoil. This measurement, however, would require its reference to another coordinate system, and the usefulness of the first as a coordinate system would be entirely destroyed. The very concepts of position and the conjugate momentum are mutually exclusive when quantities of the order of h are considered.

It is important to remember in this discussion of the methods of measuring positions and momenta that it is required that the results be such as can be used for initial conditions and for the prediction of another observation. If the position of a particle is precisely located by means of a gamma-ray microscope and a short time later the same particle happens to come into the field and be observed by another such microscope, the distance between the two positions divided by the time interval could be called the average velocity of the particle in the intervening space. It would not, however, have any great physical significance, since it would not permit a precise prediction of a future position. The reaction of the second observation would entirely destroy the significance of the first position.

4. The principle of indetermination

The generalization of the above results gives the principle of indetermination first enunciated by HEISENBERG [1927]. The principle has two aspects, which are only slightly different.

a. From an analysis of experimental procedures, in the light of known experimental results, one can conclude that an apparatus for the production of particles with a specified position and momentum will introduce such unknown interactions between the particle and the apparatus that any coordinate and its conjugate momentum will be indeterminate to the degree indicated by $\Delta p \, \Delta q \approx h$. In this form the principle rests upon an immediate experimental basis, and any satisfactory theory must be so constructed as to correspond to it.

b. On the other hand, as will be shown later, the general theory of quantum mechanics, which has been found by trial to be valid in a range exceeding that of classical mechanics, states that the interpretation of any theoretical results in terms of the classical concepts of coordinates and conjugate momenta must be expressed in statistical terms and subject to such indetermination that $\Delta p \, \Delta q \geq h/4\pi$.

These statements of the principle of indetermination enable one to formulate a statement of the range in which classical mechanics is valid. If the magnitudes of the quantities involved are such that h (6.62×10^{-27} erg sec) is negligible, the classical mechanics will give correct results. If, however, quantities of this order of magnitude must be considered, the classical results will in general be wrong and the quantum mechanics must be used. The quantity h is an "atom" of action. It is an indivisible unit of a quantity which, on the classical theory, should be capable of existing in any amount, however small. If this atom is so small that the fact of its finiteness is unimportant, the classical theory is valid. Otherwise the quantum mechanics must be used.

PROBLEM 10. Find the indetermination in the velocity of a particle when its position is determined within 0.001 cm. Find also the indetermination in the velocities of an electron and an alpha particle, respectively, when they are located between atoms 5×10^{-8} cm apart.

The above illustrations have referred to cartesian coordinates and the conjugate momenta. Other physical quantities show similar indeterminations. For quantities expressed as functions of the q's and the p's the indetermination will be shown to be proportional to the Poisson brackets. For other quantities, not known to classical mechanics, the quantum mechanical definition will involve some indetermination. Since we shall be dealing with a nonrelativistic formulation of quantum mechanics, the time is treated as an independent parameter. Nevertheless there will appear in the treatment an indetermination between energy and time such as would be expected if they were treated as canonically conjugate variables.

CHAPTER 3

FORMULATION OF WAVE MECHANICS

As the phenomena described in Chapter 1 were discovered various attempts were made to modify the current theories and views of atomic structure to accommodate them. It soon became evident that no minor changes would suffice, and in the middle of the 1920's there were formulated several superficially different mathematical systems which seemed suitable to describe the observations. Among the names associated with these developments, those of Heisenberg, Dirac and Schroedinger were especially prominent. It was soon shown that all of the theories were fundamentally equivalent and they will be treated here as aspects of one and the same quantum mechanics. For many purposes the form known as "wave mechanics" and associated with the name of E. Schroedinger is the most useful. This chapter and a few of the succeeding ones will be devoted to its formulation.

1. Momentum in wave mechanics

The discussion of the previous chapter has called attention to the difficulties with velocity as a mechanical concept in case the finiteness of h is important. Consequently it might be supposed that no use could be made of such laws of mechanics as the conservation of momentum, since in the classical theory momentum is merely a function of velocity. However, a basic accomplishment of quantum mechanics has been the establishment of a new concept of momentum, independent of the kinematic concept of velocity, and such that it is conserved. In cases in which classical mechanics is an adequate approximation the quantum-mechanical momentum reduces to the classical momentum, but in all cases an applicable conservation law can be formulated. The new concept is based on the experimental fact that a stream of "particles" is diffracted by a grating as is a train of waves. The momentum is then defined by the relationship

$$p = h/\lambda, \tag{3.1}$$

where λ is the wavelength as determined from the angle of diffraction. It is possible to produce "particles" of a specified wavelength, by the means used in optics for producing monochromatic light. In fig. 3.1 let S be the source, G the grating, and A the slit which defines the emergent beam. This slit must be wide enough not to affect appreciably the particles which pass through it, and the grating must be large enough and far enough away from both S and A to give the resolution necessary for the desired precision in λ.

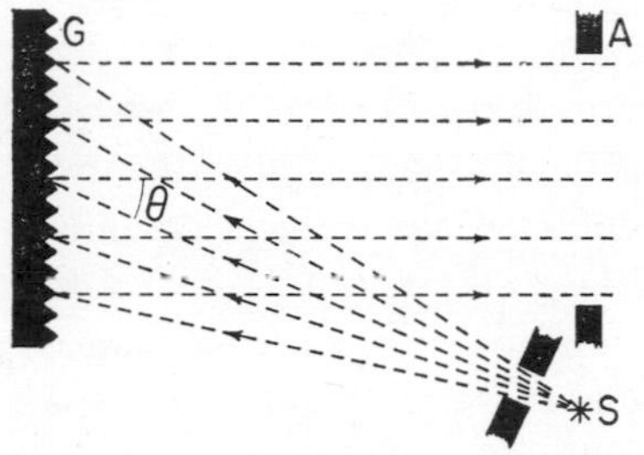

Fig. 3.1. A light beam scattered from a sufficiently large grating in a sufficiently well defined direction is a plane wave with a well defined wavelength. A "particle" passing through the same apparatus may be described by the same wavelength.

To get a precisely defined wavelength it is necessary to use an infinitely large system, just as to obtain a precise measurement of position with the gamma-ray microscope it is necessary to use light of infinitely short wavelength. These are examples of the fact that a precisely defined mechanical quantity is always an idealization. It is necessary to be able to approach the limit in principle, but it usually is not necessary to approach it very closely in practice. Although there must be no reason, inherent in the definitions of the quantities themselves, for not approaching the limit, there may be a large number of practical difficulties in the way. Before the discovery of the quantum phenomena there was no conceptual difficulty in approaching indefinitely close to the state of a particle in which both position and velocity were precisely defined; but the experimental study of the interactions involved in the necessary observations has shown this limit to be impossible. There remains the possibility, of course, that future discoveries will even more strictly circumscribe the idealizations that can be used.

In the limit in which the device in fig. 3.1 is suited for accurate specification of the momentum it becomes entirely unsuited for any specification of position, since all the dimensions of the apparatus become large. This is entirely in accord with the requirements of the principle of indetermination.

With this new concept of momentum as wavelength, the conservation law can be maintained in all interactions, and the prominent part played by the wave idea in quantum mechanics has led to the use of the term "wave mechanics" to describe it. Nevertheless, it will appear that the waves are more mathematical symbols than real waves in real space, and a too physical view of them leads to difficulties.

In the limit in which the apparatus in fig. 3.1 defines a wavelength exactly, the state of each "particle" diffracted from it can be represented by an infinite plane wave. The only properties of such a wave are its wavelength and its direction. It has no others. Similarly a "particle" diffracted by such a grating may be said to have these properties and no others, to have a momentum but no kinematical properties. The size and disposition of the apparatus are such as to obscure any information about the positions of such particles. If the size of the grating is reduced, or if other means are employed to define the particle's position, the resolving power of the system is reduced and the wavelength becomes less strictly defined. One such means is to place a shutter in front of the slit at A. If this shutter is opened for a short time, and if a particle goes through, it must have gone through while the shutter was open, so that something is known about its position. In such a case the infinite wave that represented the state of each particle must be thought of as cut off, or replaced by a finite wave train or a wave packet. This wave packet will not have a precise wavelength, nor will it have a precise position. It will be located in an extended region of space, and it can be resolved by Fourier analysis into a superposition of plane waves, each characterized by a wavelength. In such a case the momentum will not be precisely defined but will be spread over the values represented in the superposition of plane waves.

It can be shown qualitatively that such a wave packet satisfies the principle of indetermination. Let the length of the packet be Δx. This is clearly the indetermination of its position and the position of the particle whose state it represents. In order that the packet shall vanish at each end of this length on account of the interference of the superposed waves, there must be such a difference in wavelength that a phase difference of at least 2π will be set up in the distance Δx. Let the difference between the momenta of these two waves be Δp. Then

$$\frac{\Delta x}{\lambda} \approx \frac{\Delta x}{\lambda + \Delta\lambda} + 1, \tag{3.2}$$

and since $\lambda = h/p$ and $\lambda + \Delta\lambda = h/(p - \Delta p)$, it follows that $\Delta x\, \Delta p \approx h$. This analysis is very crude, both as regards the treatment of the interference and as regards the definitions of Δx and Δp, but it contains all of the essential elements, and a more precise treatment will be given later.

2. State of a system represented by a wave function

In classical mechanics the state of a system is described by a set of numbers that give the values of all the coordinates and the conjugate momenta as functions of the time. In quantum mechanics these quantities cannot all be sharply defined at the same time so that some other mode of representation is necessary. As long as attention is confined to nonrelativistic quantum mechanics, the time is treated classically and a special method of treatment is required only for the other coordinates. It is possible to speak of the state of the system at a definite time without considering the question of a possible indetermination in the time. Thus the nonrelativistic quantum mechanics is similar to the nonrelativistic classical mechanics in that the time is always the uniquely specified independent variable.

In the preceding discussion of experimental methods it has been shown possible, as an idealization, to produce particles in two kinds of states. In one the position of the particle is precisely fixed by means of a slit or by a gamma-ray microscope, and in the other the momentum (wavelength) is accurately fixed by means of a diffraction grating. In the first case nothing whatever is known about the momentum, and in the second nothing is known about the position. The existence of states between these extremes is suggested by the possibility of using a microscope with light of an intermediate wavelength, to make a rough position measurement on a particle with a prescribed momentum, without entirely obliterating the momentum. A satisfactory theory must include a method of representing such intermediate states as well as the extremes.

The general procedure is to represent the state of a system by means of a function $\psi(q_1,q_2,\ldots,q_n,t)$, in general a complex function, of the coordinates of the system. To represent the extreme cases, the suitable functions can be written down at once. For a system with only one coordinate, and in a state in which the coordinate has a precisely defined value q_0 at some given time t_0, $\psi(q,t_0) = \delta(q-q_0)$.[1] The only property of this function is that of being at

[1] The function $\delta(x)$ is defined by the properties $\delta(x) = 0$ when $x \neq 0$, and $\int_{-\infty}^{\infty} \delta(x)\,dx = 1$. Strictly speaking this is not a proper function, but it seems that its use in quantum mechanics, at least under an integral sign, does not lead to any error.

q_0, so it is admirably suited to represent a particle that has just been observed by means of a gamma-ray microscope to be at q_0. Nothing is said about the way the particle is moving, nor as to the way its position will change with the time. The only statement is that at t_0 the coordinate q has the value q_0. It is a prediction that the result of a measurement of q will be q_0, and it is equally well a description of the fact that the position has just been measured and found to be q_0. The two statements are equivalent, for two measurements of position following immediately one after the other are expected to give the same result.

For the other limiting case, in which the momentum conjugate to the single cartesian coordinate has the value p_0, the function is $\psi(q) = \exp\{(i/\hbar)p_0 q\}$.[1] Again this function is suitable for representing a particle that has been reflected at a definite angle from a grating. This function has a wavelength and nothing else. It certainly has no position, since it has the same absolute value for all values of q.

The restriction of the discussion to a system having only one coordinate makes it a little difficult to visualize just what is implied. Particularly in the case of a particle with a definite momentum because of being scattered from a grating at a suitable angle, it must be understood that the coordinate q is measured along the direction in which the particle is traveling away from the grating. An extension of these representations can be made to three dimensions without much trouble. The fact that a particle is at the point (x_0,y_0,z_0) at t_0 can be represented by

$$\psi(x, y, z, t_0) = \delta(x-x_0)\delta(y-y_0)\delta(z-z_0)$$

and the fact that it has momentum components p_x, p_y, p_z can be expressed by the function $\psi(x,y,z,t_0) = \exp\{i/\hbar)(p_x x+p_y y+p_z z)\}$.

Any function $\psi(q)$, with some unimportant exceptions, can be expressed as a superposition of elementary functions like those above. Thus

$$\psi(q) = \int_{-\infty}^{\infty} \psi(q_0)\delta(q-q_0)\, dq_0 = \int_{-\infty}^{\infty} \phi(p_0)\, e^{(i/\hbar)p_0 q}\, dp_0 . \qquad (3.3)$$

The first equality is one of the properties of the function $\delta(x)$; the second is a definition of $\phi(p_0)$. If the state of a particle is represented by this function, $\psi(q)$, it is not, in general, a state in which the coordinate q has a definite value. A knowledge of the fact that a particle is in the state represented by the function $\psi(q)$ does not imply a knowledge of a unique result to be ob-

[1] $\hbar = h/2\pi$. Since this combination is very frequently used, it is convenient to have a notation for it.

tained if the position of the particle is measured. It does, however, imply a method of producing or detecting a "particle" in a state with the properties described by $\psi(q)$. A measurement of the coordinate may be thought of as "finding" one of the functions $\delta(q-q_0)$ which represents a definite coordinate and of which $\psi(q)$ is a superposition. The situation is described by saying that the probability of finding the value q_0 for the coordinate, or of finding $\delta(q-q_0)$ as the function representing the system, is proportional to $|\psi(q_0)|^2$. After the coordinate has been measured and found to be q_0, the function $\psi(q)$ no longer represents the system. The procedure for "measuring" q must be a procedure for producing or detecting a particle at q_0. It is then properly represented only by $\delta(q-q_0)$.

Similarly the momentum of a particle in the state $\psi(q)$ is not sharply defined. If the particle were to be scattered from a diffraction grating, the result could be described by a Fourier analysis of the function $\psi(q)$. The second equality in equation (3.3) indicates this analysis. It shows that $\psi(q)$ can be made up of plane waves and that the "intensity" of each of them is proportional to $|\phi(p_0)|^2$. Thus the probability of "finding" the momentum p_0, when the particle is in the state $\psi(q)$, is $|\phi(p_0)|^2$.

The use of the term "probability" to describe the significance of a wave function is apparently necessary, but it is in one sense unfortunate. It tends to imply that the quantities concerned are merely inaccurately known. The more correct interpretation is that they are undefined or indeterminate. The significance of the probability statement is as follows: Suppose a particle is known to be in the state represented by $\psi(q)$. This corresponds to a physical situation in which the coordinate is a quite inappropriate quantity to use. The term "particle" is of limited value because no precise position can be attributed to it. If a measurement of the coordinate q is made by the gamma-ray microscope or by some other suitable means, the properties of the state of the system are not sufficient to predict exactly the result of the measurement. They are sufficient, however, to give the distribution of the results if the situation and the measurement are repeated many times.

According to classical theory one can say, "If a particle at the time t_0 has the coordinate q_0 and the momentum p_0, at the time t it will have the coordinate $q(t)$ and the momentum $p(t)$." This implies some experimental means by which a particle at the time t_0 can be "produced" with the coordinate q_0 and the momentum p_0. It implies that no unspecified forces act on the particle between t_0 and t, and it also implies that if at time t the position and momentum are measured, the results of the measurement will be $q(t)$ and $p(t)$. In classical mechanics the state of a particle is equivalent to its position and momentum. In quantum mechanics this equivalence does not hold. One can make only the statistical kind of statement about coordinates and momenta. In quantum mechanics one can say that if at $t = 0$ the system is in the state $\psi(q,0)$, at the time t it will be in the state $\psi(q,t)$ and measurements of most dynamical quantities can be only statistically predicted from this fact. However, the state is precisely defined and the physical significance of such a precise definition of the state will be emphasized later.

It is important to notice that the statistical statement cannot be tested by repeated measurements on the same particle, since after a measurement the state of the particle is no longer the same as before. There has taken place an interaction between the particle and the coordinate system, with reference to which the measurement was made, so that statements made concerning isolated systems will no longer apply. However, if it is possible to produce a large number of separate and independent particles in the state $\psi(q,t)$, and if the measurement of the coordinate q is carried out on each system, the distribution of the results of the measurement will be given by $|\psi(q)|^2$. The methods to be used in producing particles in a specified state require careful consideration, but will not be discussed here.

3. The first postulate of wave mechanics

The ideas just discussed may be incorporated into a first postulate of wave mechanics. The postulate gives the relationship between the mathematical representation of a state by means of a wave function and the physical determination of the state by means of experiments or measurement. It may be formulated as follows.

POSTULATE A. The state of a mechanical system is represented by the wave function $\psi(q_1,q_2,q_3,\ldots,q_n,t)$. This is a function of the independent cartesian coordinates of the "particles" composing the system and of the time t. It represents the state of the system in the sense defined by postulates A_1,A_2 and A_3.

POSTULATE A_1: $W(q_1,q_2,\ldots,q_n,t) = |\psi(q_1,q_2,\ldots,q_n,t)|^2$
where $W\,dq_1\ldots dq_n$ is the probability that a simultaneous measurement of the various coordinates will give results lying in the ranges q_1 to (q_1+dq_1), q_2 to (q_2+dq_2),...,q_n to (q_n+dq_n) at the time t.

The probability distribution of the momenta is also given by the function ψ through the following rule: Let

$$\phi(p_1, p_2, \ldots, p_n, t) = \\ = h^{-\frac{1}{2}n} \int \mathrm{d}q_1 \ldots \mathrm{d}q_n \, \mathrm{e}^{-(i/\hbar)(p_1 q_1 + \ldots + p_n q_n)} \, \psi(q_1, \ldots, q_n, t) \,. \tag{3.4}$$

The momentum probability distribution is given directly in terms of these functions.

POSTULATE A_2: $W(p_1,p_2,\ldots,p_n,t) = |\,\phi(p_1,\ldots,p_n,t)\,|^2$
where $W(p_1,p_2,\ldots,p_n,t)$ is defined as is $W(q_1,\ldots,q_n,t)$ above. It is important to note that although the knowledge of the state of the system, the knowledge of the function ψ, suffices to determine the probability distributions of the coordinates and momenta, the converse is not true. The wave function contains more than these probability distributions, and these distributions do not determine the function uniquely.

3.1. *Normalization of wave functions*

In order that the Postulates A_1 and A_2 shall define probabilities in the usual sense, it is necessary that the integral of $W(q,t)$ and of $W(p,t)$ over all values of the variables shall give a total probability of 1. Hence

$$\int |\,\psi(q, t)\,|^2 \,\mathrm{d}q = \int |\,\phi(p, t)\,|^2 \,\mathrm{d}p = 1 \tag{3.5}$$

is a requirement imposed on all functions used. In equation (3.5), q represents all of the cartesian coordinates, and p represents all of the conjugate momenta. This normalization can be accomplished by application of a constant multiplier in all cases in which the integrals converge; and this convergence of the integral of the square of the absolute value of the function must be required of a function if it is to be suitable for representing the state of a mechanical system. Unless otherwise indicated, it will be assumed that all functions used are normalized in this way.

As an example, consider the function $\psi(x) = N\exp(-\frac{1}{2}\alpha x^2)$. The constant N is called the normalization constant and must be given such a value that the square of the function is one. In this case

$$\int |\,\psi(x)\,|^2 \,\mathrm{d}x = N^2 \int_{-\infty}^{\infty} \mathrm{e}^{-\alpha x^2} \,\mathrm{d}x = N^2 \left(\frac{\pi}{\alpha}\right)^{\frac{1}{2}} = 1 \,. \tag{3.6}$$

From the integral it follows that

$$\psi(x) = \left(\frac{\alpha}{\pi}\right)^{\frac{1}{4}} \mathrm{e}^{-\frac{1}{2}\alpha x^2} \tag{3.7}$$

is a properly normalized function.

It will be noticed that the function $\exp\{(\mathrm{i}/\hbar)p_0 x\}$, which represents a state of precisely defined momentum, does not give a convergent integral and cannot be normalized in the usual way. Neither can $\delta(x)$. These may, however, be regarded as limiting cases of functions that can be normalized. For example

$$\mathrm{e}^{(\mathrm{i}/\hbar)p_0 x} = \lim_{\alpha\to 0} \mathrm{e}^{(\mathrm{i}/\hbar)p_0 x}\, \mathrm{e}^{-\frac{1}{2}\alpha x^2}. \tag{3.8}$$

3.2. *Functions of the momenta*

The function ϕ is as suitable as ψ for representing the state of the system, since not only is ϕ uniquely determined by ψ, but ψ can also be determined from ϕ. The Fourier inversion of (3.4) is

$$\psi(q_1, q_2, \ldots, q_n,) =$$
$$= h^{-\frac{1}{2}n} \int \mathrm{d}p_1 \ldots \mathrm{d}p_n\, \mathrm{e}^{(\mathrm{i}/\hbar)(p_1 q_1 + p_2 q_2 + \ldots + p_n q_n)}\, \phi(p_1, p_2, \ldots, p_n, t) \tag{3.9}$$

so that if the function ϕ were given it could be transformed to ψ. For many applications only the functions of the coordinates are necessary.

The state represented by the function in equation (3.7) is also represented by a particularly simple function of the momentum,

$$\phi(p) = h^{-\frac{1}{2}} \left(\frac{\alpha}{\pi}\right)^{\frac{1}{4}} \int_{-\infty}^{\infty} \mathrm{e}^{-(\mathrm{i}/\hbar)px}\, \mathrm{e}^{-\frac{1}{2}\alpha x^2}\, \mathrm{d}x$$

$$= h^{-\frac{1}{2}} \left(\frac{\alpha}{\pi}\right)^{\frac{1}{4}} \mathrm{e}^{-p^2/2\alpha\hbar^2} \int_{-\infty}^{\infty} \mathrm{e}^{-\frac{1}{2}\alpha[x+(\mathrm{i}p/\alpha\hbar)]^2}\, \mathrm{d}x$$

$$= \left(\frac{2}{h}\right)^{\frac{1}{2}} \left(\frac{\pi}{\alpha}\right)^{\frac{1}{4}} \mathrm{e}^{-p^2/2\alpha\hbar^2}. \tag{3.10}$$

The function of the momentum has, in this special case, the same form as the function of the coordinates, and the spreads of the two functions are reciprocally related to each other. If α is large, the coordinate is well determined but the momentum has a very wide spread. If α is small, the situation is reversed. It will be noticed that the function $\phi(p)$ in equation (3.10) is already normalized. The coefficient $h^{-\frac{1}{2}n}$ in equation (3.9) is inserted to provide that if $\psi(x)$ is normalized, $\phi(p)$ will also be normalized.

The fact that a state can be represented by different kinds of functions is evidence of the small amount of "physical reality" to be attributed to any of them. These "wave" functions are mathematical symbols, in terms of which it is possible to represent the state of the system and to describe correctly the observed facts. The waves are not, however, simple physical things as are sound waves. One evidence of this is the fact that they are functions of all of the coordinates of the system, rather than of just three coordinates of position in real space.

3.3. *Expectation values*

Instead of dealing with the probability distributions as given in Postulates A_1 and A_2 it is sometimes more convenient to deal with average values or, as they are called in statistical theory, expectation values. These are averages over all possible values, weighted by the probability of occurrence as given by the probability distribution. A knowledge of the expectation values of all the powers of a quantity is equivalent to a knowledge of the distribution function. The expectation value of any quantity whose statistical distribution is known can be obtained by multiplying each value by its probability and summing over all the values. In case a quantity, x, is distributed according to a simple error function, $P = A\exp(-\beta x^2)$, the expectation value of x is equal to

$$A \int_{-\infty}^{\infty} x \exp(-\beta x^2) \mathrm{d}x .$$

In this case $\langle x \rangle$[1] is zero since the probability of a given positive value is just equal to the probability of the corresponding negative value. In a symmetrical distribution such as this one, the expectation value of any odd power of the variable vanishes, and the whole distribution is characterized by the averages of the even powers.

The expectation values in a quantum-mechanical state can be determined either from the function ψ or from the function ϕ. From Postulate A_1 it follows at once that the expectation value of any power of a coordinate is given by

$$\langle q_j^s(t) \rangle = \int_{-\infty}^{\infty} \psi^*(q_1, q_2, \ldots, q_n, t) q_j^s \psi(q_1, q_2, \ldots, q_n, t) \, \mathrm{d}q_1 \ldots \mathrm{d}q_n \tag{3.11}$$

---

[1] The enclosure of a symbol such as x in brackets $\langle\ \rangle$ will be used to indicate a mean or expection value.

and from Postulate A_2 it follows at once that

$$\langle p_j^s(t)\rangle = \int_{-\infty}^{\infty} \phi^*(p_1, \ldots, p_n, t) p_j^s \phi(p_1, \ldots, p_n, t)\, \mathrm{d}p_1 \ldots \mathrm{d}p_n. \tag{3.11a}$$

These expectation values are functions of the time, as is indicated. They are the averages at a definite time, when the system is in the very definite state indicated by the wave functions. These expectation values are averages over probabilities in the sense described above and are not time averages over the motion.

If the expression (3.4) is used to express ϕ and ϕ^* in (3.11a), there results

$$\langle p_j^s\rangle = h^{-n} \int \mathrm{d}p \int \mathrm{d}q' \int \mathrm{d}q\, \mathrm{e}^{(\mathrm{i}/\hbar)p \cdot q'}\, \psi^*(q') p_j^s\, \mathrm{e}^{(-\mathrm{i}/\hbar)p \cdot q}\, \psi(q)$$

where $\mathrm{d}p$, $\mathrm{d}q'$ and $\mathrm{d}q$ are written instead of the products of all the differentials involved, and $(p.q)$ is written instead of $(p_1 q_1 + p_2 q_2 + p_3 q_3 + \ldots . p_n q_n)$. Since

$$p_j^s\, \mathrm{e}^{-(\mathrm{i}/\hbar)p \cdot q} = \left(-\frac{\hbar}{\mathrm{i}}\right)^s \frac{\partial^s}{\partial q_j^s}\, \mathrm{e}^{-(\mathrm{i}/\hbar)p \cdot q},$$

an integration by parts leads to

$$\langle p_j^s(t)\rangle = h^{-n} \left(\frac{\hbar}{\mathrm{i}}\right)^s \int \mathrm{d}p \int \mathrm{d}q' \int \mathrm{d}q\, \mathrm{e}^{(\mathrm{i}/\hbar)(p \cdot q' - p \cdot q)}\, \psi^*(q') \frac{\partial^s}{\partial q_j^s} \psi(q).$$

This integration by parts assumes that the wave functions and their derivatives vanish for the infinite values of their arguments and that the derivatives of ψ are defined at all points. Such conditions are usually fulfilled. Integrating next with respect to p_j from $p_j = -A$ to $p_j = A$ leads to expressions of the form

$$h^{-1} \int \psi^*(q')\, \mathrm{d}q_j' \int_{-A}^{A} \mathrm{d}p_j\, \mathrm{e}^{(\mathrm{i}/\hbar)p_j(q_j' - q_j)} =$$

$$= \frac{2A}{h} \int \psi^*(q')\, \mathrm{d}q_j' \frac{\sin [A(q_j' - q_j)/\hbar]}{A(q_j' - q_j)/\hbar}.$$

In the limit as $A \to \infty$ this integrand differs from zero appreciably only in the neighborhood of $q' = q$, so that $\psi^*(q')$ under the integral sign may be replaced by $\psi^*(q)$ and the integration over q' carried out without reference to this function. The integral is equal to 1 for each value of j; consequently

$$\langle p_j^s(t)\rangle = \left(\frac{\hbar}{\mathrm{i}}\right)^s \int \psi^*(q) \frac{\partial^s \psi(q)}{\partial q_j^s}\, \mathrm{d}q. \tag{3.12}$$

Similarly it can be shown that

$$\langle q_j^s(t)\rangle = \left(\frac{\hbar}{\mathrm{i}}\right)^s \int \phi^*(p)\,\frac{\partial^s\phi(p)}{\partial p_j{}^s}\,\mathrm{d}p. \tag{3.13}$$

The determination of average values from the wave function can be illustrated with the function of equation (3.7). The average value of x is clearly zero, and the average value of x^2 is given by

$$\langle x^2\rangle = \left(\frac{\alpha}{\pi}\right)^{\frac{1}{2}} \int_{-\infty}^{\infty} x^2\,\mathrm{e}^{-\alpha x^2}\,\mathrm{d}x = \frac{1}{2\alpha}. \tag{3.14}$$

The momentum averages follow directly from equation (3.10). $\langle p\rangle = 0$, and

$$\langle p^2\rangle = \frac{2}{h}\left(\frac{\pi}{\alpha}\right)^{\frac{1}{2}} \int_{-\infty}^{\infty} p^2\,\mathrm{e}^{-p^2/\alpha\hbar^2}\,\mathrm{d}p = \tfrac{1}{2}\alpha\hbar^2. \tag{3.15}$$

However, the momentum averages can also be obtained from the coordinate function $\psi(x)$ by the use of equation (3.12),

$$\langle p\rangle = \frac{\hbar}{\mathrm{i}}\left(\frac{\alpha}{\pi}\right)^{\frac{1}{2}} \int_{-\infty}^{\infty} \mathrm{e}^{-\frac{1}{2}\alpha x^2}\,\frac{\partial}{\partial x}(\mathrm{e}^{-\frac{1}{2}\alpha x^2})\,\mathrm{d}x = 0$$

and

$$\langle p^2\rangle = -\hbar^2\left(\frac{\alpha}{\pi}\right)^{\frac{1}{2}} \int_{-\infty}^{\infty} \mathrm{e}^{-\frac{1}{2}\alpha x^2}\,\frac{\partial^2}{\partial x^2}(\mathrm{e}^{-\frac{1}{2}\alpha x^2})\,\mathrm{d}x = \tfrac{1}{2}\alpha\hbar^2. \tag{3.15a}$$

The above expressions for the expectation values of the momentum are also subject to the requirement that the functions of which the derivatives are taken are such that the derivatives themselves can be expressed as Fourier integrals. The function of equation (3.7) satisfies this condition since all its derivatives are continuous. However, there are many functions widely used in wave mechanics whose higher derivatives are derivatives of discontinuous functions. In such cases, the momentum distribution must be obtained from the Fourier transform.

PROBLEM 1. Find the distribution of the coordinate x and the conjugate momentum p_x represented by the function

$$\psi(x) = \left(\frac{2\alpha}{\pi}\right)^{\frac{1}{4}} \mathrm{e}^{\mathrm{i}\omega x-\alpha(x-x_0)^2}.$$

PROBLEM 2. Find the distribution of the coordinate x and the conjugate momentum p_x represented by the function

$$\psi(x) = N\,\mathrm{e}^{\mathrm{i}\omega x-\alpha x^2}(a_0+a_1x).$$

PROBLEM 3. Find the distribution of coordinate and momentum represented by the function

$$\psi(x) = \left(\frac{\mathrm{i}}{L}\right)^{\frac{1}{2}} \mathrm{e}^{\mathrm{i}\omega x} \qquad 0 < x < L$$

$$\psi(x) = 0 \qquad x < 0,\ x > L.$$

Note that this function and its derivative are discontinuous at $x = 0$ and $x = L$ so that the usual integrals cannot be used to evaluate averages. Show that while the mean value of p can be determined from either $\psi(x)$ or $\phi(p)$, the mean value of p^2 and higher powers must be determined from $\phi(p)$.

PROBLEM 4. Find the expectation value of the momentum and the square of the momentum for a state represented by

$$\psi(x) = N\left[\sin x + \tfrac{1}{2}\mathrm{i} \sin 3x\right] \qquad 0 < x < 2\pi$$

$$\psi(x) = 0 \qquad x < 0,\ x > 2\pi.$$

3.4. *Principle of indetermination*

The method of representing the state of a system by means of a wave function guarantees the satisfaction of the principle of indetermination because the distribution of position and the distribution of momentum are related by equation (3.4), ROBERTSON [1930], SCHROEDINGER [1930].

Consider for simplicity a case of one dimension. Let the wave function be $\psi(x)$. The indetermination in x must be arbitrarily defined and the customary rms deviation may be used. Thus

$$(\Delta x)^2 = \langle (x - \langle x \rangle)^2 \rangle = \langle x^2 \rangle - \langle x \rangle^2 . \qquad (3.16)$$

According to this definition, only the mean values of x and of x^2 are necessary to evaluate the indetermination. It is not necessary to know the distribution in any more detail. The indetermination in the momentum can be similarly defined. Let

$$(x - \langle x \rangle)\psi(x) = f(x),$$

and let

$$(\hbar/\mathrm{i})\, \mathrm{d}\psi/\mathrm{d}x - \langle p \rangle \psi = g(x).$$

By the Schwartz inequality,

$$\left(\int f^* f\, \mathrm{d}x\right)\left(\int g^* g\, \mathrm{d}x\right) \geq \left|\int f^* g\, \mathrm{d}x\right|^2$$

so that

$$
\begin{aligned}
(\Delta x)^2(\Delta p)^2 &= \left[\int \psi^*(x-\langle x\rangle)^2\psi\,\mathrm{d}x\right] \\
&\quad\times\left[\int\left(-\frac{\hbar}{\mathrm{i}}\frac{\mathrm{d}\psi^*}{\mathrm{d}x}-\langle p\rangle\psi^*\right)\left(\frac{\hbar}{\mathrm{i}}\frac{\mathrm{d}\psi}{\mathrm{d}x}-\langle p\rangle\psi\right)\mathrm{d}x\right] \\
&\geq \left|\int \psi^*(x-\langle x\rangle)\left[\frac{\hbar}{\mathrm{i}}\frac{\mathrm{d}\psi}{\mathrm{d}x}-\langle p\rangle\psi\right]\mathrm{d}x\right|^2 \\
&= \left|\frac{\hbar}{\mathrm{i}}\int \psi^* x\frac{\mathrm{d}\psi}{\mathrm{d}x}\,\mathrm{d}x-\langle px\rangle\right|^2 \\
&= \left|\frac{1}{2}\frac{\hbar}{\mathrm{i}}\left[\int \psi^* x\frac{\mathrm{d}\psi}{\mathrm{d}x}\,\mathrm{d}x+\int\frac{\mathrm{d}\psi^*}{\mathrm{d}x}x\psi\,\mathrm{d}x\right]\right. \\
&\quad\left.+\frac{1}{2}\frac{\hbar}{\mathrm{i}}\left[\int \psi^* x\frac{\mathrm{d}\psi}{\mathrm{d}x}\,\mathrm{d}x-\int\frac{\mathrm{d}\psi^*}{\mathrm{d}x}x\psi\,\mathrm{d}x\right]-\langle px\rangle\right|^2.
\end{aligned}
$$

The first bracket in the last part of this expression is real, since it is the sum of a quantity and its complex conjugate. Then, because of the imaginary factor in front of it, the whole term is imaginary. The second and third quantities are real. The square of the absolute value is equal to or greater than the square of the absolute value of the imaginary term. If the wave functions vanish in the usual way at $+\infty$ and $-\infty$ and are normalized, the first bracket can be shown by partial integration to be equal to -1. Hence for any function that is properly normalized,

$$(\Delta x)^2\,(\Delta p)^2 \geq \tfrac{1}{4}\hbar^2. \tag{3.17}$$

The exact value of the factor on the right-hand side of (3.17) depends on the arbitrary definition of Δx and Δp that has been used. The essential part is the dependence on h, which shows that the theory satisfies in this respect the requirements described in the previous chapter.

It is also of importance to notice the $>$ sign in equation (3.17). The proof given above does not show that the equality can ever be attained. However, the state represented in equation (3.7) does give this minimum value of the indetermination. From equation (3.15)

$$(\Delta p)^2 = \langle p^2\rangle-\langle p\rangle^2 = \tfrac{1}{2}\alpha\hbar^2$$

and from equation (3.14)

$$(\Delta x)^2 = \langle x^2\rangle-\langle x\rangle^2 = \frac{1}{2\alpha}.$$

The product of these quantities then has the minimum value of $\frac{1}{4}\hbar^2$. Thus this state is one in which the coordinate and momentum are most accurately fixed. By suitable selection of α either x or p can be given any desired amount of precision, and the precision in the conjugate quantity will be the maximum possible.

4. Dynamical quantities other than coordinates and momenta

The previous discussion has covered the statistical distributions of the coordinates and the corresponding momenta, but these are not the only dynamical quantities in which one is interested. In classical mechanics only the coordinates and momenta need be given, since all other quantities are functions of them. In quantum mechanics, however, it is not in general possible to find the probability distribution of a function of the coordinates and momenta from the distributions of the coordinates and momenta themselves. This is because the coordinates and momenta are not statistically independent.

To deal with these other dynamical quantities the first postulate must be extended by introducing operators to represent them. Such an operator operates on a wave function and serves to permit determination of the expectation values of the various powers of the dynamical quantities. Operators will be indicated by the sans-serif type face. If Q is the operator that represents the dynamical quantity Q, then

$$\langle Q^s \rangle = \int \psi^*(q)\mathsf{Q}^s\psi(q)\,\mathrm{d}q\,. \tag{3.18}$$

This includes the cases already discussed. The operator that represents the coordinate x is "multiplication by x" so that

$$\langle x^2 \rangle = \int \psi^*(x)x^2\psi(x)\,\mathrm{d}x\,. \tag{3.18a}$$

The operator that represents the momentum conjugate to x is "$\hbar/i$ times differentation with respect to x", so that

$$\mathsf{P}_x = \frac{\hbar}{\mathrm{i}}\frac{\partial}{\partial x} \quad \text{and} \quad \langle p_x^s \rangle = \left(\frac{\hbar}{\mathrm{i}}\right)^s \int \psi^*(x)\,\frac{\partial^s\psi(x)}{\partial x^s}\,\mathrm{d}x\,. \tag{3.18b}$$

The third part of the first postulate is a prescription for the formation of the operator for any function of the coordinates and momenta.

POSTULATE A_3:

$$\mathsf{Q}(q, p) = \mathsf{Q}\left(q_i, \frac{\hbar}{\mathrm{i}} \frac{\partial}{\partial q_i}\right).$$

The rule is to form a differential operator by replacing each momentum p in the function by $(\hbar/\mathrm{i})\ \partial/\partial q_i$ while each coordinate is unchanged. This rule applies in cartesian coordinates only. The operator in any other system of coordinates can be obtained by transformation in the usual way after the operator has been formed in cartesian coordinates.

To distinguish operators from the dynamical quantities they represent, a special type face will be used.

The above rule for forming operators applies when the wave functions are expressed in terms of the coordinates. When the functions ϕ of the momenta are used, the operators must be expressed in terms of the independent variables, $\mathsf{Q}(q, p) = \mathsf{Q}(-(\hbar/\mathrm{i})\ \partial/\partial p, p)$.

The dynamical quantities of classical mechanics can always be expressed as functions of the coordinates and the momenta so that the corresponding operators can always be written down. Usually this gives a unique result, but occasionally the operator is ambiguous because the order of x_i and $\partial/\partial x_i$ is important. The operators for x_i and p_i do not commute, although classically $p_i x_i$ and $x_i p_i$ are identical. The dynamical quantity $x_i^2 p_i$ might be $x_i^2 p_i$, $x_i p_i x_i$, or $p_i x_i^2$. The corresponding operators would be different. They would be

$$\frac{\hbar}{\mathrm{i}} x_i^2 \frac{\partial}{\partial x_i};\quad \frac{\hbar}{\mathrm{i}} x_i \frac{\partial}{\partial x_i} x_i \equiv \frac{\hbar}{\mathrm{i}}\left(x_i + x_i^2 \frac{\partial}{\partial x_i}\right);\quad \frac{\hbar}{\mathrm{i}} \frac{\partial}{\partial x_i} x_i^2 \equiv \frac{\hbar}{\mathrm{i}}\left(2x_i + x_i^2 \frac{\partial}{\partial x_i}\right).$$

In such a case other considerations must be invoked to determine the proper operator to use. When two or more operator functions correspond to the same classical quantity, and also satisfy other conditions that will be discussed later, it may be considered that there are really several quantities concerned but that they coincide in the limit in which the classical theory is applicable. For quantities other than those of classical mechanics, the suitable operators must be determined in each case to fit the observed facts.

As an illustration of the use of Postulate A_3 consider the angular momentum of a particle around the origin of coordinates. The classical expression for the z component is $xp_y - yp_x$, and the corresponding operator is

$$\mathscr{L}_z = \frac{\hbar}{\mathrm{i}}\left(x \frac{\partial}{\partial y} - y \frac{\partial}{\partial x}\right). \tag{3.19}$$

This can be transformed to spherical polar coordinates, where it has the simple form

$$\mathscr{L}_z = \frac{\hbar}{\mathrm{i}} \frac{\partial}{\partial \varphi}. \tag{3.20}$$

The latter form is suitable for operating on functions expressed in polar coordinates and is an illustration of the general rule according to which the differential operator is first set up in cartesian coordinates and then transformed to any others for which it is desired to be used.

5. The equation of motion

The first postulate provides a method of representing the state of a system, and the method is such that the principle of indetermination follows immediately. Many of the characteristic features of quantum mechanics are involved in the first postulate and its consequences, but it is also necessary to know the way in which the state of the system, or the function which represents it, will change with the time. The equation for this change, or the equation of motion, constitutes the second postulate.

POSTULATE B. The wave function $\psi(q_1, q_2, \ldots, q_n, t)$ changes with the time in accordance with the Schroedinger equation

$$-\frac{\hbar}{\mathrm{i}} \frac{\partial \psi}{\partial t} = \mathsf{H}\psi \tag{3.21}$$

where H is the operator corresponding to the Hamiltonian function of the system. A similar equation holds when the function ϕ of the momenta is used, but we shall be concerned principally with the function ψ of the coordinates.

5.1. *Necessary initial conditions*

Because equation (3.21) is of the first order with respect to the time, only the value of ψ at $t = t_0$ is necessary to determine its value at all other times. The initial rate of change of ψ need not be specified. The rate of change is given uniquely in terms of ψ itself by equation (3.21).

This is a wholly satisfactory situation, since the function represents the state of the system in its entirety. In it are included all the things pertinent to the description of the state and hence all things pertinent to its change with the time. The function ψ represents not only the coordinates but the momenta as well, and hence in classical terms the rate of change of the coordinates. This may be considered as the reason for using complex functions, since a complex function consists of two real functions.

5.2. *Form of the Hamiltonian operator*

For simple systems whose Hamiltonian functions depend on the coordinates and momenta only, the Hamiltonian operators can be written down immediately from the classical functions. It is necessary only to combine the elementary operators in the manner already discussed. The only possible ambiguity will occur when the order of factors is not uniquely determined in the classical case. The operator for a coordinate does not commute with the operator for the conjugate momentum, and hence the specification of this order is important for fixing the Hamiltonian operator. In many cases such difficulty does not arise, but when it does, some additional considerations must be used to determine the correct operator.

For the case of a single particle moving in a given potential field, the Hamiltonian function is

$$H = \frac{1}{2m}(p_x^2 + p_y^2 + p_z^2) + V(x, y, z), \tag{3.22}$$

and the corresponding operator is

$$\mathsf{H} = -\frac{\hbar^2}{2m}\nabla^2 + V(x, y, z). \tag{3.22a}$$

In order that the theory be consistent it is necessary that the integrated probability of position,

$$\int W(q_1, q_2, \cdots, q_n, t)\, \mathrm{d}q$$

be a constant, independent of the time. This requirement puts a limitation on the kinds of operators which can be used for H. From the expression for W,

$$\frac{\mathrm{d}W}{\mathrm{d}t} = \frac{\partial\psi^*}{\partial t}\psi + \psi^*\frac{\partial\psi}{\partial t} = \frac{\mathrm{i}}{\hbar}[(\mathsf{H}^*\psi^*)\psi - \psi^*\mathsf{H}\psi]. \tag{3.23}$$

The equation for $\partial\psi^*/\partial t$ is obtained by taking the complex conjugate of equation (3.21). This involves taking the complex conjugate of the function ψ, changing the sign of the time derivative, and changing the sign of every i that may occur in the operator H. If the integral of equation (3.23) over all values of the coordinates is zero, the integrated probability is constant. The vanishing of this integral for an operator is one property of Hermitian operators. An operator O is Hermitian if

$$\int [(\mathsf{O}^*\psi_a^*)\psi_b - \psi_a^*\mathsf{O}\psi_b]\, dq = 0 \tag{3.23a}$$

for an arbitrary pair of functions ψ_a and ψ_b. The integral is over the whole range of the variables q. The vanishing of the integral of (3·23) is a special case of (3.23a) in which $\psi_a = \psi_b$ and the operator is H. In trying to find Hamiltonian operators to represent systems whose properties are different from those known in the classical theory or in trying to resolve the ambiguity mentioned above, only Hermitian operators need be tried.

In investigating an operator to see whether or not it is Hermitian, certain conditions must be imposed on the functions to which it is applied. The conditions satisfied by the functions in quantum mechanics are suitable vanishing at infinity. It is necessary that the function vanish sufficiently so that the integral $\int\psi_a^* q^n \psi_b\, dq$ converges for any value of n, if the system is confined to a finite region of space. Under these conditions it is easy to show that $(\hbar/\mathrm{i})\, \partial/\partial x$ is Hermitian:

$$-\frac{\hbar}{\mathrm{i}}\int_{-\infty}^{\infty} \frac{\partial\psi_a^*}{\partial x}\psi_b\, dx = -\frac{\hbar}{\mathrm{i}}\psi_a^*\psi_b\Big|_{-\infty}^{\infty} + \frac{\hbar}{\mathrm{i}}\int_{-\infty}^{\infty} \psi_a^* \frac{\partial\psi_b}{\partial x}\, dx\,.$$

Since $\psi_a^*\psi_b$ vanishes at both limits, the condition is satisfied.

Similarly it can be shown by means of partial integrations that the operator of equation (3.22) is Hermitian. The two changes in sign associated with the two partial integrations leave the sign of the second derivative unchanged in the final result.

5.3. *Computation of time derivatives and the conservation laws*

Let Q be any dynamical quantity. In general it will be a function of the coordinates, the momenta, and the time. The average value of Q at a given time t will be

$$\langle Q(t)\rangle = \int \psi^*(q, t)\mathsf{Q}\psi(q, t)\, dq\,. \tag{3.24}$$

This average is a function of the time, and its rate of change with the time will consist of two parts. The first part is the rate of change of Q itself due to its explicit dependence on the time. For many functions this will be zero, since many of the quantities in which one is interested do not depend explicitly on the time but only on the coordinates and momenta. The other part of the rate of change is due to the change with the time of the functions ψ and ψ^*, while the functional dependence of Q on the variables q does not change. Then

$$\begin{aligned}\frac{\mathrm{d}\langle Q\rangle}{\mathrm{d}t} &= \int \frac{\partial \psi^*}{\partial t} \mathsf{Q}\psi \, \mathrm{d}q + \int \psi^* \frac{\partial \mathsf{Q}}{\partial t} \psi \, \mathrm{d}q + \int \psi^* \mathsf{Q} \frac{\partial \psi}{\partial t} \mathrm{d}q \\ &= \left\langle \frac{\partial Q}{\partial t} \right\rangle + \frac{\mathrm{i}}{\hbar} \int [(\mathsf{H}^*\psi^*)\mathsf{Q}\psi - \psi^*\mathsf{QH}\psi] \, \mathrm{d}q \\ &= \left\langle \frac{\partial Q}{\partial t} \right\rangle + \frac{\mathrm{i}}{\hbar} \int \psi^*(\mathsf{HQ}-\mathsf{QH})\psi \, \mathrm{d}q . \end{aligned} \tag{3.25}$$

The last transformation is possible because H is Hermitian. If by $(\mathrm{d}\mathsf{Q}/\mathrm{d}t)$ is understood that operator whose average value gives the rate of change of the average value of Q, then

$$\frac{\mathrm{d}\mathsf{Q}}{\mathrm{d}t} = \frac{\partial \mathsf{Q}}{\partial t} + \frac{\mathrm{i}}{\hbar}(\mathsf{HQ}-\mathsf{QH}). \tag{3.26}$$

The expression $(\mathsf{HQ}-\mathsf{QH})$ is the commutator of Q and H. Equation (3.26) is the same as equation (2.9) of the previous chapter with the commutator substituted for $(\hbar/\mathrm{i})$ times the Poisson bracket of the corresponding quantities. It is quite a general result that the commutator of two operators in the quantum mechanics has the same significance as $(\hbar/\mathrm{i})$ times the Poisson bracket of the corresponding quantities in the classical theory.

It follows immediately from equation (3.26) that the average value of the Hamiltonian function is a constant if it does not contain the time explicitly.

The conservation laws of quantum mechanics can be expressed in terms of operators. If a quantity Q has an operator Q that commutes with H, and if Q does not depend explicitly on the time, the expectation value of each power of Q will be a constant. This follows because

$$\mathsf{HQ}^s - \mathsf{Q}^s\mathsf{H} = 0 \quad \text{if} \quad \mathsf{HQ}-\mathsf{QH} = 0. \tag{3.27}$$

The conservation of Q, or the constancy of Q, refers in quantum mechanics to the fact that the probability distribution of Q is a constant, regardless of whether Q has a precisely defined value or not.

In the case of a free particle the Hamiltonian operator is merely the kinetic-energy operator, $-(\hbar^2/2m)\nabla^2$. The operator for momentum in the x direction is $(\hbar/\mathrm{i})\,\partial/\partial x$. This clearly commutes with the Hamiltonian operator since the order of differentiation is nonessential. It follows that the linear momentum of a free particle is conserved in quantum mechanics as well as in classical mechanics. In classical mechanics the momentum of a free particle has a precise value that does not change with the time. In quantum mechanics a free particle has a probability distribution of momentum that does not change with the time.

PROBLEM 5. Find the average and the average squares of the coordinates (x, y, z) of a particle in a state represented by

$$\psi = Nr\,\mathrm{e}^{-\alpha r}\sin\vartheta\,\mathrm{e}^{\mathrm{i}\varphi}$$

where r, ϑ, φ are spherical polar coordinates.

PROBLEM 6. Write the operator for the energy of a free particle in cartesian coordinates and in spherical polar coordinates.

PROBLEM 7. Find the average and the average square of the angular momentum about the polar axis for the state represented by

$$\psi(r, \vartheta, \varphi) = f(r)\,\mathrm{e}^{in\varphi}$$

where n is an integer.

PROBLEM 8. Show that the total momentum of a system of particles that act on each other according to Newton's third law of motion is constant.

PROBLEM 9. Show that the total angular momentum of a system of particles acting on each other with central forces is conserved in wave mechanics.

5.4. *Hamiltonian operator in the presence of an electromagnetic field*

The force on a particle with a charge e in an electromagnetic field is given by

$$\boldsymbol{F} = e\boldsymbol{E} + e\boldsymbol{v}\times\boldsymbol{B}. \tag{3.28}$$

This follows from equation (1.1V) by integrating the force over the small volume occupied by the charged particle, with the assumption that the fields are constant in this small region. The next problem is to find the suitable Lagrangian and Hamiltonian functions for the motion of such a particle. Those functions are suitable which give the proper equations of motion. The usual Lagrangian is

$$L = T - V + e\boldsymbol{A}\cdot\boldsymbol{v} = \tfrac{1}{2}m(\dot{x}^2+\dot{y}^2+\dot{z}^2) + e\boldsymbol{A}\cdot\boldsymbol{v} - e\Phi - V' \tag{3.29}$$

where $\boldsymbol{A}$ and Φ are the values of the vector and scalar potentials at the position of the particle. V' includes whatever potential energy is not included in the electric potential Φ. The function in equation (3.29) is for a single particle in a known field of force. Additional particles would require additional terms of the same kind, as well as terms representing the interactions between the particles.

In Chapter 1 the electromagnetic field was described by means of a vector potential only. Such a description was possible in that case because only a radiation field was present. In the more general case it is necessary to use a scalar potential also. If $\boldsymbol{B} = \operatorname{curl} \boldsymbol{A}$, equation (1.1)

$$\operatorname{curl}\left(\boldsymbol{E} + \frac{\partial \boldsymbol{A}}{\partial t}\right) = 0.$$

Since the curl of this expression is zero, it can be written as the negative gradient of a scalar potential Φ. The fields are then given in terms of these potentials by

$$\boldsymbol{E} = -\operatorname{grad} \Phi - \frac{\partial \boldsymbol{A}}{\partial t} \quad \text{and} \quad \boldsymbol{B} = \operatorname{curl} \boldsymbol{A}. \tag{3.30}$$

Such relationships are sufficient to give the fields $\boldsymbol{E}$ and $\boldsymbol{B}$ uniquely in terms of the potentials $\boldsymbol{A}$ and Φ, but the converse is not true. The potentials are not uniquely determined when the fields are given. It is therefore possible to impose another condition on the potentials, and for convenience this is taken to be

$$\operatorname{div} \boldsymbol{A} = -\mu_0 \varepsilon_0 \frac{\partial \Phi}{\partial t}. \tag{3.31}$$

The potentials in equation (3.29) are usually subject to this condition, although the validity of the equation does not depend on it.

To show that equation (3.29) is the correct Lagrangian function, it is necessary to show that the correct equations of motion are obtained from it. For simplicity consider the x coordinate only:

$$\frac{\partial L}{\partial \dot{x}} = m\dot{x} + eA_x$$

$$\frac{\partial L}{\partial x} = e\left(\frac{\partial A_x}{\partial x}\dot{x} + \frac{\partial A_y}{\partial x}\dot{y} + \frac{\partial A_z}{\partial x}\dot{z}\right) - e\frac{\partial \Phi}{\partial x} - \frac{\partial V'}{\partial x}$$

$$\frac{\mathrm{d}}{\mathrm{d}t}\frac{\partial L}{\partial \dot{x}} = m\ddot{x} + e\left(\frac{\partial A_x}{\partial t} + \frac{\partial A_x}{\partial x}\dot{x} + \frac{\partial A_x}{\partial y}\dot{y} + \frac{\partial A_x}{\partial z}\dot{z}\right).$$

Lagrange's equation is then

$$m\ddot{x}+e\frac{\partial A_x}{\partial t}+e\left[\dot{y}\left(\frac{\partial A_x}{\partial y}-\frac{\partial A_y}{\partial x}\right)-\dot{z}\left(\frac{\partial A_z}{\partial x}-\frac{\partial A_x}{\partial z}\right)\right]+e\frac{\partial \Phi}{\partial x}+\frac{\partial V'}{\partial x}=0$$

whence

$$m\frac{\mathrm{d}^2x}{\mathrm{d}t^2}=eE_x+e[\boldsymbol{v}\times\boldsymbol{B}]_x-\frac{\partial V'}{\partial x}. \tag{3.32}$$

Since this is the correct equation of motion, the form of equation (3.29) is justified.

With the knowledge of the Lagrangian function the Hamiltonian function can be written. The momenta now involve the vector potential:

$$\begin{aligned} p_x &= m\dot{x}+eA_x \\ H &= p_x\dot{x}+p_y\dot{y}+p_z\dot{z}-L \\ &= \frac{1}{2m}[(p_x-eA_x)^2+(p_y-eA_y)^2+(p_z-eA_z)^2]+e\Phi+V'. \end{aligned} \tag{3.33}$$

If the external fields, and hence the potentials $\boldsymbol{A}$ and Φ, depend on the time, this function will depend on the time through them.

From this classical Hamiltonian function can be formed the Hamiltonian operator to be used in the Schroedinger equation:

$$\mathsf{H}=\frac{1}{2m}\left[\left(\frac{\hbar}{\mathrm{i}}\frac{\partial}{\partial x}-eA_x\right)^2+\left(\frac{\hbar}{\mathrm{i}}\frac{\partial}{\partial y}-eA_y\right)^2+\left(\frac{\hbar}{\mathrm{i}}\frac{\partial}{\partial z}-eA_z\right)^2\right]+e\Phi+V'. \tag{3.34}$$

In forming the squares of these binomial operator expressions it is necessary to remember that the two members do not commute since $\boldsymbol{A}$ is a function of the coordinates. The two product terms cannot generally be combined into one but must be kept separate.

5.5. *Gauge invariance*

The Hamiltonian function (3.33) and the corresponding operator contain the potentials Φ and $\boldsymbol{A}$, and yet one might expect the behavior of the system to depend only on the fields and not at all on the values of the potentials themselves. From this point of view it must be required that the results of the theory be independent of any change in the potentials which does not change the fields. The transformation

$$\boldsymbol{A}'=\boldsymbol{A}+\operatorname{grad} f,\quad \Phi'=\Phi-\frac{\partial f}{\partial t} \tag{3.35}$$

is such a transformation and is called a gauge transformation. This transformation leaves the fields unchanged with any function f, but if the relation (3.31) between the potentials is to be maintained, it is necessary that only such functions f be used for which

$$\operatorname{div} \operatorname{grad} f - \mu_0 \varepsilon_0 \frac{\partial^2 f}{\partial t^2} = 0. \tag{3.36}$$

Invariance under the transformation (3.35) is called gauge invariance.

When the quantities A' and Φ' are substituted in the operator (3.34), the resulting Schroedinger equation will be satisfied by the function

$$\psi' = \psi \, e^{(ie/\hbar)f} \tag{3.37}$$

if ψ satisfies the equation with the potentials A and Φ. The change from ψ to ψ' will not change the physical results of the theory. It is evident that the probability distribution of position will not be changed, since it depends only on the absolute value of the function. The exponential term will change the distribution of momentum, but the definition of momentum will also be changed because of the change in the vector potential, and this change will be just enough to compensate the change in distribution.

PROBLEM 10. Starting with equation (3.33) find the Hamiltonian operator for a particle of charge e in a uniform magnetic field, $\boldsymbol{B} = B\boldsymbol{k}$. Express the result in terms of the angular momentum.

6. Criteria for the validity of the theory

In order that the theory formulated in this chapter may be satisfactory three conditions must be satisfied.

1. The theory must be self-consistent. This requirement has already been used in the formulation to show that the Hamiltonian operator in the Schroedinger equation must be Hermitian. Similarly under all circumstances the conclusions drawn from one part of the theory must not contradict those drawn from another part.

2. The conclusions drawn must approach those of the classical theory in the limit in which h can be neglected and in cases in which the classical theory is known to be correct. Chapter 5 will be devoted to showing that the classical mechanics is a special case of the wave mechanics.

3. The results must be correct in cases in which the classical theory does not apply. Unless this were true, the theory would be useless. Various ap-

plications will bring out this point. In the course of these applications the interpretations of the paradoxes mentioned in Chapter 1 should become clear, as well as the application of the theory to a wide variety of other phenomena. However, it is not to be expected that all problems can be handled. In the first place the theory is nonrelativistic and cannot be expected to give correct results in cases in which velocities comparable with the velocity of light are present. In the second place, although it is a generalization of ordinary mechanics to include the quantum of action h, it still is limited in its region of applicability. These limits cannot be precisely stated since no more inclusive theory has been developed, but they are suggested in some applications.

CHAPTER 4

CHARACTERISTIC STATES OR EIGENSTATES

Many of the features peculiar to quantum mechanics are inherent in the method of representing the state of a physical system, but there has been left open thus far the question of how it is possible to associate the wave function to be used as the initial state with the physical apparatus or the material system under consideration. How is it possible to know the state of a system, or to prepare a system in a prescribed state?

The things that can be done experimentally must be described in classical terms. They are usually described as measurements of positions, or wavelengths. They may sometimes be more properly regarded as specifications. The position of a "particle" which passes through a slit is taken to be the position of the slit. The wavelength of a "particle" diffracted from a grating is taken to be given by the properties of the grating and the angle of diffraction. Thus properties ascribed to electrons must really be properties of some large scale equipment in which the electrons are functioning.

The analysis of the previous chapters has pointed out the general fact that no apparatus can be devised, operating in the realm of the known laws of physics, which at the same time can fix precisely two canonically conjugate quantities. Only one such quantity can be regarded as established by any conceivable arrangement of measuring devices. Consequently, the best possible specification of the "state" of a system is the precise specification, or measurement, of a maximum set of quantities, no two of which are canonically conjugate. The wave function to represent such a situation is an eigenfunction or a characteristic function. These two terms will be used interchangeably.

1. Definition of an eigenstate

An eigenstate, or a characteristic state, is a state such that one or more dynamical quantities have precise values. The function χ which represents such a state is a characteristic function, or an eigenfunction, of the operators corresponding to the dynamical quantities.

A characteristic feature of quantum mechanics is the assertion that certain dynamical quantities, such as angular momentum, can have only discrete values and these are called the eigenvalues, or the characteristic values, of the dynamical quantity. The statement that a certain quantity has a precisely defined value is clear for such quantities. For quantities with a continuum of eigenvalues, the statement involves a limiting process.

Consider a system of one dimension only, and call the coordinate x. The function $\delta(x-x_0)$ is characteristic of the quantity x and represents a state in which x has the value x_0. Similarly the function $\exp\{(\mathrm{i}/\hbar)p_0 x\}$ represents a state characteristic of the momentum, in which the momentum has the value p_0.

If a gamma-ray microscope using gamma rays of infinitely short wavelength is available to "measure" the "position" of the particle, it may be determined that $x = x_0$. Immediately after such a measurement the system will be in a state represented by the function $\chi = \delta(x-x_0)$.

When a stream of particles is reflected from a diffraction grating in a direction corresponding to the wavelength λ, each one passing through the defining slit will be represented by the function $\chi = \exp\{(\mathrm{i}/\hbar)\boldsymbol{p}_0 \cdot \boldsymbol{r}\}$, where $\boldsymbol{p}_0$ has the direction of the scattered beam and a magnitude equal to h/λ.

In general, a system in a characteristic state can be produced by an apparatus which selects those systems with the prescribed characteristic values of the quantity in question. Exact values of these quantities are, of course, limits, as in classical mechanics, but since the limit can be approached as closely as is desired, it is convenient to consider such a limit as a possible situation.

The general equation for a characteristic function is

$$\mathsf{Q}\chi = Q_0\chi \tag{4.1}$$

where Q is the operator corresponding to the dynamical quantity of which the state represented by χ is characteristic. Q_0 is a constant. It is a characteristic value of the operator Q and is the value of the dynamical quantity Q when the system is in the state χ.

It is clear that in the state χ, the expectation value of the n'th power of Q will be the n'th power of the expectation value of Q.

For many operators Q, there are only certain discrete values of Q_0 for which equation (4.1) will have admissible solutions, solutions that are single valued and quadratically integrable. Hence this formulation of the theory represents the observed fact that some quantities appear experimentally to be "quantized". There are only certain discrete values which can be ascribed to a physical system. An ideal experiment to measure Q precisely will give a definite result, and since this fact must be representable by a characteristic function, it follows that only certain values of Q_0 can be obtained as a result of such a measurement. This is a definite statement, derived from the theory, as to a physical fact; and this is one of the first of the facts to be discovered which led to the formulation of Quantum Mechanics. In his work described in Chapter 1, Planck made the assumption that the normal vibrations of the electromagnetic field could have only discrete values of the energy, $E = nh\nu$. These will be shown later to be related to the characteristic values of the energy for such a system. The discrete amounts of energy which atoms have been found to possess are the characteristic values of the energy for the mechanical system constituting the atom.

Early in the study of spectroscopy it was found that the component, in a given direction, of the angular momentum of an atom can have only certain discrete values. In Chapter 3 it was demonstrated that the operator in spherical polar coordinates, for the component of the angular momentum around the polar axis, is $(\hbar/\mathrm{i})\,\partial/\partial\varphi$. The equation for the functions characteristic of this quantity is

$$\frac{\hbar}{\mathrm{i}}\frac{\partial\chi}{\partial\varphi} = L_z\chi. \tag{4.2}$$

The solutions which satisfy the necessary conditions are

$$\chi(r, \vartheta, \varphi) = f(r, \vartheta)\, \mathrm{e}^{im\varphi}, \tag{4.3}$$

where $L_z = m\hbar$ and m is an integer. This result indicates that only integral multiples of $\hbar$ are to be permitted as values of the angular momentum, a conclusion which is in agreement with many well known experiments.

The study of eigenstates is of particular importance because any experimental arrangement which completely defines the state of a physical system must be described by an eigenstate.

If the operator Q is Hermitian, the characteristic functions for different values of Q_0 are orthogonal to each other. To show this, consider the two equations

$$\mathrm{Q}\chi_\mathrm{a} = Q_\mathrm{a}\chi_\mathrm{a}$$

$$\mathrm{Q}^*\chi_\mathrm{b}^* = Q_\mathrm{b}^*\chi_\mathrm{b}^*$$

where the subscripts a and b refer to two different functions corresponding to the two different characteristic values Q_a and Q_b. Multiply the first equation by χ_b^*, the second by χ_a, and integrate the difference over the whole range of the variables:

$$\int [\chi_b^* Q \chi_a - (Q^* \chi_b^*)\chi_a] \, dv = (Q_a - Q_b^*) \int \chi_b^* \chi_a \, dv = 0. \tag{4.4}$$

The first term of the equation is zero because the operator is Hermitian. The equality of the second term to zero shows the orthogonality of the functions if $Q_a \neq Q_b^*$. Since $\chi_a^* \chi_a$ is always positive, $Q_a = Q_a^*$ and the characteristic values of such an operator are always real.

The property of orthogonality makes characteristic functions especially useful. Other states can be expanded in a series of such orthogonal functions, and the coefficients can be determined. If an arbitrary state ψ is expanded as a series of functions characteristic of the quantity Q, the coefficients in the expansion give the probability distribution of the quantity Q in the state ψ. To show this, let

$$\psi = \sum a_n \chi_n$$

where χ_n is the characteristic function associated with the characteristic value Q_n. Operating on both sides of this equation with the operator Q, multiplying by ψ^*, and integrating gives

$$\int \psi^* Q \psi \, dv = \sum_n a_n Q_n \int \sum_s a_s^* \chi_s^* \chi_n \, dv = \sum_n | a_n |^2 Q_n . \tag{4.5}$$

The left-hand side of (4.5) is the average value of Q, while the right-hand side is a sum of terms each of which is a possible value of Q multiplied by its probability.

2. Energy Eigenfunctions

States in which an isolated physical system has a definite energy are very much used because the energy is such an important constant of the motion. A system in an energy characteristic state will stay in that state until it is disturbed. Actual atomic systems and other systems of charged particles are not quite isolated because they are coupled to the electromagnetic field; but for many purposes this coupling is so weak it can be neglected, and the description of the behavior of the system in terms of energy states is useful as a close approximation.

Since for an isolated system H is the operator representing the energy, the functions representing the energy states satisfy the equation

$$-\frac{\hbar}{\mathrm{i}}\frac{\partial \chi}{\partial t} = \mathsf{H}\chi = E\chi. \tag{4.6}$$

The last two terms are equal because χ is an energy characteristic state, and the equality of the first and last terms shows that

$$\chi(q, t) = u(q)\,\mathrm{e}^{-(\mathrm{i}/\hbar)Et}, \tag{4.7}$$

where $u(q)$ is a function of the coordinates only. This function $u(q)$ also satisfies the second equality in (4.6). Hence it follows that the characteristic states of a single particle, moving in a field of force whose potential energy is $V(x, y, z)$, can be obtained from the equation

$$\mathsf{H}u = -\frac{\hbar^2}{2m}\nabla^2 u + Vu = Eu$$

where m is the mass of the particle. This is usually transformed to read

$$\nabla^2 u + \frac{2m}{\hbar^2}(E-V)u = 0, \tag{4.8}$$

and is often known as Schroedinger's equation with the time eliminated.

Energy states are often called stationary states because the change with the time is all included in the exponential, as shown in equation (4.7). The quantity $\chi^*\chi$ is a constant in time, and the probability distributions of all quantities are constant. From this it follows that a system with a precisely defined energy cannot be followed through a motion in space.

A general solution of the Schroedinger equation of motion can be written in terms of the energy characteristic states. Let

$$\psi(x, y, z, t) = \sum_n a_n u_n(x, y, z)\,\mathrm{e}^{-(\mathrm{i}/\hbar)E_n t} \tag{4.9}$$

where the $u_n(x, y, z)$ are the functions associated with the energies E_n. The $u_n(x, y, z)$ form a complete orthogonal set, so that by suitable selection of the coefficients a_n the function $\psi(x, y, z, t)$ can be expressed at any given time. Furthermore the sum can be shown by substitution to be a solution of the Schroedinger equation.

There are relatively few cases in which characteristic functions can be written in closed form. In most cases some method of approximation must be used to express them, but for the cases of a harmonic oscillator and a particle in an inverse-square field the energy characteristic functions can be given exactly. These two cases are of great use as the basis of approximation treatments for other problems, and we shall give them special attention.

3. The one-dimensional harmonic oscillator

As an example of a system for which the energy characteristic functions can be obtained exactly, consider a particle free to move in one dimension and attracted toward the origin with a force proportional to the distance. The Schroedinger equation for the energy functions is

$$\frac{\mathrm{d}^2u}{\mathrm{d}x^2}+\frac{2m}{\hbar^2}(E-ax^2)u=0. \tag{4.10}$$

This has the form of equation (4.8) with only a single second derivative, since the motion is considered to be in one dimension only. To simplify this let

$$\frac{2m}{\hbar^2}E=\varepsilon,\quad \frac{2m}{\hbar^2}a=\alpha^2,\quad \alpha^{\frac{1}{2}}x=\xi,\quad u=v\,\mathrm{e}^{-\frac{1}{2}\xi^2}. \tag{4.11}$$

Then

$$\frac{\mathrm{d}^2v}{\mathrm{d}\xi^2}-2\xi\frac{\mathrm{d}v}{\mathrm{d}\xi}+\left(\frac{\varepsilon}{\alpha}-1\right)v=0. \tag{4.11a}$$

If $(\varepsilon/\alpha)-1=2n$, where n is a positive integer, this is the equation for the Hermite polynomials of order n. If $(\varepsilon/\alpha)-1\neq 2n$, the solution for large values of ξ approaches e^{ξ^2} so that $u\to \mathrm{e}^{\frac{1}{2}\xi^2}$. Such a function cannot represent a state of the oscillator, since the integral of its square does not converge. The requirement of integrability thus determines the values of ε for which a suitable solution can be found. If $\varepsilon=(2n+1)\alpha$,

$$E=(n+\tfrac{1}{2})\hbar\sqrt{2a/m}=(n+\tfrac{1}{2})h\nu. \tag{4.12}$$

The quantity $(1/2\pi)\sqrt{(2a/m)}=\nu$ is the frequency with which the classical oscillator would vibrate. The expression for the energy in (4.12) differs from that first assumed by Planck in his treatment of black-body radiation by having $\frac{1}{2}h\nu$ as the minimum value instead of 0. Since, however, the changes in energy are in steps of hv, Planck's law of black-body radiation can be obtained from this expression with the exception that a constant, $\frac{1}{2}hv$, is added to the average energy of each mode of vibration.

When the condition (4.12) is satisfied, the solution of equation (4.9) is

$$u_n = \frac{1}{(2^n n!)^{\frac{1}{2}}} \left(\frac{\alpha}{\pi}\right)^{\frac{1}{4}} \mathrm{e}^{-\frac{1}{2}\alpha x^2} H_n(\alpha^{\frac{1}{2}} x) \tag{4.13}$$

where H_n is the Hermite polynomial of the indicated argument. Some properties of these polynomials are given in the Appendix.

A general solution of the Schroedinger equation for the oscillator can be written as an infinite series of these energy characteristic states:

$$\psi(x, t) = \sum_{n=0}^{\infty} a_n \chi_n(x, t) = \sum_{n=0}^{\infty} a_n u_n \, \mathrm{e}^{-2\pi \mathrm{i}(n+\frac{1}{2})vt}. \tag{4.14}$$

If the function is known at some time, say $t = 0$, its value at any time t can be determined from equation (4.14). The constants a_n can be obtained from the known function at $t = 0$ by using the orthogonality of the u_n.

It is frequently important to classify states according to their parity. The differential equation (4.10) is unchanged when x is replaced by $-x$, and so if $u(x)$ is a solution, $u(-x)$ is also a solution. These two solutions can differ only by a multiplicative factor since the solutions of (4.10) are not degenerate. It follows from a repetition of the argument that the multiplicative factor can be only 1 or -1. Each function (4.13) is either unchanged or is multiplied by -1 when x is replaced by $-x$.

The Hermite polynomials are even or odd as n is even or odd, and so the function u_n has the parity of n.

PROBLEM 1. Find the average and the average square of the coordinate and the momentum for the energy characteristic states of an oscillator. Show that the product of the indeterminations is $(n+\frac{1}{2})\hbar$.

PROBLEM 2. Find the averages of the kinetic and the potential energies for an oscillator whose total energy is $(n+\frac{1}{2})hv$.

PROBLEM 3. Find the probability that an oscillator in its state of lowest energy will have a value of its coordinate greater than the maximum for a classical oscillator of the same energy.

PROBLEM 4. An oscillator is in such a state that at $t = 0$ the probability that its energy is $\frac{3}{2}hv$ is $\frac{1}{2}$ and the probability that it is $\frac{5}{2}hv$ is $\frac{1}{2}$. Show that a large number of functions can be written which satisfy this condition.

PROBLEM 5. Write a wave function that satisfies the conditions of the above problem and at the same time makes the average of x equal to zero at $t = 0$. Find the average value of x as a function of the time.

4. Degenerate states and the two-dimensional oscillator

It is frequently the case that to a given value Q' of the dynamical quantity Q there corresponds more than one characteristic function. This constitutes degeneracy. Degeneracy does not occur in bounded one-dimensional problems, but in more than one dimension it is quite common and is an important phenomenon. In a degenerate case the one quantity Q is insufficient to specify the state of the system uniquely. For this purpose some other quantities must also be given precisely defined values. In general, as many quantities must be specified as there are independent coordinates, but these quantities cannot be selected at random. They must be such that their operators commute.

The situation in a degenerate case can be illustrated by the problem of the two-dimensional oscillator. This is a particle moving in the $x-y$ plane and attracted toward the origin with a force proportional to the distance. The equation for the energy states of the system is

$$\frac{\partial^2 u}{\partial x^2} + \frac{\partial^2 u}{\partial y^2} + \frac{2m}{\hbar^2}(E - ax^2 - ay^2)u = 0. \tag{4.15}$$

This equation is separable and can be satisfied by a product of solutions of the two equations

$$\frac{d^2 v}{dx^2} + \frac{2m}{\hbar^2}(E_x - ax^2)v = 0$$

$$\frac{d^2 w}{dy^2} + \frac{2m}{\hbar^2}(E_y - ay^2)w = 0$$

provided that $E = E_x + E_y$. These equations are separately of the same form as equation (4.8) so that a solution of (4.15) is

$$u_{s,t}(x, y) = \frac{1}{(2^{s+t}s!t!)^{\frac{1}{2}}}\left(\frac{\alpha}{\pi}\right)^{\frac{1}{2}} e^{-\frac{1}{2}\alpha(x^2+y^2)} H_s(\alpha^{\frac{1}{2}}x)H_t(\alpha^{\frac{1}{2}}y) \tag{4.16}$$

with

$$E_{s,t} = (s+\tfrac{1}{2})h\nu + (t+\tfrac{1}{2})h\nu = (s+t+1)h\nu.$$

The energy is specified by the sum $(s+t)$, but to specify the state uniquely it is necessary to specify s and t separately. The two quantities E and E_x will specify the state, but E alone will not. For each value of E there are several characteristic states, each with a different E_x, and for each value of E_x there are many states with different values of E.

In cases of degeneracy the characteristic value equation will be separable in more than one system of coordinates. The two-dimensional oscillator can be treated in polar as well as in cartesian coordinates, PAULING and WILSON [1935]. In plane polar coordinates the equation for u is

$$\frac{\partial^2 u}{\partial r^2} + \frac{1}{r}\frac{\partial u}{\partial r} + \frac{1}{r^2}\frac{\partial^2 u}{\partial \vartheta^2} + \frac{2m}{\hbar^2}(E - ar^2)u = 0. \tag{4.17}$$

This equation can be satisfied by $u(r, \vartheta) = f(r)\,\mathrm{e}^{im\vartheta}$,if m is a positive or negative integer to insure the single-valuedness of the function, and $f(r)$ satisfies the equation

$$\frac{\mathrm{d}^2 f}{\mathrm{d}r^2} + \frac{1}{r}\frac{\mathrm{d}f}{\mathrm{d}r} + \left(\varepsilon - \alpha^2 r^2 - \frac{m^2}{r^2}\right) f = 0. \tag{4.17a}$$

The quantities ε and α are as defined in equation (4.11). Equation (4.17a) can be transformed to permit a power series solution by means of the substitution $f = g\,\mathrm{e}^{-\frac{1}{2}\rho^2}$ with $\rho = \alpha^{\frac{1}{2}} r$. Then

$$\frac{\mathrm{d}^2 g}{\mathrm{d}\rho^2} - \left(2\rho - \frac{1}{\rho}\right)\frac{\mathrm{d}g}{\mathrm{d}\rho} + \left(\frac{\varepsilon}{\alpha} - 2 - \frac{m^2}{\rho^2}\right) g = 0. \tag{4.18}$$

This equation has a singularity at the origin, but it is a nonessential singularity, and the solution can be developed in a power series around it. Let $g = \sum_s b_s \rho^s$. Substituting in the differential equation and equating the coefficients of each power of ρ to zero gives

$$[m^2 - (s+2)^2] b_{s+2} = \left(\frac{\varepsilon}{\alpha} - 2(s+1)\right) b_s. \tag{4.19}$$

This two-term recursion formula shows that a series can be formed of either odd or even powers of ρ. In either case $b_{|m|-2}$ will be zero, and the series having the same parity as m will begin with $\rho^{|m|}$. The other series will contain an infinite number of negative powers and will not give a usable solution.

If $\varepsilon = 2\alpha(n+1)$, where n is an integer with the same parity as m, the series will terminate with ρ^n. If ε has any other value, the infinite series solution will not be suitable for a characteristic function. The function g is thus a polynomial and can be designated by the two integers, or quantum numbers, n and m. The final result is

$$u_{n,m}(r, \vartheta) = N_{n,m} g_{n,m}(\alpha^{\frac{1}{2}} r)\, e^{-\frac{1}{2}\alpha r^2 + im\vartheta} \tag{4.20}$$

with $E = (n+1)h\nu$.

A few of the $g_{n,m}$ given by the recursion formula are

$$g_{0,0} = 1, \qquad g_{1,-1} = \alpha^{\frac{1}{2}} r, \qquad g_{1,1} = \alpha^{\frac{1}{2}} r,$$

$$g_{2,-2} = \alpha r^2, \qquad g_{2,0} = 1 - \alpha r^2 \qquad g_{2,2} = \alpha r^2.$$

These functions must be multiplied by the normalization constant $N_{n,m}$ in order that the functions $u_{n,m}(r, \vartheta)$ may give the coordinate probability distribution directly.

Again in this treatment the total energy does not uniquely specify the function, but the number m is needed in addition. The angular momentum is $m\hbar$ so that the simultaneous specification of the energy and the angular momentum is possible and will completely define the function and the state of the oscillator.

Another important property of degenerate systems is illustrated by the fact that the functions $u(r, \vartheta)$ are linear combinations of the functions $u(x, y)$ that are eigenfunctions of the energy with the same eigenvalue. The probability distribution of E_x and E_y, when the state is represented by $u(r, \vartheta)$, is given by the coefficients of the linear combination. Similarly the distribution of angular momentum, when the system is in a state $u_{s,t}(x, y)$ can be obtained from an expansion of this function in terms of the $u_{n,m}(r, \vartheta)$.

The parity of the functions $u_{s,t}$ is clearly just the parity of $(s+t)$. When the function is expressed in polar coordinates, the parity is determined by a reversal in direction of r, which corresponds to replacing ϑ by $\vartheta + \pi$. Such a transformation changes the wave function $u_{n,m}(r, \vartheta)$ by $e^{im\pi}$ so that the function has the parity of m. Since n and m have the same parity, this is identical with the parity of the function $u_{s,t}$.

PROBLEM 6. Show that if equation (4.1) is satisfied the quantity Q has a definite value.

PROBLEM 7. Show that if $\varepsilon \neq (2n+1)\alpha$, $u \to e^{\frac{1}{2}\alpha x^2}$ is a solution for the one-dimensional harmonic oscillator.

PROBLEM 8. If at the time $t = 0$ the function describing the state of an oscillator is

$$\psi(x, 0) = (2b/\pi)^{\frac{1}{4}}\, e^{-b(x-x_0)^2},$$

find it at a later time t. To do this consider the expansion of the function in a series of energy eigenfunctions. Show that the probability distribution oscillates back and forth somewhat like the corresponding classical oscillator and that it returns to its original form every period, although the wave function requires two full periods to return to its original form.

PROBLEM 9. Find the values of the normalization constants $N_{n,m}$ for $n = 0, 1, 2$.

PROBLEM 10. Show that

$$u_{0,0}(r, \vartheta) = u_{0,0}(x, y),$$

$$u_{1,1}(r, \vartheta) = \frac{1}{\sqrt{2}} u_{1,0}(x, y) + \frac{\mathrm{i}}{\sqrt{2}} u_{0,1}(x, y),$$

$$u_{1,-1}(r, \vartheta) = \frac{1}{\sqrt{2}} u_{1,0}(x, y) - \frac{\mathrm{i}}{\sqrt{2}} u_{0,1}(x, y).$$

PROBLEM 11. Expand the functions $u_{n,m}(r, \vartheta)$ as linear combinations of the functions $u_{s,t}(x, y)$ for $n = 2$. Find the probability of each value of E_x and E_y for each of these states.

PROBLEM 12. Show that the operator for angular momentum commutes with the operator for energy in this problem.

5. Measurements in quantum mechanics

In Chapter 3 two postulates were formulated. The first referred to the representation of the "state" of a mechanical system at a time t by means of a wave function, and the interpretation of the wave function in terms of possible observations or measurements. The second referred to the way in which such a wave function changes with the time. Although mention has been made several times of the lack of any "reality" to be ascribed to the wave function and attention has been called to its symbolic nature, there still remains the necessity of clearly associating the wave function, with which calculations are made, and the physical situations to which it is intended to apply. Several remarks are in order.

It is a characteristic of wave mechanics that a wave function does not describe an electron or group of electrons, it describes a particular physical situation which is described in terms of an electron or group of electrons. Hence different experiments will require different wave functions. The arrangements for an experiment and the results of it must be described in classical language. We have no other suitable language. But we must remember that the language applies to the experiment rather than to a supposed object on which the experiment is carried out.

The previous discussion has treated the wave function at $t = 0$ as arbitrary, but well defined. The wave function, of course is subject to certain boundary conditions and conditions of integrability. This initial wave function defines a probability distribution of the momenta and the position coordinates, but these probability distributions do not define the wave function. To be defined, the wave function must be an eigenfunction of some dynamical observable, and the experimental arrangement to measure, or to select, a value of this observable is described by saying that at $t = t_0$ the system under consideration has the value $0'$ of the observable. In one dimension, only one observable and functions of it can be specified experimentally. The canonically conjugate quantity is completely undefined. In n dimensions there must be defined n independent quantities whose operators commute. These constitute a "physically complete" set of commuting quantities and the state of a system is completely defined when it is a simultaneous eigenstate of these quantities.

The Schroedinger equation then gives the wave function at a later time t. This later wave function is also an eigenfunction of a physically complete set of commuting quantities. They will not be the same quantities as those fixed at $t = t_0$, but they can be identified from the properties of the wave function. The Schroedinger equation then gives the probability distribution of these and all other observables at the time t. The wave mechanics, then, may be regarded as a procedure for predicting from a complete set of precise measurements at $t = t_0$, all possible probability distributions at a later time t.

Of course, as in classical mechanics, a precise measurement of dynamical quantities is a limiting case. Few attainable arrangements will give complete precision. However, the approach to this limiting case is not hindered by the fact that h is finite. In the case of quantities with a continuum of eigenvalues, such as cartesian coordinates and momenta, the approach to precision is always hampered experimentally. In the cases of discrete eigenvalues, such as those of angular momentum, systems with one eigenvalue can be isolated.

Most experimental arrangements are not suitable for attaining the ultimate accuracy needed to define a pure state. When this lack of accuracy is important, recourse must be had to statistical mechanics and the treatment of mixed states. These are dealt with in Chapter 17.

CHAPTER 5

THE CLASSICAL APPROXIMATION

For the wave mechanics as formulated in Chapter 3 to be useful, it must include the classical mechanics as a special case. Every problem that can be treated satisfactorily by classical mechanics must have a quantum-mechanical solution, essentially identical with the classical solution. In such cases the classical treatment may be preferred if it is simpler; and in every case it should be possible to give an estimate of the error to be expected in using the classical approximation. This chapter will be devoted to showing briefly that such is the case.

1. Ehrenfest's theorem

EHRENFEST [1927] showed that if the probability distribution of the position of a particle is essentially localized in a small volume, the average values of the coordinates will change in a manner very similar to the change of the precisely defined coordinates in the classical theory. Thus the classical theory gives an approximate description of the motion.

In the special case of a system of particles subject to conservative forces the proof is simple. From equation (3.26) it follows that

$$\frac{\mathrm{d}\langle x_j\rangle}{\mathrm{d}t} = \frac{\mathrm{i}}{\hbar}\int \psi^*(\mathsf{H}\mathsf{x}_j-\mathsf{x}_j\mathsf{H})\psi\,\mathrm{d}v = \frac{\hbar}{im_j}\int \psi^*\frac{\partial\psi}{\partial x_j}\,\mathrm{d}v = \frac{\langle p_j\rangle}{m_j}. \quad (5.1)$$

Thus the connection between the average momentum of a particle and the rate of change of its average coordinate is rigorously the same as in classical mechanics, and this relationship is subject to no restriction whatever. However the average coordinate may not be a very significant description of the location of the particle if the indetermination is large.

A second differentiation of $\langle x_j\rangle$ leads to

$$\begin{aligned}\frac{d^2\langle x_j\rangle}{dt^2} = \frac{1}{m_j}\frac{d\langle p_j\rangle}{dt} &= \frac{\hbar}{im_j}\int\left[\frac{\partial\psi^*}{\partial t}\frac{\partial\psi}{\partial x_j} + \psi^*\frac{\partial^2\psi}{\partial x_j\partial t}\right]dv \\ &= \frac{1}{m_j}\int\left[\psi^*\mathsf{H}\frac{\partial\psi}{\partial x_j} - \psi^*\frac{\partial}{\partial x_j}(\mathsf{H}\psi)\right]dv \qquad (5.2) \\ &= -\frac{1}{m_j}\int\psi^*\frac{\partial V}{\partial x_j}\psi\, dv.\end{aligned}$$

Equation (5.2) shows that the acceleration of the average coordinate, or of the "center of mass" of the probability distribution, is given by the average value of the negative of the derivative of the potential energy, or the average value of the force. If the probability distribution is confined to a region in which the force does not change too much, and this is the significance of the requirement that the wave function differs from zero only in a small volume, the content of equation (5.2) is much the same as that of Newton's second law of motion. Of course the question must be considered as to how long the probability distribution will remain essentially in a small region even if it starts in such a way. This time will be found to depend on the particular form of the wave function involved and on the Hamiltonian function, and not on the probability distribution only.

2. Motion of a free particle

The content of Ehrenfest's theorem, as well as a method of handling the Schroedinger equation of motion, can be illustrated by considering the case of a free particle in one dimension. At the time $t = 0$ let the state of the particle be represented by the function

$$\psi(x, 0) = (\alpha/\pi)^{\frac{1}{4}} \exp\{-\tfrac{1}{2}\alpha x^2 + i\beta x\}. \qquad (5.3)$$

This function represents a situation in which the average coordinate is zero, and the dispersion can be made very small by taking α large. Also, the average momentum is $\beta\hbar$, and the dispersion of the momentum is inversely proportional to that of the coordinate. The function $\psi(x,0)$ is also an eigenfunction of the operator

$$\mathsf{A} = \hbar^2\alpha^2 x^2 + (p - \hbar\beta)^2. \qquad (5.3a)$$

It may be difficult to specify a procedure for producing particles in a state specified by the operator A, but at least A has a well defined significance in terms of a coordinate and the conjugate momentum. Even without a precise specification for its preparation, the state represented in (5.3) may be regarded as typifying a particle whose position and momentum are as well defined as it is possible to define them under the restrictions of quantum mechanics.

In the previous chapter it was shown how a general solution of Schroedinger's equation can be written as a sum of energy-characteristic solutions, each with an exponential time dependence. In the case of a free particle, the energy-characteristic states are also characteristic of momentum, since the energy is a function of the momentum only. Furthermore, all energy values are possible, so that the sum must be replaced by an integral. Thus the general solution of the Schroedinger equation for a free particle in one dimension is

$$\psi(x, t) = h^{-\frac{1}{2}} \int_{-\infty}^{\infty} \mathrm{d}p\, \phi(p) \exp\left\{\frac{\mathrm{i}}{\hbar}\left(px - \frac{p^2 t}{2m}\right)\right\}. \tag{5.4}$$

The term $p^2/2m$ is the energy of a state whose momentum is p, and $\exp\{(\mathrm{i}/\hbar)px\}$ gives the coordinate dependence of the wave function for such a state. To evaluate the function $\phi(p)$ it is necessary to use the function at $t = 0$. From the Fourier integral theorem, equation (3.4), it follows that

$$\begin{aligned}\phi(p) &= h^{-\frac{1}{2}} \int_{-\infty}^{\infty} \left(\frac{\alpha}{\pi}\right)^{\frac{1}{4}} \mathrm{d}x \exp\left\{-\alpha x^2 + \mathrm{i}\beta x - \frac{\mathrm{i}}{\hbar} px\right\} \\ &= \left(\frac{2}{h}\right)^{\frac{1}{2}} \left(\frac{\pi}{\alpha}\right)^{\frac{1}{4}} \exp\left\{-\frac{1}{2\alpha}\left(\frac{p}{\hbar} - \beta\right)^2\right\}.\end{aligned} \tag{5.5}$$

The integration in equation (5.4) can then be carried out with the result that

$$\begin{aligned}\psi(x, t) = (\pi\alpha)^{-\frac{1}{4}} \left(\frac{\alpha m}{m + \mathrm{i}\alpha\hbar t}\right)^{\frac{1}{2}} \exp\Bigg[-\frac{\alpha}{2[1 + (\alpha^2\hbar^2 t^2/m^2)]} \times \\ \times \left\{\left(x - \frac{\beta\hbar}{m} t\right)^2 - \mathrm{i}\left(\frac{\alpha\hbar t}{m} x^2 + \frac{2\beta}{\alpha} x + \frac{\beta\hbar t}{\alpha m}\right)\right\}\Bigg].\end{aligned} \tag{5.6}$$

Equation (5.6) gives the state of the system at the time t, and from it a number of conclusions can be drawn.

1. The average momentum, at any time t, is $\beta\hbar$ and is independent of the time. This is in accord with Ehrenfest's theorem when the potential energy is constant.

2. The average coordinate is $(\beta\hbar/m)t$, and hence it changes just as would the coordinate of a classical particle of mass m and momentum $\beta\hbar$.

3. The distribution of momentum is independent of the time in accordance with the result of Problem 8, Chapter 3. The momentum is a constant of the motion in case of a free particle.

4. The indetermination in position increases with the time. The function at the time $t = 0$ represents the minimum indetermination, as was shown in Problem 1, Chapter 3. It then increases quadratically with the time. If the function at a time $t<0$ were taken as the initial state, the indetermination would decrease for a time to its minimum value and then increase again.

From these results it follows that insofar as the indetermination in position and momentum can be neglected, classical mechanics is quite adequate to describe the motion of a free particle, but if the accuracy of observation becomes so great that the dispersion must be considered, quantum mechanics must be used.

PROBLEM 1. Show that in the presence of a magnetic field the average velocity is equal to $\langle(\boldsymbol{p}-e\boldsymbol{A}/m)\rangle$ when $\boldsymbol{p}$ is the three dimensional vector momentum, $\boldsymbol{A}$ is the vector potential and the $\langle\ \rangle$ signifies the quantum mechanical average.

PROBLEM 2. Show that Ehrenfest's theorem holds for the case of a force due to a magnetic field.

PROBLEM 3. Illustrate Ehrenfest's theorem for an initial state such as that in Problem 8 of Chapter 4.

3. Expansion in powers of $\hbar$

Classical mechanics is that limit of quantum mechanics in which $\hbar$ is negligible. This suggests that an expansion of solutions of the Schroedinger equation in powers of $\hbar$ might well give the classical case as a zero-order approximation and permit an estimate of its error. It turns out to be more satisfactory to expand the logarithm. Hence let

$$\psi(q, t) = \exp\left\{\frac{\mathrm{i}}{\hbar}\left[S_0 + \frac{\hbar}{\mathrm{i}}S_1 - \hbar^2 S_2 + \ldots\right\}\right] \tag{5.7}$$

where the S_j are functions of the coordinates and the time. The trouble with this expansion is that it does not converge. The approximation does not become better as more terms are taken. Nevertheless for a given number of terms there are regions in which the approximation is fairly good.

For illustration take the case of a single particle moving in a field of force, and substitute (5.7) into the corresponding differential equation. The terms independent of $\hbar$ must themselves satisfy the equation, and this condition leads to

$$-\frac{\partial S_0}{\partial t} = \frac{1}{2m}\left[\left(\frac{\partial S_0}{\partial x}\right)^2 + \left(\frac{\partial S_0}{\partial y}\right)^2 + \left(\frac{\partial S_0}{\partial z}\right)^2\right] + V(x, y, z). \qquad (5.8)$$

This equation is identical with equation (2.16) where the function S represents a transformation from the state of the system at time t_0 to the state at at time t. Such a transformation can be applied to a wide variety of initial states and so its appearance in wave mechanics is consistent with the necessary indetermination in the initial conditions.

If S_0 in equation (5.8) is such that $S_0 = S - Et$, with S independent of the time, the equation becomes equivalent to (2.17). Such a condition has a definite meaning in classical mechanics where E is the energy of the system. Its use then restricts the applicability of the transformation to the different initial conditions corresponding to the energy E. In wave mechanics, likewise, this condition on S_0 implies that the function $\psi(q, t)$ in equation (5.7) is one representing a definite energy, and is an energy eigenfunction. The function S in classical mechanics leads directly to a solution of the mechanical problem. The classical trajectories are perpendicular to the lines of constant S, since the momenta are obtained by taking the derivatives of S. In quantum mechanics the average vector momentum would be obtained by taking grad ψ, which would involve grad S. Thus, in this approximation, the possible classical trajectories and the possible directions of the quantum-mechanical momentum coincide.

The fact that equation (5.8) contains only the squares of the derivatives shows that S_0 will be real in the region in which $V < E$, that is, in the region in which the particle would move according to the classical mechanics. In this region ψ will be a complex oscillating function. Outside of this region, where $V > E$, and where according to the classical mechanics the particle could never be found, S_0 will be complex, and ψ will contain a real exponential function of the coordinates. This must be a decreasing exponential since the increasing exponential would not satisfy the conditions for a wave function. From this it follows that although the probability of finding a particle in a region in which $E < V$ does not vanish, it rapidly becomes negligible as one gets farther from the region within which the motion would be confined by the classical laws.

The terms in the differential equation which contain the first power of $\hbar$, after (5.7) has been substituted, lead to

$$-\frac{\partial S_1}{\partial t} = \frac{1}{2m}\left[\nabla^2 S_0 + 2\left(\frac{\partial S_0}{\partial x}\frac{\partial S_1}{\partial x} + \frac{\partial S_0}{\partial y}\frac{\partial S_1}{\partial y} + \frac{\partial S_0}{\partial z}\frac{\partial S_1}{\partial z}\right)\right]. \tag{5.9}$$

The function S_1 gives a modulation of the oscillating function in the region of the classical motion and provides a small oscillating term outside of it. If the quantities S_2, S_3 are neglected, the solution will have approximate validity in much of the range of the variables. There will be regions, however, in which the solution will not be at all significant, and great care must be taken in assuming a connection between solutions approximately correct in different regions.

If a solution of equation (5.9) is found in which $\partial S_1/\partial t = 0$, the approximation will be to an energy-characteristic function with the energy E. In this case S_1 will be real in the region of the classical motion and will provide a modulation of the oscillations given by S_0. Only in the case of a precisely defined energy is the region of the classical motion sharply defined.

With an energy eigenfunction the probability of position is constant with time, but even when $\partial S_1/\partial t \neq 0$ equation (5.9) guarantees the conservation of total probability to the approximation in which the form (5.7) using only S_0 and S_1 is useful. Since

$$W(x, y, z, t) = \psi^*\psi = \exp\left\{\frac{i}{\hbar}(S_0 - S_1^*) + (S_1 + S_1^*)\right\}$$

and

$$-\frac{\partial W}{\partial t} = \left\{-\frac{i}{\hbar}\left(\frac{\partial S_0}{\partial t} - \frac{\partial S_0^*}{\partial t}\right) + \left(\frac{\partial S_1}{\partial t} + \frac{\partial S_1^*}{\partial t}\right)\right\}\exp\left\{\frac{i}{\hbar}(S_0 - S_0^*) + (S_1 + S_1^*)\right\}$$

$$= \operatorname{div}\left\{\frac{1}{2m}\left[\operatorname{grad}(S_0 + S_0^*)\right] e^{S_1 + S_1^*}\right\}, \tag{5.10}$$

the term $\{(1/2m)\,[\operatorname{grad}(S_0 + S_0^*)] \exp(S_1 + S_1^*)\}$ can be regarded as a flux of probability density. Equation (5.10) shows that a decrease of probability density at one point is accompanied by a flow away from that point. The integral of the right hand side of (5.10) over the whole volume is zero since the wave function must vanish at infinity.

4. Application to the harmonic oscillator

The application of the above analysis to the harmonic oscillator in one dimension, compared with the exact solution in the previous chapter, is instructive in showing some of the connection between classical and quantum mechanics. For this case equation (5.8) becomes

$$\frac{\partial S_0}{\partial x} = [2m(E-ax^2)]^{\frac{1}{2}} \text{ and } S_0 = \int [2m(E-ax^2)]^{\frac{1}{2}}\,\mathrm{d}x. \quad (5.11)$$

S_0 has two values depending on the sign of the square root which is taken. In the classical theory these two values represent motions in opposite directions, and the motion can change direction only at the end points where $E-ax^2 = 0$. In the quantum theory both are required, to give the general solution. To this approximation

$$\psi(x) = A\,\mathrm{e}^{(i/\hbar)\,|S_0|} + B\,\mathrm{e}^{-(i/\hbar)\,|S_0|}. \quad (5.12)$$

In the classical theory the momentum is given by $\partial S_0/\partial x$, and in quantum theory as well the derivative of ψ with respect to x involves $\partial S_0/\partial x$. In the classical theory the solution has no significance in regions in which E—$ax^2 < 0$, while in the quantum theory the imaginary value of S_0 gives a valid approximation solution which is

$$\psi(x) = A'\mathrm{e}^{-(1/\hbar)S_0} + B'\mathrm{e}^{(1/\hbar)S_0}. \quad (5.13)$$

It must not be hastily concluded that A' and B' in (5.13) are identical with A and B in (5.12), for the approximation is not valid at the point $E = ax^2$ and the connection between the approximations in the two regions must be determined by a more detailed discussion of the differential equation. The only solution which can give an integrable function is that for which the exponential decreases, so either A' or B' must be zero. The corresponding values of A and B must then be determined accordingly.

The equation for S_1 in the case of a stationary state follows immediately from equation (5.9) to be

$$\frac{\mathrm{d}S_1}{\mathrm{d}x} = -\frac{\mathrm{d}^2S_0/\mathrm{d}x^2}{2(\mathrm{d}S_0/\mathrm{d}x)} = \frac{ax}{2(E-ax^2)} \quad (5.14)$$

whence $S_1 = -\frac{1}{4}\log(E-ax^2)$. From this it follows that the approximation wave function will contain as a factor $(E-ax^2)^{-\frac{1}{4}}$ and the probability density will be proportional to $(E-ax^2)^{-\frac{1}{2}} = (2/m)^{\frac{1}{2}}/v$. In the classical theory the time spent by a particle in a given element $\mathrm{d}x$ is inversely proportional to its velocity so that the two theories agree to this extent.

In fig. 5.1 the value of $(E-ax^2)^{-\frac{1}{2}}$ is plotted as a function of x for a number of energy states. The probability density as computed from the exact solution of the Schroedinger equation is also given. The agreement can be seen to improve as the quantum number increases. An increase in quan-

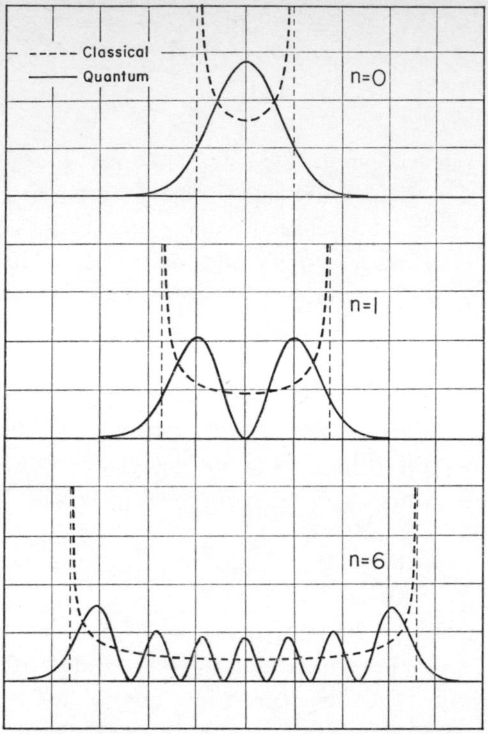

Fig. 5.1. The probability distribution of the position of a simple harmonic oscillator in an energy eigenstate approaches a corresponding time distribution of a classical oscillator as the energy is taken large compared with $\hbar\omega$.

tum number is equivalent in this respect to a decrease in $\hbar$, since the product $nh\nu$ appears in the energy. As n increases, or $\hbar$ decreases, the wavelength decreases until it would take very precise measurements to do anything more than to evaluate the average over several wavelengths. This average agrees closely with the classical function.

PROBLEM 4. Compute the wavelength, near the center of its oscillation, associated with the motion of an oscillator in an energy eigenstate and show that the number of such wavelengths between the classical limits of oscillation is independent both of the frequency and of $\hbar$.

CHAPTER 6

MOTION OF A PARTICLE IN A CENTRAL FIELD

The motion of a particle in a Coulomb field can be given an exact treatment in terms of energy-characteristic functions, both in spherical and in parabolic coordinates. The more general problem, with an arbitrary central field, can be treated in spherical polar coordinates, and the determination of the energy-characteristic functions can be reduced to the solution of an ordinary differential equation in the radius. Such solutions are of considerable importance, not only because of the value of exact solutions for the hydrogen and ionized helium atoms, but also because they provide a useful basis for the approximate solution of many more complicated problems.

1. The problem of two bodies

The general problem of two bodies which attract or repel each other with a force depending only on the distance between them can be reduced to the central-field problem by separation of the motion of the center of mass from the motion about this center. The process is entirely analogous to the corresponding separation in classical mechanics, and the possibility of carrying it out is associated with the definition of the total linear momentum of the pair of bodies.

The differential equation for the energy-characteristic states in this case is

$$\begin{aligned}\frac{1}{2m_1}\left(\frac{\partial^2 U}{\partial x_1^2}+\frac{\partial^2 U}{\partial y_1^2}+\frac{\partial^2 U}{\partial z_1^2}\right)+\frac{1}{2m_2}\left(\frac{\partial^2 U}{\partial x_2^2}+\frac{\partial^2 U}{\partial y_2^2}+\frac{\partial^2 U}{\partial z_2^2}\right)+\\ +\frac{1}{\hbar^2}\{E_0-V[(x_1-x_2)^2+(y_1-y_2)^2+(z_1-z_2)^2]\}U=0,\end{aligned} \tag{6.1}$$

where the subscripts 1 and 2 designate quantities which characterize the first and second particles, respectively. The coordinates of the center of mass are

$$X = \frac{m_1 x_1 + m_2 x_2}{m_1 + m_2}, \quad Y = \frac{m_1 y_1 + m_2 y_2}{m_1 + m_2}, \quad Z = \frac{m_1 z_1 + m_2 z_2}{m_1 + m_2}$$

and corresponding relative coordinates are

$$\xi = x_1 - x_2, \qquad \eta = y_1 - y_2, \qquad \zeta = z_1 - z_2.$$

In terms of these coordinates, equation (6.1) becomes

$$\frac{1}{M}\left(\frac{\partial^2 U}{\partial X^2} + \frac{\partial^2 U}{\partial Y^2} + \frac{\partial^2 U}{\partial Z^2}\right) + \left(\frac{1}{m_1} + \frac{1}{m_2}\right)\left(\frac{\partial^2 U}{\partial \xi^2} + \frac{\partial^2 U}{\partial \eta^2} + \frac{\partial^2 U}{\partial \zeta^2}\right) + + \frac{2}{\hbar^2}[E_0 - V(\xi^2 + \eta^2 + \zeta^2)]U = 0. \tag{6.2}$$

This equation can be satisfied by the product

$$U(X, Y, Z, \xi, \eta, \zeta) = U_c(X, Y, Z)U_r(\xi, \eta, \zeta)$$

if

$$\nabla^2 U_c + \frac{2M}{\hbar^2} E_t U_c = 0 \tag{6.3a}$$

and

$$\nabla^2 U_r + \frac{2\mu}{\hbar^2}[E_0 - E_t - V(\xi^2 + \eta^2 + \zeta^2)]U_r = 0. \tag{6.3b}$$

In these equations $M = m_1 + m_2$ and $\mu = m_1 m_2/(m_1 + m_2)$.

The solutions of equation (6.3a) represent the uniform translation of the center of mass, and the solutions of (6.3b) represent the motion about this center. Since the system as a whole is free, the solutions of (6.3a) are characteristic of momentum as well as of the kinetic energy of translation E_t. The product function U will be characteristic of the total energy of the whole system if both equations (6.3a) and (6.3b) are satisfied, and the sum of E_t and the energy of the relative motion will be equal to E_0. For this reason the energy of the relative motion is written as $E_0 - E_t$ in equation (6.3b). The whole problem is degenerate in this respect. There can be many solutions representing the same total energy E, but in which this energy is differently distributed between translation and relative motion.

PROBLEM 1. Write the solutions of equation (6.3a), and show that they exist for all values of E_t. Form such a combination of those solutions that at $t = 0$ the probability distribution of the position of the center of mass shall be an error function. What must be the indetermination in the energy in order that the center of mass of a hydrogen atom may be located with an r.m.s. indetermination of 10^{-9} cm?

2. The central-field problem

Equation (6.3b) is just the equation for a particle of mass μ moving in a central field. The variables can be separated in spherical polar coordinates. In these coordinates the equation is

$$\frac{1}{r^2}\frac{\partial}{\partial r}\left(r^2\frac{\partial U}{\partial r}\right)+\frac{1}{r^2\sin\vartheta}\frac{\partial}{\partial\vartheta}\left(\sin\vartheta\frac{\partial U}{\partial\vartheta}\right)+\frac{1}{r^2\sin^2\vartheta}\frac{\partial^2 U}{\partial\varphi^2}+$$

$$+\frac{2\mu}{\hbar^2}[E-V(r)]U=0. \tag{6.4}$$

Here E has been written for E_0-E_t, and the subscript has been dropped from the U. To effect the separation, let

$$U(r,\vartheta,\varphi)=R(r)Y(\vartheta,\varphi.) \tag{6.5}$$

Substitution in (6.4), division by RY, and multiplication by r^2 gives

$$\frac{1}{R}\frac{\mathrm{d}}{\mathrm{d}r}\left(r^2\frac{\mathrm{d}R}{\mathrm{d}r}\right)+\frac{1}{Y\sin\vartheta}\frac{\partial}{\partial\vartheta}\left(\sin\vartheta\frac{\partial Y}{\partial\vartheta}\right)+\frac{1}{Y\sin^2\vartheta}\frac{\partial^2 Y}{\partial\varphi^2}+$$

$$+\frac{2\mu}{\hbar^2}(E-V)r^2=0. \tag{6.5a}$$

In this equation the first and last terms are functions of r only, and the other terms are functions of the angles only. Thus each of these two parts must be a constant. Let the constant to which the function of r is equal be C. $R(r)$ must satisfy the equation

$$\frac{1}{r^2}\frac{\mathrm{d}}{\mathrm{d}r}\left(r^2\frac{\mathrm{d}R}{\mathrm{d}r}\right)+\left[\frac{2\mu}{\hbar^2}(E-V)-\frac{C}{r^2}\right]R=0. \tag{6.6}$$

The function $Y(\vartheta,\varphi)$ can also be separated. Let

$$Y(\vartheta,\varphi)=\Theta(\vartheta)\Phi(\varphi) \tag{6.7}$$

then

$$\frac{\sin\vartheta}{\Theta}\frac{\mathrm{d}}{\mathrm{d}\vartheta}\left(\sin\vartheta\frac{\mathrm{d}\Theta}{\mathrm{d}\vartheta}\right)+C\sin^2\vartheta+\frac{1}{\Phi}\frac{\mathrm{d}^2\Phi}{\mathrm{d}\varphi^2}=0. \tag{6.7a}$$

The first two terms are functions of ϑ only, and the last term is a function of φ only. Let the first two equal m^2. Then

$$\frac{\mathrm{d}^2\Phi}{\mathrm{d}\varphi^2} + m^2\Phi = 0 \tag{6.8a}$$

and

$$\frac{1}{\sin\vartheta}\frac{\mathrm{d}}{\mathrm{d}\vartheta}\left(\sin\vartheta\frac{\mathrm{d}\theta}{\mathrm{d}\vartheta}\right) + \left(C - \frac{m^2}{\sin^2\vartheta}\right)\theta = 0. \tag{6.8b}$$

The solution of (6.8a) can be written down immediately. When normalized it is

$$\Phi_m = \frac{1}{\sqrt{2\pi}}\,\mathrm{e}^{\mathrm{i}m\varphi}. \tag{6.9}$$

In order that this function be single-valued, it is necessary for m to be a real integer, positive, negative, or zero. This requirement restricts the suitable solutions of the differential equation to a discrete set.

With m an integer, if $C = l(l+1)$, where l is a positive integer such that $l \geq |m|$, equation (6.8b) is satisfied by the associated Legendre polynomials with cos ϑ as the argument, $P_l^m(\cos\vartheta)$. If C is not equal to $l(l+1)$, the solutions of the equation are not finite and quadratically integrable in the range of the variable ϑ. If the normalized associated Legendre polynomial is designated by Π_l^m, it follows that

$$Y_l^m = \frac{1}{\sqrt{2\pi}}\,\Pi_l^m(\cos\vartheta)\,\mathrm{e}^{\mathrm{i}m\varphi}. \tag{6.10}$$

From the above analysis it is evident that the solutions of the central-field problem in spherical polar coordinates are not characteristic of the energy alone, but may be specified in addition by two integers, l and m, which are called quantum numbers. These quantum numbers are connected with the characteristic values of the angular momentum, as will be shown in the next section. For reasons which are to be found in the historical development of spectroscopy, l is called the azimuthal quantum number and m the magnetic quantum number.

3. Angular momentum in a central field

It follows from equation (6.9), and from the results of Problem 7 of Chapter 3, that the functions obtained are eigenfunctions of the angular momentum about the polar axis and that the eigenvalue is $m\hbar$. It might at first be thought desirable to select functions characteristic of vector angular momentum itself, but this is impossible. It is impossible because the operators for different components of the angular momentum do not commute. A precise specification of one component precludes the precise specification of the others. Nevertheless, the operator for each component commutes with the operator for the square of the absolute magnitude, or with the operator for the sum of the squares of the three components. The operator for this quantity is

$$\mathsf{L}^2 = -\hbar^2 \left[\left(y\frac{\partial}{\partial z} - z\frac{\partial}{\partial y}\right)^2 + \left(z\frac{\partial}{\partial x} - x\frac{\partial}{\partial z}\right)^2 + \left(x\frac{\partial}{\partial y} - y\frac{\partial}{\partial x}\right)\right]. \quad (6.11)$$

If this operator is transformed to spherical polar coordinates, the equation for an eigenfunction of the quantity L^2 can be written as

$$\frac{1}{\sin\vartheta}\frac{\partial}{\partial\vartheta}\left(\sin\vartheta\frac{\partial\chi}{\partial\vartheta}\right) + \frac{1}{\sin^2\vartheta}\frac{\partial^2\chi}{\partial\varphi^2} = -\frac{L_0^2}{\hbar^2}\chi. \quad (6.12)$$

Equation (6.12) is identical with equation (6.7a) if $L_0^2/\hbar^2 = C$, which shows that the eigenvalues of L^2 are $l(l+1)\hbar^2$. Thus the function of the angles, Y_l^m, is characteristic of the square of the total angular momentum with the characteristic value $l(l+1)\hbar^2$ and also of the angular momentum around the polar axis with the characteristic value $m\hbar$.

The physical significance of the requirement that $l \geq |m|$ is now apparent, since one component of the angular momentum cannot be greater than the magnitude of the whole. The different values of m are often pictured as corresponding to the different possible orientations of the vector $\boldsymbol{L}$ with reference to the polar axis. Out of the $2l+1$ different functions for a given value of l it is possible to form linear combinations representing an integral value of the component of the angular momentum in any one direction. Such properties as this are of much importance in the classification of the possible stationary states of atoms.

PROBLEM 2. Show that the operator in polar coordinates for the component of the angular momentum about the x axis is

$$\mathsf{L}_x = -\frac{\hbar}{\mathrm{i}}\left(\sin\varphi\frac{\partial}{\partial\vartheta} + \cot\vartheta\cos\varphi\frac{\partial}{\partial\varphi}\right). \quad (6.13)$$

PROBLEM 3. Show that the function

$$Z = \frac{1}{2}Y_1^1 + \frac{1}{\sqrt{2}}Y_1^0 + \frac{1}{2}Y_1^{-1}$$

is characteristic of the square of the angular momentum with the characteristic value $2\hbar^2$ and also of the angular momentum about the x axis with the value $\hbar$. Find the corresponding linear combinations for the other characteristic values of the angular momentum around the x axis.

PROBLEM 4. Show that

$$\mathsf{L}_x\mathsf{L}_y - \mathsf{L}_y\mathsf{L}_x = -\frac{\hbar}{\mathrm{i}}\mathsf{L}_z \tag{6.14}$$

and in general that

$$\mathbf{L}\times\mathbf{L} = -\frac{\hbar}{\mathrm{i}}\mathbf{L}. \tag{6.14a}$$

PROBLEM 5. If a particle moving in a central field is in such a state that the value of the square of the angular momentum is $2\hbar^2$, and if the component about the z axis has one of its characteristic values, find the distribution of the values of the component about the x axis.

PROBLEM 6. Show that

$$\begin{aligned}\mathsf{L}_x + \mathrm{i}\mathsf{L}_y &= \frac{\hbar}{\mathrm{i}}\,\mathrm{e}^{\mathrm{i}\varphi}\left(\mathrm{i}\frac{\partial}{\partial\vartheta} - \cot\vartheta\,\frac{\partial}{\partial\varphi}\right),\\ \mathsf{L}_x - \mathrm{i}\mathsf{L}_y &= -\frac{\hbar}{\mathrm{i}}\,\mathrm{e}^{-\mathrm{i}\varphi}\left(\mathrm{i}\frac{\partial}{\partial\vartheta} + \cot\vartheta\,\frac{\partial}{\partial\varphi}\right).\end{aligned} \tag{6.15}$$

4. Parity in a central field

Since the potential energy is a function of r only, equation (6.4) is unchanged when the direction of the radius vector is reversed. This involves replacing ϑ by $\pi-\vartheta$ and φ by $\pi+\varphi$. The usual argument shows that the wave function $U(r, \vartheta, \varphi)$ will be either odd or even, or at least can be written so as to be either odd or even.

The change of φ to $\pi+\varphi$ changes the function by $\mathrm{e}^{\mathrm{i}m\pi}$, and the change of ϑ to $\pi-\vartheta$ changes the argument of P_l^m from $(\cos\vartheta)$ to $(-\cos\vartheta)$. These polynomials, as shown in the Appendix, have the parity of $l-m$, so the whole function has the parity of l.

In the simple cases thus far considered, the one- and two-dimensional oscillator, and the single particle in a central field, the parity has always corresponded to that of some other quantum number. In such cases the parity is of little use in classifying the states. For atoms with more than one electron, however, and particularly for atomic nuclei, the parity provides a useful classification of states in addition to the classification in terms of angular momentum.

5. The energy in a Coulomb field

To complete the determination of the energy-characteristic states, it is necessary to find solutions of equation (6.6) which satisfy the boundary conditions. These solutions will depend upon the form of $V(r)$ and must ordinarily be obtained by approximation methods. In the special case in which $V(r) = -Ze^2/4\pi\varepsilon_0 r$, the solution has been extensively investigated, and the energy values have been determined. The characteristic energies are divided into two classes, those greater than zero and those less than zero. The negative energies represent cases in which the particles are bound together, while the positive energies represent states in which the particles have enough energy to fly apart. These two cases must be considered separately.

5.1. *Negative energies*

To make the equation appear less complicated let

$$\frac{2\mu}{\hbar^2}E = -\frac{1}{r_0^2} \quad \text{and} \quad \frac{2\mu}{\hbar^2}\frac{Ze^2}{4\pi\varepsilon_0} = \frac{2D}{r_0}.$$

Then equation (6.6) takes the form

$$\frac{1}{r^2}\frac{\mathrm{d}}{\mathrm{d}r}\left(r^2\frac{\mathrm{d}R}{\mathrm{d}r}\right) + \left[\frac{2D}{r_0 r} - \frac{l(l+1)}{r^2} - \frac{1}{r_0^2}\right]R = 0. \tag{6.16}$$

The singular points of this equation are at zero and at infinity, and it is possible to find at once the behavior of the solutions in those regions. When $r \to 0$, the dominating term in the bracket is $-l(l+1)/r^2$. If the other terms are neglected, the equation is satisfied by r^l or r^{-l-1}. Only the solution with the positive exponent can be used, since the square of the other would not be integrable. As $r \to \infty$ the dominating term is the constant $(-1/r_0^2)$. Neglecting the others, the equation is satisfied by e^{r/r_0} and by e^{-r/r_0}. The general theory of differential equations shows that if R is a solution which approaches r^l as $r \to 0$, it will approach a linear combination, $\alpha \exp(r/r_0) + \beta \exp(-r/r_0)$ as $r \to \infty$. The solution which is finite near zero will, in general, be infinite as r becomes infinite. There exist, however, special cases in which the coefficient α is zero so that the function vanishes at infinity also.

To find the condition for the occurrence of such a special case, let

$\rho = 2r/r_0$, and let $R = \rho^l \mathrm{e}^{-\frac{1}{2}\rho} v(\rho)$. Substitution of these forms into equation (6.16) leads to

$$\frac{\mathrm{d}^2 v}{\mathrm{d}\rho^2} + \left[\frac{2(l+1)}{\rho} - 1\right]\frac{\mathrm{d}v}{\mathrm{d}\rho} + [D-(l+1)]\frac{v}{\rho} = 0. \tag{6.17}$$

This equation has a nonessential singularity at the origin, so a solution can be obtained as a power series,

$$v = \sum_{s=0}^{\infty} a_s \rho^s .$$

The recursion formula obtained by substitution of this power series in the equation is

$$[s+2(l+1)](s+1)a_{s+1} = (s+l+1-D)a_s . \tag{6.18}$$

For large values of s and ρ this recursion formula gives a function which increases as e^{ρ} and prevents the factor $\mathrm{e}^{-\frac{1}{2}\rho}$ from causing the solution to vanish. But in the special case in which D is an integer the series will terminate. If $D = n'+l+1$, the series will terminate with the term $a_{n'}\,\rho^{n'}$. When the series is merely a polynomial, the exponential factor causes the function to vanish for large values of ρ.
Thus the condition for the existence of a solution which satisfies the boundary conditions is that

$$D = \frac{\mu}{\hbar^2}\frac{Ze^2 r_0}{4\pi\varepsilon_0} = n'+l+1 = n$$

or

$$E = -\frac{\mu Z^2 e^4}{8\varepsilon_0^2 h^2}\frac{1}{n^2}. \tag{6.19}$$

This is the equation for the energy levels of a hydrogen atom. It leads to the famous Balmer formula for the wavelengths of the lines of the atomic spectrum of hydrogen. The emergence of this formula from the Schroedinger equation in 1925 was a major factor in the rapid acceptance of "wave mechanics" as the correct description of electron behavior.

The polynomial solution of equation (6.17) is the $(2l+1)$th derivative of the $(n+l)$th Laguerre polynomial whose properties are described in the Appendix. If this is designated by L_{2l+1}^{n+l}, the energy-characteristic solutions of the problem have the radial factor

$$R(\rho) = N\rho^l L_{n+l}^{2l+1}(\rho)\,\mathrm{e}^{-\frac{1}{2}\rho} \tag{6.20}$$

where N is the normalization factor. It is important to remember that the variable ρ is related to the radius vector by the quantity r_0, which depends upon the energy. Thus the scale used for describing this radial function is different for different energies.

5.2. *Positive energies*

For positive energies r_0 will be imaginary, so let $r_0 = \mathrm{i}s_0$. The solutions of equation (6.16) will then be $\exp(\pm \mathrm{i}r/s_0)$ as $r \to \infty$. Neither of these solutions becomes infinite and so neither can be excluded for that reason. The linear combination of them corresponding to the solution r^l at the origin is a suitable eigenfunction and all positive values of E are eigenvalues of the energy. Of course such a function is not quadratically integrable but it has the same properties as the functions for a free particle, and its form imposes no restriction on the energy. If it is necessary to treat the function as normalized it may be multiplied by $\mathrm{e}^{-\alpha r}$ and α allowed to approach zero at the end of the calculation.

The power series based on the recursion formula (6.18) provides two solutions, v and v^*, depending on whether $-\mathrm{i}D$ or $\mathrm{i}D$ is inserted. These are not of a great deal of use except near the origin, and other forms of solution are treated in the Appendix. However something of the nature of the solutions as $r \to \infty$ can be seen from the differential equation.

The solution corresponding to r^l must be real for all values of r, since the differential equation contains no imaginary coefficients. Hence

$$R = Nr^l(v\,\mathrm{e}^{\mathrm{i}r/s_0} + v^*\,\mathrm{e}^{-\mathrm{i}r/s_0}). \tag{6.21}$$

Something of the nature of v and v^* can be learned from a consideration of the differential equation (6.17) as $\rho \to \pm\mathrm{i}\,\infty$. If the second derivative is neglected, and only the leading term in the coefficient of the first derivative is considered, the equation is approximated by

$$\frac{\mathrm{d}v}{\mathrm{d}\rho} = -\,[\mathrm{i}\,|D| + l+1]\,\frac{v}{\rho}. \tag{6.22}$$

This suggests the approximate solutions

$$\log v = -\,[\mathrm{i}\,|D| + l+1]\log\rho + \text{const.} \tag{6.23}$$

and the corresponding complex conjugate. It must be remembered that ρ is an imaginary number and has opposite signs in v and v^*. If ρ is replaced by $\mp 2r/\mathrm{i}s_0$, and the necessary powers of i are taken into the constants,

$$R \approx \frac{\text{const}}{r} A \exp\left\{\mathrm{i}\left[|D| \log\frac{r}{s_0} - \frac{r}{s_0}\right]\right\} + B \exp\left\{-\mathrm{i}\left[|D| \log\frac{r}{s_0} - \frac{r}{s_0}\right]\right\}. \tag{6.24}$$

This approximate solution shows an important property of the Coulomb force. The particle really never gets away from the influence of the center of attraction. The attraction, contained in D, influences the motion for all values of r. The energy never becomes entirely kinetic.

PROBLEM 7. In the case of an atom of hydrogen in the state of lowest energy find the average distance of the electron from the nucleus.

PROBLEM 8. In the case of an atom of hydrogen in the state of lowest energy find the probability that the electron will be found at a distance from the nucleus greater than its energy would permit on the classical theory.

PROBLEM 9. Treat the problem of a free particle in spherical coordinates, and show that the radial factor of an energy-characteristic solution is a spherical Bessel function.

PROBLEM 10. Treat the problem of the three dimensional harmonic oscillator in spherical coordinates.

Chapter 7

METHODS OF APPROXIMATION

In only a few cases can the Schroedinger equation be solved exactly. This is similar to the situation in Newtonian mechanics where Newton's equations can be solved in closed form for only a relatively few simple problems. In most cases an approximation method must be used to obtain an estimate of the desired results.

The methods of approximation in wave mechanics may be considered as falling into two classes. There are the methods for approximating the characteristic values, or the eigenvalues, and the corresponding eigenfunctions of various operators. Then there are the methods for approximating the change of an initial wave function with the time. A number of such methods will be described and illustrated in this chapter. The recent literature on theoretical or mathematical physics contains many modifications of the methods treated here and numerous others adapted to special purposes.

1. The use of discontinuous potentials

In some cases the actual potential energy can be replaced with sufficient accuracy by a function which is constant except for a small number of discontinuities. The energy eigenvalues and eigenfunctions can frequently be determined for such a potential and provide an approximation to those for the original problem.

As an example consider the case of a particle moving in one dimension, which will be taken as the x axis. Let the forces involved be such that no force acts on the particle except in the neighborhood of $x = \pm a$. In these regions let there be a large force of attraction toward the origin acting over such a distance that

$$-\int_{a-\Delta}^{a+\Delta} F \, dx = W.$$

Thus the potential energy at the origin may be taken to be zero, and at large distances from the origin will be W, as illustrated in fig. 7.1. If the force is sufficiently concentrated in two small regions, an approximation to this situation can be obtained by setting $V(x) = 0$ for $-a < x < a$ and $V(x) = W$

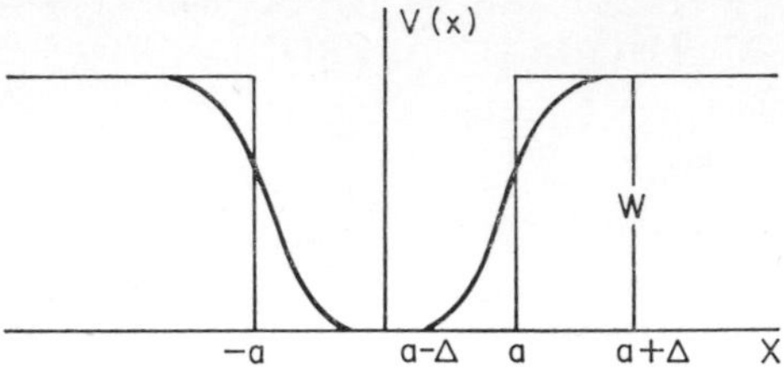

Fig. 7.1. A potential energy function which changes rapidly in certain regions can be approximated by a discontinuous function for which the energy eigenvalue equation can be solved graphically or numerically.

for $x > a$ and $x < -a$. With this potential function the energy states can be found.

Between $x = -a$ and $x = a$ the differential equation is

$$\frac{d^2u}{dx^2} + \frac{2m}{\hbar^2} Eu = 0 \tag{7.1}$$

while for $x < -a$ and $x > a$,

$$\frac{d^2u}{dx^2} + \frac{2m}{\hbar^2}(E-W)u = 0. \tag{7.1a}$$

Consider first the case in which $E < W$. The solution of (7.1) is then

$$u = \alpha\, e^{igx} + \beta e^{-igx} \tag{7.2a}$$

and of (7.1a) is

$$u = \alpha\, e^{-\gamma x} + \beta e^{\gamma x} \tag{7.2b}$$

where both g and γ are real.

Equations (7.2a) and (7.2b) are forms into which solutions in the respective regions can be put. There remains the problem of determining the values of the constants so as to represent the same solution in all regions. Since the differential equation is of the second order, the value of the function and the value of its first derivative are necessary to specify a solution at a given point. These quantities will be continuous along a single solution. Only the second derivative will be discontinuous at the points of discontinuity of the potential. A solution over the whole x-axis can be obtained by matching the forms given in (7.2a) and (7.2b), and adjusting the constants so the first derivatives are continuous, across the discontinuities in potential. Such a solution will be the limit approached by a solution for the continuous potential as the size of the region in which the potential changes approaches zero. The requirement to be imposed is then that the functions and their derivatives must be continuous at the points $x = \pm a$. This requirement is necessary to insure that one is dealing with the "same" solution in all three regions.

The above conditions lead to four equations. In addition it must be required that the functions do not become infinite at infinity, so that the coefficients of the increasing real exponentials must be set equal to zero. If the subscript 1 refers to the solutions for $x < -a$ and the subscript 2 refers to the solutions for $x > a$, $\alpha_1 = \beta_2 = 0$. And

$$\begin{aligned} \beta_1 \, e^{-\gamma a} &= \alpha \, e^{-iga} + \beta \, e^{iga} \\ \gamma\beta_1 \, e^{-\gamma a} &= ig\alpha \, e^{-iga} - ig\beta \, e^{iga} \\ \alpha_2 \, e^{-\gamma a} &= \alpha \, e^{iga} + \beta \, e^{-iga} \\ -\gamma\alpha_2 \, e^{-\gamma a} &= ig\alpha \, e^{iga} - ig\beta \, e^{-iga} \end{aligned} \tag{7.3}$$

where the constants without subscripts refer to the central region. These four simultaneous equations for the determination of the four constants have a solution different from zero only when they are compatible. The condition for compatibility is

$$\tan 2ga = \frac{2g\gamma}{g^2 - \gamma^2}. \tag{7.4}$$

Solutions of this transcendental equation can be obtained by graphical or numerical means and give the values of E which are the characteristic values of the energy. The corresponding solutions of (7.3) then give the energy-characteristic solutions of the problem.

Figure 7.2 shows $\tan 2ga$ and $2g\gamma/(g^2 - \gamma^2)$ plotted as a function of $g^2 = (2m/\hbar^2)E$ for the case in which $(2m/\hbar^2)W = 10 \text{ cm}^{-2}$ and $a = 2$ cm. The two curves intersect at those values of g^2 which represent possible characteristic values of the energy. In case m is taken as the electron mass, $W = 6.1 \times 10^{-27}$ erg. This is not a typical case, but the same curves can be used for some other cases by changing the units. If $W = 6.1 \times 10^{-15}$ erg and $a = 2 \times 10^{-6}$ cm, the curves of fig. 7.2 are correct when the abscissas are read in units of 10^{12}. The maximum value of g^2 is then $10 \times 10^{12} = 10^{13}$ per square centimeter.

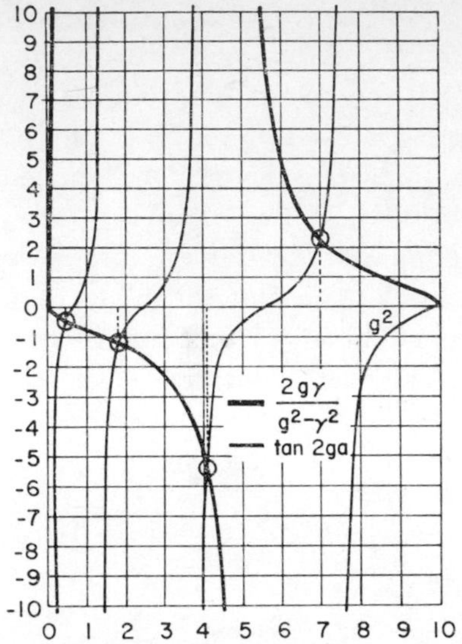

Fig. 7.2. Graphical solution of the transcendental equation for the energy eigenvalues of a particle confined between two discontinuities in potential.

When equation (7.4) is satisfied, the equations (7.3) lead to

$$\begin{aligned} \beta &= \pm\alpha \\ \beta_1 &= (e^{-iga} \pm e^{iga})\, e^{\gamma a}\alpha \\ \alpha_2 &= (e^{iga} \pm e^{-iga})\, e^{\gamma a}\alpha \end{aligned} \tag{7.5}$$

and the normalization condition gives

$$|\,\alpha\,|^2 = \left(4a + \frac{2}{\gamma} \pm \frac{2}{g}\sin 2ga \pm \frac{2}{\gamma}\cos 2ga\right)^{-1}. \tag{7.5a}$$

The upper signs are associated with the lowest value of the energy and with the odd-numbered energy values. The second, fourth, etc., take the lower roots.

PROBLEM 1. Work out an approximate value for the energy of the lowest state when $E \ll W$. For the lowest state, $2ga$ will be near π.

PROBLEM 2. Show that when $W/E \to \infty$, the wave functions vanish at $x = \pm a$ and are zero for $x > a$ and $x < -a$. Compute the energy values and the corresponding functions for this case.

PROBLEM 3. Evaluate the probability that the particle of problem 1 will be found in a region into which, according to classical mechanics, it has not enough energy to penetrate.

PROBLEM 4. Work out the energy-characteristic states for a particle enclosed in a rectangular box (three dimensions) with impenetrable walls, i.e., infinitely high potential energy outside of it.

For the second case, in which $E > W$, the solutions will all be of the form (7.2a), and since this never becomes infinite, there will be six constants to be evaluated. The four equations obtained by matching the solutions will not suffice to determine all of these, so it is necessary to classify the solutions by some arbitrary assignment of values. This can be done by letting $\alpha_2 = 0$ for one solution and $\beta_2 = 0$ for the other. There result two independent solutions, characteristic of momentum in the region $x > a$. One corresponds to positive and the other to negative momentum in this region. The four equations suffice to determine the relationships between the remaining five constants for all values of the energy. Hence an energy-characteristic function can be found for any energy, and all energies are permitted.

PROBLEM 5. Work out the solutions for the case $E > W$.

2. W.K.B. approximation in one dimension

The solution of the Schroedinger equation for energy eigenstates in one dimension has been given a great deal of attention because of its mathematical interest as well as because of its frequent application. The problem requires some very careful mathematical analysis which will be omitted here. For the details there are numerous treatments. A good one is given by MATHEWS and WALKER [1964].

Let the potential energy be an arbitrary function of x and let the energy of the desired eigenfunction be E. Then the x axis will contain two kinds of regions, those for which $(E-V) > 0$ and those for which $(E-V) < 0$. The usual case of interest to which the W.K.B. method is applied contains just one region where $(E-V) > 0$ surrounded on both sides by regions where $(E-V) < 0$.

The equation to be solved is

$$\frac{d^2u}{dx^2} + K(E-V)u = 0 \tag{7.6}$$

where $K = 2m/\hbar^2$. When $(E-V)$ does not vary too rapidly, an approximate solution is given by

$$u = [K(V-E)]^{\frac{1}{4}} \left\{ a \exp\left\{ \int_0^x [K(V-E)]^{\frac{1}{2}} \, dx \right\} + \right.$$
$$\left. + \, b \exp\left\{ - \int_0^x [K(V-E)]^{\frac{1}{2}} \, dx \right\} \right\} \tag{7.7}$$

where a and b are arbitrary constants. Differentiation of (7.7) shows that

$$\frac{\mathrm{d}^2 u}{\mathrm{d}x^2} = \left\{K(V-E) - \tfrac{1}{4}(V-E)^{-1}\frac{\mathrm{d}^2 V}{\mathrm{d}x^2} + \tfrac{5}{16}(V-E)^{-2}\left(\frac{\mathrm{d}V}{\mathrm{d}x}\right)^2\right\} u. \quad (7.8)$$

Hence equation (7.6) is satisfied to the extent that the last two terms on the right side of (7.8) are negligible. If $(V-E)$ is large enough the changes in V are not important, but if $(V-E)$ is small the error gets large unless V is nearly constant. Hence it is clear that in the neighborhood of $V = E$, the approximate solution (7.7) is no good at all.

Although equation (7.7) is valid for both kinds of regions, $V \gg E$ and $V \ll E$, it has quite different behaviors in the two cases. If $V > E$ the exponentials are real and one of them increases without limit. On the other hand, when $V < E$ the exponents are imaginary and the function is oscillating.

Let x_0 be a value of x for which $E = V$ and let $E < V$ for $x < x_0$. In this region the solution may be written

$$u_1 = [K(V-E)]^{-\frac{1}{4}} \left[a_1 \exp\left\{\int_x^{x_0} [K(V-E)]^{\frac{1}{2}}\,\mathrm{d}x\right\} + \right.$$

$$\left. + \, b_1 \exp\left\{-\int_x^{x_0} [K(V-E)]^{\frac{1}{2}}\,\mathrm{d}x\right\}\right]. \quad (7.9\text{a})$$

Unless $a_1 = 0$ the function will increase without limit as $x \to -\infty$ and will be unsuitable for use as an eigenfunction.

For $x > x_0$ the solution may be written

$$u_0 = [K(E-V)]^{-\frac{1}{4}} \left[a_0 \exp\left\{\mathrm{i}\int_{x_0}^{x} [K(E-V)]^{\frac{1}{2}}\,\mathrm{d}x\right\} + \right.$$

$$\left. + \, b_0 \exp\left\{-\mathrm{i}\int_{x_0}^{x} [K(E-V)]^{\frac{1}{2}}\,\mathrm{d}x\right\}\right]. \quad (7.9\text{b})$$

However the relation between (a_1, b_1) and (a_0, b_0) is yet to be determined in order that (7.9a) and (7.9b) shall be approximations to the same exact solution of (7.6). When discontinuous potentials are used, the two kinds of solutions can be connected by matching the functions and their first derivatives at the point of discontinuity in the potential. For the case now under consideration such a procedure is impossible because the approximate forms are not good approximations in the region near $x = x_0$.

Without giving the analysis, it is sufficient here to state that if $a_1 = 0$, the relation between a_0 and b_0 is such that

$$u_0 = A[K(E-V)]^{-\frac{1}{4}} \cos\left\{\int_{x_0}^{x} [K(E-V)]^{\frac{1}{2}}\, dx - \tfrac{1}{4}\pi\right\}. \tag{7.9c}$$

The inference does not hold in the reverse direction, for (7.9c) is only an approximation, and a very slight change in the phase factor $\frac{1}{4}\pi$ will bring in some of the other exponential in (7.9a) which will eventually dominate the solution where $x < x_0$.

If the potential energy after dropping below E at $x = x_0$ rises again through E at $x = x_1$, similar considerations can be applied to connect solutions on the two sides of this point. The equality of the two forms between x_0 and x_1 requires that

$$\cos\left\{\int_{x_0}^{x} [K(E-V)]^{\frac{1}{2}}\, dx - \tfrac{1}{4}\pi\right\} = \pm\cos\left\{\int_{x}^{x_1} [K(E-V)]^{\frac{1}{2}}\, dx - \tfrac{1}{4}\pi\right\}$$
$$= \pm\cos\left\{-\int_{x}^{x_1} [K(E-V)]^{\frac{1}{2}}\, dx + \tfrac{1}{4}\pi\right\}. \tag{7.10}$$

Equation (7.10) can be satisfied if

$$\int_{x_0}^{x} [K(E-V)]^{\frac{1}{2}}\, dx + \int_{x}^{x_1} [K(E-V)]^{\frac{1}{2}}\, dx =$$
$$= \int_{x_0}^{x_1} [K(E-V)]^{\frac{1}{2}}\, dx = (n+\tfrac{1}{2})\pi. \tag{7.11}$$

This is then the condition for the existence of energy eigenvalues, and from this they can be approximately determined.

The above process of determining the energy values can be easily illustrated in the case of the harmonic oscillator. In this particular case the result is exact. Let $V = ax^2$. Equation (7.11) then gives

$$(n+\tfrac{1}{2})\pi = K^{\frac{1}{2}} \int_{x=-(E/a)^{\frac{1}{2}}}^{x=(E/a)^{\frac{1}{2}}} (E-ax^2)^{\frac{1}{2}}\, dx = \tfrac{1}{2}\left(\frac{K}{a}\right)^{\frac{1}{2}} \pi E$$

which gives

$$E = \left(\frac{4a}{K}\right)^{\frac{1}{2}} (n+\tfrac{1}{2}).$$

PROBLEM 6. Find approximate energy levels for a particle attracted toward the origin with a force proportional to the third power of the distance.

PROBLEM 7. Show that if $V = a\,|\,x\,|^{s}$, approximate allowed values of the energy are given by

$$E_n = (n+\tfrac{1}{2})^{2s/(s+2)} N_s - 2s \left[\frac{ha}{(32m)^s}\right]^{1/(s+2)}$$

where

$$N_s = \int_0^1 (1 - t^s)^{\frac{1}{2}}\, dt.$$

3. Perturbation method for eigenfunctions

Perturbation methods are often used quite effectively for the approximate calculation of eigenvalues and the corresponding eigenfunctions. Although the methods can be used with almost any operator, they will be described here in terms of the energy operator, the Hamiltonian function.

The methods are based on the fact that the eigenfunctions of most operators form a complete set in terms of which an arbitrary function can be expanded. The problem is to find the coefficients in the expansion after having found a form of expansion which shows some reasonable tendency to converge.

Consider a system whose Hamiltonian operator can be expanded in terms of a parameter α so that

$$\mathsf{H} = \mathsf{H}_0 + \alpha \mathsf{H}_1 + \alpha^2 \mathsf{H}_2 + \dots \tag{7.12}$$

and H_0 is an operator whose characteristic values and functions are known. The perturbation procedure then gives results that are strictly valid as α approaches zero, but in many cases are useful even when α is fairly large.

Let the characteristic values of H_0 be $E_n^{(0)}$ and the corresponding functions be $u_{n,l}$. The subscript l in addition to n indicates the possibility of degeneracy; to each n and $E_n^{(0)}$ there may belong a number of different functions designated by values of l. These will be understood to be orthogonal to each other, since they can always be taken so. Now the energy eigenfunctions, $\mu_{n\lambda}$, of H are of course different from those of H_0; however, they can always be taken to be a linear combination of the complete set of the eigenfunction $u_{n,l}$ of H_0. This expansion naturally falls into two subgroups: those eigenfunctions $U_{n,l}$ that are needed to describe an eigenstate of H when $\alpha = 0$, and those that depend upon the value of α. In the same way the eigenvalues $E_{n,\lambda}$ of H can be expanded in the eigenvalues of H_0. Thus

$$\begin{aligned} U_{n,\lambda} &= \sum_{m,l} a_{n,\lambda;m,l} U_{m,l} \\ &= \sum_{l} a_{n,\lambda;n,l} U_{n,l} + \sum_{m \neq n} a_{n,\lambda;m,l} U_{m,l} \\ &= \sum_{l} a_{n,\lambda;n,l} U_{n,l} + \alpha V^{(1)}_{n,\lambda} + \alpha^2 V^{(2)}_{n,\lambda} + \dots \end{aligned} \tag{7.13a}$$

where the last expansion is possible since all $a_{n,\lambda:m,l}$ will vanish when α is zero. In the same way

$$E_{n,\lambda} = E^{(0)}_n + \alpha E^{(1)}_{n,\lambda} + \alpha^2 E^{(2)}_{n,\lambda} + \dots . \tag{7.13b}$$

The subscript λ designates the different states whose energies approach $E^{(0)}_n$ as α approaches zero. There will be as many different values of λ as there are of l.

If the expansions (7.13) are substituted into the equation

$$(\mathsf{H} - E_{n,\lambda}) U_{n,\lambda} = 0, \tag{7.14}$$

and the coefficients of the different powers of α are separately equated to zero, a series of equations is obtained which can be solved successively. The solutions give the coefficients of the powers of α in (7.13). The first two equations are

$$(\mathsf{H}_0 - E^{(0)}_{n,\lambda}) V^{(1)}_{n,\lambda} = \sum_l a_{n,\lambda;n,l} (E^{(1)}_{n,\lambda} - \mathsf{H}_1) u_{n,l}, \tag{7.15a}$$

$$(\mathsf{H}_0 - E^{(0)}_n) V^{(2)}_{n,\lambda} = (E^{(1)}_{n,\lambda} - \mathsf{H}_1) V^{(1)}_{n,\lambda} + \sum_l a_{n,\lambda;n,l} (E^{(2)}_{n,\lambda} - \mathsf{H}_2) u_{n,l}. \tag{7.15b}$$

Recalling that the $V^{(1)}_{n,\lambda}$ are linear combinations of all those $u_{m,l}$ with $m \neq n$ then it follows that the right-hand side must be orthogonal to all of the $u_{n,l}$ for the n in question. This requirement leads to a series of homogeneous algebraic equations for the $a_{n,\lambda;n,l}$; and the condition for the compatibility of these equations gives the set of values $E^{(1)}_{n,\lambda}$. For each value of $E^{(1)}_{n,\lambda}$ there will be a set of a $a_{n,\lambda;n,l}$ designated by a value of λ. The linear combinations $\sum_l a_{n,\lambda;n,l} u_{n,l}$ are called the zero-approximation solutions of the problem. They are those solutions of the unperturbed problem toward which the solutions of the actual problem approach as $\alpha \to 0$. Frequently the zero-approximation solutions, and the first-approximation corrections to the energy $E^{(1)}_{n,\lambda}$, are adequate to give the desired information.

To illustrate the procedure up to this point, consider the two-dimensional harmonic oscillator in cartesian coordinates, and add to the Hamiltonian function a small perturbation, $\gamma(x+y)^2$. This represents an additional force parallel to the line $x = +y$ and proportional to the distance from the line

$x = -y$. Classically such an additional force would mean that the oscillator now has two normal frequencies and that the curves of constant potential energy in the x-y plane are ellipses whose principal axes make angles of 45° with the coordinate axes. The "perturbation parameter" in terms of which the solution is to be expanded, is γ, and the series (7.12) contains only two terms.

The state with $n = 0$ is nondegenerate in the unperturbed problem and does not exhibit the behavior to be illustrated. Hence take the state with $n = 1$. As shown in Chapter 4 there are two orthogonal states with the energy $2\hbar\omega$, and hence two values of l. From Chapter 4 the solutions of the unperturbed problem may be written as $u_{n,l}$:

$$u_{1,0}(x, y) = \left(\frac{\alpha}{2\pi}\right)^{\frac{1}{2}} \exp\{-\tfrac{1}{2}\alpha(x^2+y^2)\}H_0(\sqrt{\alpha}x)H_1(\sqrt{\alpha}y),$$

$$u_{1,1}(x, y) = \left(\frac{\alpha}{2\pi}\right)^{\frac{1}{2}} \exp\{-\tfrac{1}{2}\alpha(x^2+y^2)\}H_1(\sqrt{\alpha}x)H_0(\sqrt{\alpha}y). \tag{7.16}$$

In these equations the first subscript on the left-hand side represents the $(s+t)$ of Chapter 4, and the second represents s. The first subscript represents the total energy as does the n in the polar coordinate functions. The second, however, does not represent angular momentum but the energy along the x axis.

These functions are to be inserted in the right side of (7.15a). The whole expression is then multiplied by $u^*_{1,0}$, integrated over x and y, and equated to zero. This is the condition that the right-hand side be orthogonal to $u_{1,0}$, and it gives one equation for the $a_{n,\lambda;n,l}$. The same process with $u^*_{1,1}$ gives another. The two equations are

$$[E^{(1)}_{1,\lambda} - \langle 1, 0 \mid \mathsf{H}_1 \mid 1, 0\rangle]a_{1,\lambda;1,0} - \langle 1, 0 \mid \mathsf{H}_1 \mid 1, 1\rangle a_{1,\lambda;1,1} = 0$$

$$-\langle 1, 1 \mid \mathsf{H}_1 \mid 1, 0\rangle a_{1,\lambda;1,0} + [E^{(1)}_{1,\lambda} - \langle 1, 1 \mid \mathsf{H}_1 \mid 1, 1\rangle]a_{1,\lambda;1,1} = 0 \tag{7.17}$$

where

$$\langle n, l \mid \mathsf{H}_1 \mid n', l'\rangle = \int_{-\infty}^{\infty}\int u^*_{n,l}\mathsf{H}_1 u_{n',l'}\, dx\, dy.$$

The determinant of the coeffients of equations (7.17) is a quadratic function of $E^{(1)}_{n,\lambda}$. When equated to zero it gives two possible values of the energy correction, which are designated by two values of λ. Because of the symmetry of H_1, in this particular case:

$$\langle 1, 0 \mid \mathsf{H}_1 \mid 1, 0\rangle = \langle 1, 1 \mid \mathsf{H}_1 \mid 1, 1\rangle$$

and the solutions of the quadratic equation reduce to

$$E_{1,0}^{(1)} = \langle 1,0 \mid \mathsf{H}_1 \mid 1,0\rangle + \langle 1,0 \mid \mathsf{H}_1 \mid 1,1\rangle,$$
$$E_{1,1}^{(1)} = \langle 1,0 \mid \mathsf{H}_1 \mid 1,0\rangle - \langle 1,0 \mid \mathsf{H}_1 \mid 1,1\rangle. \tag{7.18}$$

Here the second subscript on E is a value of λ. The two solutions are arbitrarily designated as 0 and 1. When these values of $E_{1,\lambda}$ are used in equations (7.17), the ratios of the coefficients $a_{1,\lambda;1,s}$ are given immediately. If it is then required that the resulting linear combinations be normalized, the $a_{1,\lambda;1,s}$ are the following:

$$a_{1,0;1,0} = \frac{1}{\sqrt{2}}, \qquad a_{1,0;1,1} = \frac{1}{\sqrt{2}},$$
$$a_{1,1;1,0} = \frac{1}{\sqrt{2}}, \qquad a_{1,1;1,1} = \frac{-1}{\sqrt{2}}. \tag{7.19}$$

The zero-approximation functions are then

$$u_{1,0}^{(0)} = \frac{1}{\sqrt{2}}u_{1,0} + \frac{1}{\sqrt{2}}u_{1,1},$$
$$u_{1,1}^{(0)} = \frac{1}{\sqrt{2}}u_{1,0} - \frac{1}{\sqrt{2}}u_{1,1}. \tag{7.20}$$

These are solutions of the unperturbed problem rather than of the total problem, but they are those solutions which are changed only a little by the small perturbation.

PROBLEM 8. Carry through the above analysis for the energy level with $n = 2$.

PROBLEM 9. Consider a charged particle moving in a Coulomb field and a uniform magnetic field. Treat the magnetic field as a perturbation. First let the magnetic field be parallel to the axis of the polar coordinates, and treat the case of an arbitrary azimuthal quantum number. Then treat the case of $l = 1$ when the field makes the angle θ with the polar axis. In each case find the zero-order functions and the first-order correction to the energy.

After the zero-approximation functions and the first approximation energy have been determined, the right side of equation (7.15a) is known and one can proceed to the evaluation of $V_{n,\lambda}^{(1)}$. For this purpose it can be expanded in a series of the original unperturbed functions. Let

$$V_{n,\lambda}^{(1)} = \sum_{n',l'}' a_{n,\lambda;n',l'}\, u_{n',l'}\,. \tag{7.21}$$

The prime on the summation sign indicates that the case $n' = n$ is to be omitted from the sum. If this series is substituted into (7.15a) and both sides are multiplied by $u^*_{n',l'}$ and integrated over the variables, it follows that

$$a_{n,\lambda;n',l'} = \frac{\sum_l a_{n,\lambda;n,l}\langle n', l' \mid \mathsf{H}_1 \mid n, l\rangle}{E_n^{(0)} - E_{n'}^{(0)}}\,. \tag{7.22}$$

Since $n' \neq n$, there is no difficulty with a zero in the denominator.

The evaluation of the expansion coefficients by means of equation (7.22) provides the first-approximation correction to the wave function. This can then be used in (7.15b), which can be treated in a similar fashion to get a second approximation. The question of the convergence of this process is one we shall not consider. In most cases the complication increases very rapidly as one goes to higher approximations, and it is customary to stop as soon as the desired effects first appear.

PROBLEM 10. Work out a few of the terms in the first-order correction to the wave function for the case of the two-dimensional oscillator treated above.

PROBLEM 11. Consider a one-dimensional harmonic oscillator in the state $n = 0$ subjected to a small perturbation energy αx. Develop a perturbation procedure for this case of no degeneracy, parallel to that discussed in the text for the case of degenerate states, and develop the first-order corrections to the eigenfunction and to the energy.

4. General expansion in a series of eigenfunctions

The eigenfunctions of a complete set of simultaneous dynamical quantities constitute a complete set for the expansion of other functions describing the same physical system. If the functions are selected so as to be normalized and orthogonal, the determination of the coefficients is formally very simple. Consider, for example, a set of functions V_n characteristic of the operator Q with the characteristic values Q_n. If then it is desired to find the solution of

$$(\mathsf{H} - E)u = 0 \tag{7.23}$$

let

$$u = \sum_n a_n V_n\,. \tag{7.23a}$$

This expansion will converge if the set of functions is complete for the expansion of the desired function u. Substitution of this series in (7.23), multiplication by V_m^* and integration gives

$$\sum_n a_n \int V_m^* \mathsf{H} V_n \, d\tau - E a_m = 0. \tag{7.24}$$

An equation of this type can be written for every value of m, and the a's can, in principle, be determined by solving this set of simultaneous equations. These equations are homogeneous and are compatible for only certain values of E, the eigenvalues of H in equation (7.23).

However, the system of equations is frequently infinite, and the practical determination of the solutions may be no simpler than the treatment of equation (7.23) by some other means. Nevertheless, by judicious selection of the set of functions V_n it is sometimes possible to learn something about the functions u and the eigenvalues E.

a. If the operator H is not too greatly different from Q it may be set equal to (Q+R). Then equation (7.24) will take the form

$$\sum_n a_n \int V^* \mathsf{R} V_n \, d\tau + (Q_m - E) a_m = 0. \tag{7.24a}$$

If the Q_m are all different, the zero order approximation to the eigenvalues is given by the diagonal terms of the determinant of the coefficients in the equations represented by (7.24a):

$$E_m = Q_m + \int V_m^* \mathsf{R} V_m \, d\tau. \tag{7.25}$$

The perturbation procedure described earlier is a systematic approach to the approximate solution of these equations when R is small compared with Q.

If two or more diagonal terms are equal, the situation is the one of degeneracy described above. In that case the off diagonal terms cannot all be neglected. Let

$$Q_m + \int V_m^* \mathsf{R} V_m \, d\tau = Q_{m'} + \int V_{m'}^* \mathsf{R} V_{m'} \, d\tau$$

for example. The two equations to be solved for a zero order approximation are then

$$\left\{E - Q_m - \int V_m^* \mathsf{R} V_m \, d\tau\right\} a_m - \int V_m^* \mathsf{R} V_{m'} \, d\tau \, a_{m'} = 0,$$

$$-\int V_{m'}^{*}\mathsf{R}V_{m}\,\mathrm{d}\tau\, a_{m} + \left\{E - Q_{m'} - \int V_{m'}^{*}\mathsf{R}V_{m'}\,\mathrm{d}\tau\right\} a_{m'} = 0.$$

The two solutions give the zero order energy and function as in the perturbation procedure.

b. If the operator Q of which the V_m are eigenfunctions is Hermitian, and if it commutes with H, a great many of the integrals in (7.24) will vanish. Since

$$\int V_{m}^{*}\mathsf{H}\mathsf{Q}V_{n}\,\mathrm{d}\tau = Q_{n}\int V_{m}^{*}\mathsf{H}V_{n}\,\mathrm{d}\tau = \int V_{m}^{*}\mathsf{Q}\mathsf{H}V_{n}\,\mathrm{d}\tau = Q_{m}\int V_{m}^{*}\mathsf{H}V_{n}\,\mathrm{d}\tau \tag{7.26}$$

the integral $\int V_{m}^{*}\mathsf{H}V_{n}\,\mathrm{d}\tau$ is zero unless $Q_n = Q_m$. Thus the equations represented by (7.24) can be separated into smaller sets, each one of which corresponds to a single Q_n. The solutions of each set are thus also eigenfunctions of Q, and the determinant which must be made to vanish by suitable selection of E is reduced to a series of squares on the diagonal. However, each square may still be infinite and there may be an infinite number of them.

PROBLEM 12. Show by the above method that the introduction of a change in the potential energy as a function of the radius will not change the angular dependence of the energy functions in the central-field problem. This is equivalent to showing that the solutions of the perturbed as well as of the unperturbed problem can be divided into groups characterized by the quantum numbers describing the angular momentum.

5. The variation method

The Schroedinger equation for any energy-characteristic state of a conservative system is the differential equation for a variation problem. The quantity to be given a stationary value is the average of the Hamiltonian function, and the auxiliary condition is that the function u must be normalized. That such is the case can be shown very simply for a single particle in one dimension. The extension to the general case is complicated but not otherwise more difficult.

The integral is

$$\langle H\rangle = \int_{-\infty}^{\infty}\left[-\frac{\hbar^2}{2m}u^{*}\frac{\mathrm{d}^2u}{\mathrm{d}x^2} + u^{*}V(x)u\right]\mathrm{d}x, \tag{7.27}$$

with the auxiliary condition that

$$\int_{-\infty}^{\infty} u^* u \, dx = 1.$$

The second derivative can be integrated once, and the variation problem, including the auxiliary condition, can be formulated as follows:

$$\delta \int_{-\infty}^{\infty} \left(\frac{\hbar^2}{2m} \frac{du^*}{dx} \frac{du}{dx} + u^* V u - \lambda u^* u \right) dx = 0. \tag{7.28}$$

The differential equation for u is then

$$\frac{\hbar^2}{2m} \frac{d^2 u}{dx^2} - Vu + \lambda u = 0 \tag{7.29}$$

and the equation for u^* is similar. The value of λ is just the value of $\langle H \rangle$ obtained by putting a solution of (7.29) into (7.27).

The importance of the variation problem lies in the fact that it may be attacked by direct methods and so sometimes the solutions of the differential equation can be obtained from the variation problem rather than proceeding in the reverse order. This is usually practicable for only the lowest of the energy levels, since for the others a number of additional auxiliary conditions must be introduced which very much complicate things. The general procedure for solving the variation problem directly is merely to try a variety of functions in the integral (7.27). That function which gives the lowest value of $\langle H \rangle$ is considered to be the best approximation to the true function. At least the value of $\langle H \rangle$ which it gives is an upper bound to the true energy value as is also shown by the following argument:

Let $\langle \alpha \mid \mathsf{H} \mid \alpha \rangle = \int \psi_\alpha^* \mathsf{H} \psi_\alpha \, d\tau$, where α represents the parameters in the assumed wave function ψ_α. Let the exact wave function for the ground state be ψ_0 and the other eigenfunctions of H be ψ_n. ψ_0 and the other ψ_n form a complete orthogonal set, and since ψ_0 is the ground state, $E_n \geq E_0$ for all n. ψ_α can be expanded in terms of the ψ_n so let

$$\psi_\alpha = \sum_n a_n(\alpha) \psi_n .$$

Then

$$\begin{aligned}
\langle \alpha \mid \mathsf{H} \mid \alpha \rangle &= \sum_{m,n} a_m^*(\alpha) \langle m \mid \mathsf{H} \mid n \rangle a_n(\alpha) \\
&= \sum_{m,n} a_m^* a_n E_n \delta_{m,n} = \sum_n |a_n|^2 E_n \\
&= |a_0|^2 E_0 + \sum_{n>0} |a_n|^2 E_n \\
&= E_0 + \sum_{n>0} |a_n|^2 (E_n - E_0) > E_0 .
\end{aligned}$$

The above proof assumes no degeneracy, which simplifies the notation. It also makes use of the normalization of all functions so that

$$\sum_{n=0}^{\infty} | a_n^2 | = 1 .$$

PROBLEM 13. Assume as the solution for the ground state of the one-dimensional oscillator the function $u = A\,e^{-\alpha x^2}$, and find that value of α which makes the energy integral a minimum.

PROBLEM 14. With the approximation function of the above problem, find an approximate value for the lowest energy of a particle attracted toward the origin with a force proportional to the third power of the distance.

6. Development of wave function with time

The general problem in quantum mechanics, as in classical mechanics, is to predict the state of a system at the time t in terms of its state at a previous time t_0. The Schroedinger equation gives this change with the time in terms of the Hamiltonian operator describing the system. Various methods have been devised for approximating solutions of this equation. They each have advantages for special purposes. The one described here is often called the method of the variation of constants.

Let the initial state of the system be an eigenstate of the operator Q. This is no limitation since any state is an eigenstate of some operator, and in particular, any state produced by an arrangement of physical apparatus is an eigenstate of that apparatus. Then at any time

$$\psi(r, t) = \sum_n A_n(t) V_n(r), \tag{7.30}$$

where $\mathsf{Q} V_n = Q_n V_n$. Whatever form the function $\psi(r, t)$ takes at the time t, it can be expanded in this kind of a series. The Schroedinger equation then leads to

$$-\frac{\hbar}{\mathrm{i}} \sum_n \frac{\mathrm{d}A_n}{\mathrm{d}t} V_n = \sum_n A_n \mathsf{H} V_n, \tag{7.31}$$

where H is the Hamiltonian operator of the system; and the orthogonality of the functions V_n leads to the conclusion that

$$-\frac{\hbar}{\mathrm{i}}\frac{\mathrm{d}A_m}{\mathrm{d}t} = \sum_n A_n \int V_m^* \mathsf{H} V_n \,\mathrm{d}\tau$$

$$= \sum_n A_n \int V_m^* (\mathsf{H}-\mathsf{Q})\, V_n \,\mathrm{d}\tau + Q_m A_m \tag{7.32}$$

$$= (Q_m + W_m)A_m + \sum_{n\neq m} A_n \int V_m^* (\mathsf{H}-\mathsf{Q}) V_n \mathrm{d}\tau,$$

where

$$W_m = \int V_m^* (\mathsf{H}-\mathsf{Q}) V_m \,\mathrm{d}\tau.$$

If H and Q are identical, equation (7.32) has a very simple form and no perturbation treatment is necessary. If H and Q are widely different, a perturbation treatment is not very useful, but if $(\mathsf{H}-\mathsf{Q})$ is relatively small an approximate solution of equations (7.32) can be useful.

It is then convenient to define $\mathsf{H}-\mathsf{Q} = \mathsf{W}$ so that W may be regarded as the perturbing operator and

$$W_m = \int V_m^* (\mathsf{H}-\mathsf{Q}) V_m \,\mathrm{d}\tau = \int V_m^* \mathsf{W} V_m \,\mathrm{d}\tau.$$

It is important to notice that H is the Hamiltonian operator of the total system under consideration. Q may be the Hamiltonian operator of a portion of the system or it may merely be some other operator, not too far different from H.

Because of the presence of the term $(Q_m + W_m)A_m$ in equation (7.32), it is convenient to define a new set of coefficients

$$a_n(t) = A_n(t) \exp\left\{+\frac{\mathrm{i}}{\hbar}(W_n + Q_n)t\right\} = A_n \exp\left\{\frac{\mathrm{i}}{\hbar} H_n t\right\}. \tag{7.33}$$

With the definitions of W_n and Q_n it follows that

$$H_n = \int V_n^* \mathsf{H} V_n \,\mathrm{d}\tau. \tag{7.33a}$$

Equation (7.32) can then be rewritten in terms of the a_m as

$$\frac{\mathrm{d}a_m}{\mathrm{d}t} = -\frac{\mathrm{i}}{\hbar}\sum_{n\neq m} a_n \exp\left\{-\frac{\mathrm{i}}{\hbar}(H_n - H_m)t\right\} \int V_m^* \mathsf{W} V_n \,\mathrm{d}\tau. \tag{7.34}$$

If W is zero, or is a function of Q, equation (7.34) shows that the coefficients a_m are constants, since

$$\int V_m^* \mathsf{W} V_n \, \mathrm{d}\tau = 0 \text{ for } m \neq n .$$

The states $V_n(r)$ are then already stationary states. This is the reason for the convenience of the definition in equation (7.33). A similar conclusion follows if H commutes with Q, for then the only terms appearing on the right side of (7.34) are those for which $Q_n = Q_m$, and there may be only a few such terms.

To proceed with the perturbation solution of equation (7.34) let the initial state be V_0. Then $A_0(0) = a_0(0) = 1$, and all other A_m and a_m are zero at $t = 0$. The initial rates of change of the coefficients are given by

$$\frac{\mathrm{d}a_m}{\mathrm{d}t} = -\frac{\mathrm{i}}{\hbar} \exp\left\{-\frac{\mathrm{i}}{\hbar}(H_0 - H_m)t\right\} \int V_m^* \mathsf{W} V_0 \, \mathrm{d}\tau \tag{7.35}$$

of which the integral is

$$a_m = \frac{\langle m \mid \mathsf{W} \mid 0 \rangle}{H_0 - H_m} \left[\exp\left\{-\frac{\mathrm{i}}{\hbar}(H_0 - H_m)t\right\} - 1\right]. \tag{7.36}$$

In this expression

$$\langle m \mid \mathsf{W} \mid 0 \rangle = \int V_m^* \mathsf{W} V_0 \, \mathrm{d}\tau . \tag{7.36a}$$

This expression is clearly valid for only a short time, but for this short time the expansion (7.30) leads to

$$\psi(r, t) \approx \exp\left\{-\frac{\mathrm{i}}{\hbar} H_0 t\right\} \left\{V_0 + \sum_{n \neq 0} \frac{\langle n \mid \mathsf{W} \mid 0 \rangle}{H_0 - H_n} \times \right.$$
$$\left. \times \left[1 - \exp\left\{-\frac{\mathrm{i}}{\hbar}(H_n - H_0)t\right\}\right] V_n \right\}. \tag{7.37}$$

Because of the short lived validity of this relationship, the exponential may be expanded to give

$$\psi(r, t) \approx \exp\left\{-\frac{\mathrm{i}}{\hbar} H_0 t\right\} \left\{V_0 - \frac{\mathrm{i}}{\hbar} \sum_{n \neq 0} \langle n \mid \mathsf{W} \mid 0 \rangle t V_n \right\}. \tag{7.37a}$$

If only these terms were included in the summation the function would not remain normalized. However the departure from normalization would be represented by a term proportional to t^2 and would be compensated in the next order of approximation.

It is not difficult to show that the exact solution of equations (7.34) does conserve the normalization. If equation (7.34) is multiplied by a_m^* and the complex conjugate equation for the rate of change of a_m^* is multiplied by a_m, the sum gives the rate of change of $|a_m|^2$:

$$\begin{aligned}\frac{\mathrm{d}}{\mathrm{d}t}|a_m|^2 &= a_m^* \frac{\mathrm{d}a_m}{\mathrm{d}t} + \frac{\mathrm{d}a_m^*}{\mathrm{d}t} a_m \\ &= -\frac{\mathrm{i}}{\hbar}\sum_{n\neq m}\left\{a_m^* a_n \exp\left\{-\frac{\mathrm{i}}{\hbar}(H_n-H_m)t\right\}\langle m|\mathrm{W}|n\rangle\right. \\ &\qquad \left.-a_m a_n^* \exp\left\{+\frac{\mathrm{i}}{\hbar}(H_n-H_m)t\right\}\langle n|\mathrm{W}|m\rangle\right\}. \end{aligned} \tag{7.38}$$

The sum of these expressions over all values of m gives zero, so

$$\sum_m |a_m|^2 = 1$$

at all times for the exact solution of the set of equations (7.34).

To proceed to the second approximation one inserts in equation (7.34) first order solutions (7.36). For a_0 this gives

$$\frac{\mathrm{d}a_0}{\mathrm{d}t} = -\frac{\mathrm{i}}{\hbar}\sum_{n\neq 0}\frac{\langle n|\mathrm{W}|0\rangle}{H_0-H_n}\left[\exp\left\{-\frac{\mathrm{i}}{\hbar}(H_0-H_n)t\right\}-1\right]\times$$

$$\times\exp\left\{-\frac{\mathrm{i}}{\hbar}(H_n-H_0)t\right\}\langle 0|\mathrm{W}|n\rangle \approx -\frac{\mathrm{i}}{\hbar}\sum_{n\neq 0}|\langle n|\mathrm{W}|0\rangle|^2\, t \tag{7.38a}$$

after expanding the exponential. This equation can be integrated, either before or after the expansion, to give

$$a_0 \approx 1 - \frac{1}{2\hbar^2}\sum_{n\neq 0}|\langle n|\mathrm{W}|0\rangle|^2\, t^2. \tag{7.38b}$$

From equation (7.36)

$$a_n = -\frac{\mathrm{i}}{\hbar}\langle n|\mathrm{W}|0\rangle\, t \tag{7.39}$$

so that to terms in t^2 the normalization is preserved.

For the derivatives of a_n other than a_0 one inserts in equation (7.34) the first order expressions for a_n. Thus

$$\frac{\mathrm{d}a_n}{\mathrm{d}t} = -\frac{\mathrm{i}}{\hbar} a_0(0) \exp\left\{-\frac{\mathrm{i}}{\hbar}(H_0-H_n)t\right\} \langle m \mid \mathsf{W} \mid 0\rangle$$

$$-\frac{\mathrm{i}}{\hbar} \sum_{\substack{m\neq n\\ m\neq 0}} \frac{\langle m \mid \mathsf{W} \mid 0\rangle}{H_0-H_m} \left[\exp\left\{-\frac{\mathrm{i}}{\hbar}(H_0-H_n)t\right\}\right.$$

$$\left. - \exp\left\{-\frac{\mathrm{i}}{\hbar}(H_m-H_n)t\right\}\right] \langle n \mid \mathsf{W} \mid m\rangle \tag{7.40}$$

which gives

$$a_n \approx a_0 \langle n \mid \mathsf{W} \mid 0\rangle \frac{\exp\left\{-\frac{\mathrm{i}}{\hbar}(H_0-H_n)t\right\} - 1}{H_0-H_n} \times$$

$$\times \sum_{\substack{m\neq n\\ m\neq 0}} \frac{\langle n \mid \mathsf{W} \mid m\rangle \langle m \mid \mathsf{W} \mid 0\rangle}{H_0-H_m} \times$$

$$\times \left[\frac{\exp\left\{-\frac{\mathrm{i}}{\hbar}(H_0-H_n)t\right\} - 1}{H_0-H_n} - \frac{\exp\left\{-\frac{\mathrm{i}}{\hbar}(H_m-H_n)t\right\} - 1}{H_m-H_n}\right]. \tag{7.41}$$

If these exponentials are expanded to include the first non-vanishing power of t

$$a_n \approx -\frac{\mathrm{i}}{\hbar} \langle n \mid W \mid 0\rangle t - \frac{1}{2\hbar^2} \sum_{\substack{m\neq n\\ m\neq 0}} \langle n \mid W \mid m\rangle \langle m \mid W \mid 0\rangle t^2. \tag{7.42}$$

The above analysis has tacitly assumed that all of the quantities H_n are distinct and sufficiently separated that the denominators do not become unduly small. The extensive use of such perturbation methods has led to the development of a terminology which is often useful in describing as well as suggesting the perturbation process, but whose limited significance also needs to be kept in mind. The expressions for $\mid A_n(t) \mid^2$ obtained from equations (7.33) and (7.42) are often described as giving the probability that the system has "jumped" during the time t from the initial state V_0, in which it started, to the state V_m. Such a statement needs to be used with care to be properly understood. The following things may be said.

a. The wave function has changed continuously from

$$\psi(r, 0) = V_0(r) \text{ at } t = 0$$

to

$$\psi(r, t) \approx \left[1 - \frac{1}{2\hbar^2} \sum_{n \neq 0} |\langle n \mid \mathrm{W} \mid 0\rangle|^2 t^2\right] \exp\left\{-\frac{i}{\hbar} H_0 t\right\} V_0(r)$$

$$- \sum_{n \neq 0} \left[-\frac{i}{\hbar} \langle n \mid \mathrm{W} \mid 0\rangle t - \frac{1}{2\hbar^2} \sum_{\substack{m \neq n \\ m \neq 0}} \langle n \mid \mathrm{W} \mid m\rangle \langle m \mid \mathrm{W} \mid 0\rangle t^2\right] \times$$

$$\times \exp\left\{-\frac{i}{\hbar} H_n t\right\} V_n(r) \qquad (7.43)$$

at the time t. There has been no discontinuity in the wave function.

b. At $t = 0$ the function ψ was an eigenfunction of the operator Q with the eigenvalue Q_0. At the time t it is an eigenfunction of the operator which Q has become by virtue of the equation (7.26) of Chapter 3.

c. The justification for the use of the term "jumped" is that if an apparatus is used which indicates a value of Q at the time t, it may give any one of the values Q_n with a relative probability given by equation (7.37a). There are no values in between. In case $\langle n \mid \mathrm{W} \mid 0\rangle = 0$ it is said the system cannot "jump" from v_0 to V_n directly, that such a "transition" is forbidden. However such a transition can be made by way of the states V_m with $m \neq n$.

In the time development of a mechanical system there are usually several quantities that are invariant, or conserved. If there are no external forces the energy is conserved. This means that the distribution of energy over its possible values (eigenvalues) does not change with the time. The mean values of all the powers of the energy must be the same for the state given in equation (7.42) as for the state V_0. In such a computation the "intermediate" or "virtual" states, designated by m in equations (7.40)–(7.42), do not appear except in the evaluation of the coefficients a_n.

The above results all indicate probabilities of "transition" that increase with the second or higher powers of the time. This is a result of the assumption that the values H_n are well separated and well separated from H_0. For the idea of a "transition probability" to be most applicable the probability must increase linearly with the time. This occurs when a number of states V_m have values of H_m very close together and overlapping H_0. In such a situation one may be interested in $\Sigma_m \mid a_m \mid^2$ and it can be evaluated from the form of (7.37), if the quantity $|\langle m \mid \mathrm{W} \mid 0\rangle|^2$ changes slowly enough with m when H_m is near H_0 that it can be taken outside the summation sign. Then

$$\sum_m | a_m(t) |^2 \approx 4 | \langle m | \mathsf{W} | 0 \rangle |^2 \sum_m \frac{\sin^2 \{(H_0 - H_m)t/2\hbar\}}{(H_0 - H_m)^2} \, . \tag{7.43}$$

If the states V_m are such that a density function $\rho(H_m)$ can be used to describe their distribution as a function of H_m, the sum can be replaced by an integral and

$$\sum_m \left| a_m(t) \right|^2 \approx \frac{2\pi}{\hbar} \left| \langle m | \mathsf{W} | 0 \rangle \right|^2 \rho(H_m) t \, . \tag{7.44}$$

The coefficient of t is then designated as the total probability per unit time that the system will have jumped from V_0 to some one of the states for which $H_m \approx H_0$.

CHAPTER 8

A GENERAL FORMULATION OF QUANTUM MECHANICS

The preceding chapters have been devoted to the wave mechanics associated with the names of Schroedinger and de Broglie. For many problems in atomic and molecular physics in which the dynamical quantities of interest correspond to similar quantities in classical mechanics, it is a convenient method of describing physical situations. However, there are other forms of quantum mechanics which can be applied to such problems and can also deal more naturally with dynamical variables having no classical counterparts, such as electron spin. One of these, the matrix mechanics of Heisenberg, antedated the wave mechanics and is often useful because of its close analogy in form with classical dynamics.

The various forms of quantum mechanics are entirely equivalent. One or another may be chosen for its convenience in application to a particular problem, but the choice is largely a matter of taste. There will be developed in this chapter a general notation and terminology, due to Dirac, which is applicable to most problems and which, by its somewhat abstract nature, emphasizes the statements already made as to the symbolic nature of the wave functions of Schroedinger's mechanics. In this notation, the equations of quantum mechanics can be written in close analogy with those of classical mechanics, and the nature of the relationship between the mathematics and the associated physical situations helps to relegate to the background questions as to whether an electron "*is*" a wave or a particle.

1. Vectors in abstract space

A vector in three dimensions is a familiar concept in physics and in mathematics. It can be pictured as an arrow whose length and direction are both significant. Many physicists like to think in such terms. It can also be defined as a set of three numbers, the vector components, which transform in a prescribed way when the coordinates, or basic unit vectors, are rotated. The idea is easily generalized to more than three dimensions and eventually to an infinity of dimensions. It can be further generalized to the case in which the dimensions are not distinct, or discrete. Thus a function $f(x)$ may be regarded as a vector if each value of x is regarded as a "direction" in the vector space, and $f(x)$ is the length of the component in the "direction" designated by x. Such a generalization, of course, requires attention to questions of convergence.

The "state" of a mechanical system can be represented by such a "vector". In the cases already treated by wave mechanics the state vectors (wave functions) are vectors in a space of infinitely many dimensions. There are cases, however, in which the dimensions, although infinite in number, are discrete. There are other cases in which only a finite number of dimensions need to be considered.

The state of a system of particles in classical mechanics is represented by a point in the phase space and such a point can be regarded as defining a vector. The point is the end of the vector from the origin. In quantum mechanics, however, the vector is in a more general space whose nature will be discussed in detail, and whose properties are those of the vectors defined in it.

The vectors in such an abstract space can be added together by adding the corresponding components. They can be multiplied by numbers, real or complex. The results of such addition and multiplication by numbers will be other vectors in the same space. Such vectors will be called "ket" vectors and designated by the symbol $| A\rangle$. The latter is used to indicate which ket vector is meant. In most applications to quantum mechanics the length of the vector is not significant. $| A\rangle$, $2 | A\rangle$, $e^{i\alpha} | A\rangle$, $- | A\rangle$ all represent the same state. It is the "direction" which is important.

The multiplication of vectors requires particular attention. In the usual three dimensional space the scalar product and the vector product (sometimes called the inner product and the outer product) are defined. To any vector space there is a "dual" space, and products are defined of a vector in the dual space and one in the primary space. For the familiar cartesian vectors the dual space is identical with the primary space so the distinction is often ignored, but when nonorthogonal coordinates are used the distinction again becomes important.

In the study of crystal lattices there is the lattice of the crystal and the reciprocal lattice which is dual to it. A vector of the reciprocal lattice has the dimension of an inverse length, so that the scalar product of a vector of the reciprocal lattice with a vector of the crystal lattice is a number without dimensions. Such a product is the sum of products of a component of the vector in the dual space with the corresponding component of the vector in the primary space. The sum is over all of the components.

In the wave mechanics the corresponding product is a scalar product of such a pair of dual vectors each in a space of infinitely many dimensions. One cannot sum over the different "directions" since they are not discrete. The scalar product of two such vectors is given by an integral $\int f^*(x)g(x)\,\mathrm{d}x$. The use of the complex conjugate symbolizes the vector in the dual space.

In Dirac's formulation of quantum mechanics it is necessary to define a space dual to the space of the ket vectors. Vectors in this dual space are called "bra" vectors and are designated by the symbol $\langle B \,|$. The scalar product of $\langle B \,|$ with $|\, A\rangle$ is designated by $\langle B \,|\, A\rangle$ and is a number, which may be real or complex. If $\langle B \,|\, A\rangle = 0$ for all possible ket vectors $|\, A\rangle$, then $\langle B \,| = 0$. Similarly, if $\langle B \,|\, A\rangle = 0$ for all $\langle B \,|$, then $|\, A\rangle = 0$.

If a ket vector $|\, A\rangle$ represents a certain quantum mechanical state, the same state can be represented by a bra vector, and to indicate the correspondence the bra vector will be designated by the same letter, $\langle A \,|$. However, $c \,|\, A\rangle$ will correspond to $c^*\langle A \,|$, where c^* is the complex conjugate of the number c. Since multiplication by complex numbers is permitted, the "components" of both ket and bra vectors may be complex numbers, and the vectors themselves may be regarded as complex quantities. But since a bra and a ket cannot be added together to obtain a meaningful result, the usual methods of separating a complex number into its real and imaginary parts are not applicable. Dirac has used the term "complex imaginary" to refer to the relationship between bra vectors and the corresponding ket vectors. The scalar product of a bra with a ket vector is, in general, a complex number and

$$\langle B \,|\, A\rangle = \langle A \,|\, B\rangle^* . \tag{8.1}$$

Consequently $\langle A \,|\, A\rangle$ is real and is greater than zero unless $|\, A\rangle = 0$.

In three dimensional vector analysis one introduces a triad of unit vectors in terms of which any vector can be expressed. It is often convenient to take these as unit vectors along the axes of a cartesian coordinate system. Similarly, it is desirable to introduce a complete set of ket vectors and bra vectors in terms of which an arbitrary vector can be expressed. A complete set of basic vectors will have as many members as there are dimensions in the vector space. This dimensionality has nothing to do with the dimensionality of the mechanical system under consideration, but is the number of different (linearly independent) states in which the system might be found. A particle moving in one dimension of physical space may be represented by a ket vector in a non-denumerably infinite dimensional abstract space, i.e. one whose components are $\psi(x)$.

It is usually convenient to take the basic vectors as orthogonal. Two vectors are orthogonal if the scalar product of the bra of one with the ket of the other is zero. Thus $| n\rangle$ and $| m\rangle$ are orthogonal if, and only if,

$$\langle m | n\rangle = \langle n | m\rangle = 0. \tag{8.2}$$

If the basic vectors are discrete, an arbitrary vector can be expressed as a sum of them with suitable coefficients. For example,

$$| P\rangle = \sum_n p_n | n\rangle, \tag{8.3a}$$

where the sum may be finite or infinite. If the dimensions of the space are not discrete, each basic vector can be identified by a value of a continuous variable x, and the expansion of an arbitrary ket has the form

$$| P\rangle = \int \mathrm{d}x\, p(x) | x\rangle. \tag{8.3b}$$

In order that expressions such as (8.3a) and (8.3b) have a well defined meaning it is necessary that a convention be adopted as to the length of a ket or a bra, even though the length has no meaning in the representation of the state of a mechanical system. The usual "normalization" is to require that

$$\langle n'' | n'\rangle = \delta(n'', n') \tag{8.4}$$

when the kets are distinct, and

$$\int \langle x'' | x'\rangle \,\mathrm{d}x' = \int \delta(x''-x')\,\mathrm{d}x' = 1 \tag{8.4a}$$

when they are not distinct. The orthogonality of $| x''\rangle$ and $| x'\rangle$ means that $\langle x'' | x'\rangle = 0$ when $x'' \neq x'$. With the normalization as indicated in equations (8.4) and (8.4a), the property of orthogonality makes possible the direct evaluation of the p_n and the $p(x)$ in equations (8.3a) and (8.3b). If equation (8.3a) is multiplied on the left by $\langle m |$, there results

$$\langle m | P\rangle = \sum_n p_n \langle m | n\rangle = p_m. \tag{8.3c}$$

Similarly equation (8.3b) leads to

$$\langle x'' | P\rangle = \int \mathrm{d}x\, p(x)\langle x'' | x\rangle = p(x''). \tag{8.3d}$$

The last equality follows because $\langle x'' \mid x \rangle = 0$ except when $x = x''$, so that $p(x)$ can be taken outside the integral as $p(x'')$. The quantity $\langle x'' \mid x \rangle$ is equivalent to Dirac's delta function $\delta(x''-x)$, which is defined by equations (8.4a) and (8.3d).

Corresponding to the set of orthogonal basic vectors in the ket space is a set of similarly orthogonal vectors in the bra space, as implied in the above discussion.

PROBLEM 1. Show that the coefficients p_n in equation (8.3a) are the complex conjugates of the coefficients in the expansion

$$\langle P \mid = \sum_n p'_n \langle n \mid .$$

PROBLEM 2. Evaluate the scalar product of two vectors $\mid A \rangle$ and $\mid B \rangle$ when each is expressed in terms of the discrete set of basic vectors $\mid n \rangle$, and when each is expressed in terms of the continuous set of basic vectors $\mid x' \rangle$.

2. Linear operators

Thus far the only combination of bra and ket vectors discussed has been the scalar product, with the bra written to the left of the ket. Such a product is a complex number. However, it is also possible to give a meaning to a ket vector written to the left of a bra. Such a combination is to be interpreted as a linear operator. $\mid A \rangle \langle B \mid$ can be applied to the ket vector $\mid P \rangle$ and the distributive law maintained if

$$\mid A \rangle \langle B \parallel P \rangle = \mid A \rangle \{\langle B \mid P \rangle\} = \{\langle B \mid P \rangle\} \mid A \rangle . \qquad (8.5)$$

The operator $\mid A \rangle \langle B \mid$ changes the vector $\mid P \rangle$ into a vector in the direction of $\mid A \rangle$ and with a length (complex) equal to $\langle B \mid P \rangle$ times the length of $\mid A \rangle$. Such a combination of vectors is called a dyadic and is often used in elementary three dimensional analysis.

A dyadic can also operate to the left on a bra vector

$$\langle P \parallel A \rangle \langle B \mid = \{\langle P \mid A \rangle\} \langle B \mid . \qquad (8.6)$$

In general, the operator $\mid A \rangle \langle B \mid$ operating on $\langle P \mid$, which is the complex imaginary of $\mid P \rangle$, gives $\langle P \mid A \rangle \langle B \mid$ which is *not* the complex imaginary of $\langle B \mid P \rangle \mid A \rangle$ obtained by operating on $\mid P \rangle$. However, there is another operator, in this case $\mid B \rangle \langle A \mid$, which operates on $\langle P \mid$ to give $\langle P \mid B \rangle \langle A \mid$ which is the complex imaginary of $\langle B \mid P \rangle \mid A \rangle$. *Such an operator is called the adjoint of the first* and is indicated by a dagger. Thus

$$(|A\rangle\langle B|)^\dagger = |B\rangle\langle A|. \tag{8.7}$$

If an operator is equal to its adjoint, it is called "self adjoint" or "Hermitian", and such operators are used to represent real physical quantities in quantum mechanics. This definition of hermitian operators is equivalent to that used in Chapter 3.

A general operator cannot be expressed as a single dyadic but may be regarded as a linear combination of dyadics, with a possible number of terms equal to the square of the number of dimensions (possibly infinite) in the space.

If the simple dyadic of equation (8.5) is applied to the ket vector $|P\rangle$, and the scalar product of the result is taken with the bra vector $\langle Q|$, the result is a "bra-ket", or bracket expression, whose meaning is uniquely defined.

$$\begin{aligned}\langle Q|\{|A\rangle\langle B|P\rangle\} &= \{\langle Q|A\rangle\langle B|\}|P\rangle \\ &= \langle Q|\{|A\rangle\langle B|\}|P\rangle = \langle Q|\alpha|P\rangle\end{aligned} \tag{8.8}$$

where the single letter α is used to designate the dyadic $|A\rangle\langle B|$.

When the ket and bra vectors are expressed in terms of a complete set of basic vectors, the operators can be expressed in the same way. If the ket vector $|A\rangle$ and the bra vector $\langle B|$ are expressed in terms of the set of basic ket vectors $|n\rangle$ and the conjugate imaginary bra vectors $\langle n|$,

$$|A\rangle\langle B| = \sum_n a_n|n\rangle\sum_m b_m^*\langle m| = \sum_{n,m} a_n b_m^*|n\rangle\langle m|. \tag{8.9}$$

When a number of dyadics expressed in this way are added together to give a general operator, the result is expressed as a sum of the dyadics $|n\rangle\langle m|$ and has the form

$$\alpha = \sum_{n,m} \alpha_{n,m}|n\rangle\langle m|. \tag{8.10}$$

The $\alpha_{n,m}$ are complex numbers, and if the operator is Hermitan, $\alpha_{n,m} = \alpha_{m,n}^*$.

In equation (8.10) the operator α is expressed as a sum of dyadics, each multiplied by a complex number. *The complex numbers themselves form a matrix, a matrix representation of α on the basis formed by the ket vectors $|n\rangle$ and their associated bra vectors.*

The addition of two dyadics, or two general operators, gives merely that dyadic or operator which produces the same result as the sum of the results of the two operators separately. This addition clearly obeys the usual laws of algebraic addition. However, the multiplication of operators implies their application one after the other, and such multiplication is not, in general commutative.

$$\{|A\rangle\langle B|\}\{|C\rangle\langle D|\}|P\rangle = \langle D|P\rangle\{|A\rangle\langle B|\}|C\rangle$$
$$= \langle D|P\rangle\langle B|C\rangle|A\rangle. \tag{8.11a}$$

On the other hand

$$\{|C\rangle\langle D|\}|A\rangle\langle B|\}|P\rangle = \langle B|P\rangle\langle D|A\rangle|C\rangle. \tag{8.11b}$$

The algebra of linear operators in this kind of a vector space is a noncummutative algebra and careful attention must be paid to this point in algebraic manipulations.

The lack of commutation in operator multiplication is obvious also when the matrix representation of the operators is used. Following equation (8.10) let

$$\alpha = \sum_{n,m} \alpha_{n,m}|n\rangle\langle m| \text{ and } \beta = \sum_{r,s} \beta_{r,s}|r\rangle\langle s|,$$

$$\alpha\beta = \sum_{n,m}\sum_{r,s} \alpha_{n,m}\beta_{r,s}|n\rangle\langle m|r\rangle\langle s| = \sum_{n,s}\{\sum_m \alpha_{n,m}\beta_{m,s}|n\rangle\langle s|\}. \tag{8.12}$$

The sum $\sum_m \alpha_{n,m}\beta_{m,s}$ is the (n, s) element of the matrix product of the two matrix representations of α and β, it is an example of the rule for multiplication of matrices.

The same rule of matrix multiplication can be used to describe the application of an operator to a ket vector if the ket vector is represented as a column matrix. If

$$|P\rangle = \sum_n p_n|n\rangle \tag{8.13}$$

the p_n can be considered as a column matrix. The operator then gives another column matrix according to the rule

$$\alpha|P\rangle = \sum_{n,m} \alpha_{n,m}|n\rangle\langle m|\sum_s p_s|s\rangle$$
$$= \sum_{n,m} \alpha_{n,m}p_m|n\rangle = \sum_n\{\sum_m \alpha_{n,m}p_m\}|n\rangle. \tag{8.13a}$$

$\sum_m \alpha_{n,m}p_m$ is an example of the rule for multiplying a column matrix by a square matrix to give another column matrix. Similarly a bra vector can be represented as a row matrix and multiplied into the square matrix representing an operator to give another row matrix.

$$\langle P \mid \alpha = \sum_s p_s^* \langle s \mid \sum_{n,m} \alpha_{n,m} \mid n \rangle \langle m \mid$$

$$= \sum_m \{\sum_n p_n^* \alpha_{n,m}\} \langle m \mid . \tag{8.14}$$

PROBLEM 3. Show from the definition of Hermitian operators that the matrix representation $\alpha_{n,m}$ of such an operator satisfies the equation

$$\alpha_{n,m} = \alpha_{m,n}^* .$$

3. Eigenvectors and eigenvalues

A particular class of ket vectors of special importance in quantum mechanics is the class of eigenvectors. These are vectors such that the application of a given operator merely multiplies the vector by a constant. The dyadic $2 \mid A \rangle \langle A \mid$, for example, has $\mid A \rangle$ for an eigenvector since

$$2 \mid A \rangle \langle A \parallel A \rangle = 2 \mid A \rangle ,$$

with the multiplying factor 2. This factor is called the eigenvalue of the operator and of the eigenket. If one has the operator

$$\mathsf{O} = a \mid A \rangle \langle A \mid + \, b \mid B \rangle \langle B \mid ,$$

where $\mid A \rangle$ and $\mid B \rangle$ are orthogonal, both $\mid A \rangle$ and $\mid B \rangle$ are eigenkets with the eigenvalues a and b respectively. In general one may write

$$\alpha \mid A \rangle = a \mid A \rangle , \tag{8.15}$$

if $\mid A \rangle$ is an eigenket of α with the eigenvalue a.

The complex imaginary of an eigenket $\mid A \rangle$ is an eigenbra of the same operator if the operator is a "normal operator". A normal operator is one which commutes with its adjoint. The product of an operator and its adjoint is Hermitian. However, the effect of operating on the bra is to multiply it by the complex conjugate of the eigenvalue by which the ket is multiplied. If the eigenvalues are real, a normal operator multiplies a bra vector by the same eigenvalue as that by which it multiplies a ket.

If $\alpha \mid A \rangle = a \mid A \rangle$, then $\langle A \mid \alpha^\dagger = \langle A \mid a^*$.

Then

$$\langle A \mid \alpha^\dagger \alpha \mid A \rangle = \{\langle A \mid \alpha^\dagger\} \{\alpha \mid A \rangle\} = \{\langle A \mid \alpha\} \{\alpha^\dagger \mid A \rangle\} .$$

As is evident from the above illustrations, every ket vector is an eigenket of some operator, in particular of the dyadic formed from the ket and its complex imaginary bra. However, the important kets are those which are eigenkets of operators having some physical significance.

If an operator can be represented as a sum of dyadics composed of the orthogonal basic ket vectors each with its complex imaginary bra vector and multiplied by a constant, it has the form

$$\alpha = \sum_n a_n \mid n\rangle \langle n \mid .$$

The eigenvectors of this operator are just the basic kets and bras in terms of which it is expressed, and the a_n are the eigenvalues.

If $\mid A\rangle$ is an eigenket of an operator α it is also an eigenket of any function of α. If two operators commute, it is possible to find ket vectors that are simultaneously eigenkets of both commuting operators. If, however, two operators do not commute it may be possible to find some exceptional ket vectors that are simultaneous eigenkets of both operators, but it is not possible to find a mathematically complete set of simultaneous eigenkets.

PROBLEM 4. Show that the eigenvalues of an Hermitian operator are real. An Hermitian operator is equal to its adjoint.

PROBLEM 5. Show that eigenkets corresponding to different eigenvalues of the same operator are orthogonal to each other.

PROBLEM 6. Show from their matrix representations that two operators do not necessarily commute.

PROBLEM 7. Show that if $\alpha^2 = 1$ and $\mid P\rangle$ is any ket vector, $(\alpha-1) \mid P\rangle$ and $(\alpha+1) \mid P\rangle$ are eigenvectors of α.

PROBLEM 8. Let the operator satisfy the equation

$$\alpha^3 - 6\alpha^2 + 11\alpha - 6 = 0.$$

Show that $(\alpha-2)(\alpha-3) \mid P\rangle$, $(\alpha-1)(\alpha-3) \mid P\rangle$ and $(\alpha-1)(\alpha-2) \mid P\rangle$ are eigenvectors of α unless they are zero. Show also that the arbitrary vector $\mid P\rangle$ is a linear combination of the three eigenvectors.

PROBLEM 9. If it is assumed that the eigenkets of an operator form a complete set, show that it is possible to find a complete set of simultaneous eigenkets of two commuting operators. To do this express the eigenkets of one operator as a linear combination of eigenkets of the other.

4. Relationship between the mathematics and the physical situation

The three preceding sections have outlined a system of abstract vectors and linear operators which is self contained and of mathematical interest. In developing a physical theory, however, it is necessary to be very clear as to the relationship between the mathematics and the physics. For the special case of wave mechanics this relationship was expressed by means of the two basic postulates of Chapter 3. In this more general formulation, the postulates must be reworded.

In the wave mechanics the state of a system is represented by a wave function. Few restrictions are normally placed on the nature of the function used to represent a state. It is expected to be single valued and quadratically integrable, although this latter requirement is relaxed for functions characteristic of position and of momentum. However the functions found to be eigenfunctions of simple operators are also continuous. Such functions are descriptive of "pure" states. This point is worth some emphasis. A pure state is a state defined as precisely as possible and thus distinguished from a state for which statistical mechanics must be used. In classical mechanics a pure state of a system of particles is one in which the coordinate and the momentum of each particle are precisely specified. The basic reason for quantum mechanics is that such a specification is experimentally impossible. A quantum mechanical pure state then is one in which a complete set of "noninterfering" quantities is precisely specified. In a one dimensional system there is only one such quantity. It may be position, or momentum, or energy, or some other function of such quantities, but only one such quantity can be independently specified.

A wave function, then, is an eigenfunction of a "physically" complete set of operators representing independent physical quantities. This imposes no very serious limitation on the function used, for almost any function of real variables can be shown to be an eigenfunction of some Hermitian operator.

These properties of wave functions can be translated into corresponding statements about ket and bra vectors in the following postulate.

POSTULATE A_1, *A pure state of a physical system is represented by a ket vector, and its conjugate imaginary bra vector, which are eigenvectors of a set of commuting operators representing a physically complete set of independent physical quantities.* Such a set contains only half of the usual set of canonical quantities.

The necessity for quantum mechanics can be based on Heisenberg's observation that the canonically conjugate quantities of classical mechanics are experimentally incompatible. A physical arrangement which can be described by a precise value for one member of a conjugate pair permits no value to be assigned to the other member. In wave mechanics this fact is represented by the selection of specific operators to represent cartesian coordinates and momenta. More generally only the commutation rule need be established. Hence we have another part of the first postulate:

POSTULATE A_2. *Operators representing classically defined physical quantities do not necessarily commute, but their commutator is proportional to the Poisson bracket of the classical quantities*:

$$[\alpha, \beta] = \alpha\beta - \beta\alpha = -\frac{\hbar}{i}[\alpha, \beta] = -\frac{\hbar}{i}\sum_m \left[\frac{\partial\alpha}{\partial q_m}\frac{\partial\beta}{\partial p_m} - \frac{\partial\alpha}{\partial p_m}\frac{\partial\beta}{\partial q_m}\right]. \quad (8.16)$$

For quantities, such as electron spin, that are not classical quantities, the commutation rules of the representative operators must be determined to fit the experimental observations. For cartesian coordinates and momenta equation (8.16) leads simply to

$$[q_s, q_r] = [p_s, p_r] = 0, \qquad [q_s, p_r] = i\,\hbar\delta(r, s).$$

The information contained in a state vector about physical properties of the system, other than those represented by operators of which the vectors are eigenvectors, must be the subject of a third part of the first postulate. For each physically complete set of commuting operators there is a mathematically complete set of eigenvectors in terms of which any vector may be expressed. If the set of operators is $\xi_1, \xi_2, \ldots, \xi_n$, the eigenkets may be designated by

$$\xi_1', \xi_2', \xi_3', \ldots, \xi_n'.$$

The ξ_i' indicates one of the eigenvalues of the linear operator ξ_i. The notation can be shortened so that the whole physically complete set of commuting operators is designated by the single letter ξ and the corresponding set of eigenvalues by ξ'. The bra and ket vectors selected in this way are orthogonal so that

$$\langle\xi'' \mid \xi'\rangle = \delta(\xi'', \xi').$$

This form will be used both with discrete and with continuous values of ξ'. For continuous values the above expression implies an integral. The eigenvalues will be represented by primes. Thus ξ' and ξ'' are eigenvalues of the operator which represents the physical quantity ξ.

With a complete set of normalized and orthogonal bra and ket vectors, each one of which is an eigenvector of a physically complete set of commuting operators, any state vector may be specified by a set of numbers:

$$| A\rangle = \sum_n a_n \,| \xi^{(n)}\rangle, \tag{8.17a}$$

$$\langle A | = \sum_n a_n^* \,\langle \xi^{(n)} |. \tag{8.17b}$$

The sum is over all possible sets of eigenvalues $\xi^{(n)}$, and when the eigenvalues form a continuum the sum becomes an integral. The set of complex numbers a_n or the complex function $a(\xi')$, is a *representation* of the state $| A\rangle$. It is the ξ representation. Multiplying equation (8.17a) on the left by $\langle \xi^{(m)} |$ and (8.17b) on the right by $| \xi^{(m)}\rangle$ gives

$$\langle \xi^{(m)} | A\rangle = a_m, \tag{8.18a}$$

$$\langle A | \xi^{(m)}\rangle = a_m^*. \tag{8.18b}$$

The bracket (bra-ket) $\langle \xi^{(m)} | A\rangle$ is a number, and the set of numbers for all values of (m) is the representative of $| A\rangle$ in the ξ representation. The values of the numbers depend both on the state $| A\rangle$ and the set of basic vectors $| \xi^{(n)}\rangle$. The number $\langle \xi^{(n)} | A\rangle$ is the projection of $| A\rangle$ on the direction of $| \xi^{(n)}\rangle$. It may be called the component of $| A\rangle$ in the direction $| \xi^{(n)}\rangle$.

In these terms the first postulate may be completed by

POSTULATE A_3. Let $W_n = | \langle \xi^{(n)} | A\rangle |^2 = | \langle A | \xi^{(n)}\rangle |^2$. *$W_n$ is the probability that a system in state $| A\rangle$ will be found to be in state $| \xi^{(n)}\rangle$.* Stated more explicitly in terms of physical operations (measurements), W_n is the probability that a simultaneous measurement of the quantities $\xi_1, \xi_2, \ldots, \xi_s$ on a system known to be in state $| A\rangle$ will give the results $\xi_1^{(n)}, \xi_2^{(n)}, \ldots, \xi_s^{(n)}$. The statement that the system is in the state $| A\rangle$ means that a simultaneous determination of the physically complete set of quantities $A_1, A_2, \ldots, A_s$ has been made to find, or to put, the system in the state

$$| A_1', A_2', \ldots, A_s'\rangle = | A\rangle.$$

Postulate A_3 indicates the way in which an experimental arrangement designed to indicate the various values of the ξ_i will respond to a situation described by the eigenvalues $A_1', A_2', \ldots, A_n'$ of the quantities $A_1, A_2, \ldots, A_n$.

5. Relationship with the wave function notation

Since $\langle \xi' | A\rangle$ is a set of numbers dependent on the value of ξ' it can be written as a function of ξ' and called $\psi_A(\xi')$. This form is especially useful when ξ' represents a continuum of values such as the values of a Cartesian coordinate. It is also, of course, a function of A, where A may represent a set of discrete numbers such as those designating the energy eigenkets of an harmonic oscillator, or may itself also represent a continuum. The set of numbers $\langle \xi' \mid A \rangle$ is thus a matrix of a rather general kind, since the rows and the columns may represent quite different kinds of quantities.

The ket vector $\mid A \rangle$ may be expressed in terms of the ket vectors $\mid \xi' \rangle$ as a basis:

$$\mid A \rangle = \int \langle \xi' \mid A \rangle \mid \xi' \rangle \, \mathrm{d}\xi' = \int \psi_A(\xi') \mid \xi' \rangle \, \mathrm{d}\xi' . \tag{8.18}$$

The function $\psi_A(\xi')$ is thus a representation of the ket vector $\mid A \rangle$ on the basis formed by the vectors $\mid \xi' \rangle$. It can be regarded as a column matrix.

It is also possible to establish a notation, which is sometimes useful, by expressing a ket vector as the result of operating on a "ket without a label" by a function of a set of commuting operators. Thus

$$\mid A \rangle = \int \psi_A(\xi') \mid \xi' \rangle \, \mathrm{d}\xi' = \psi_A(\xi) \rangle . \tag{8.18a}$$

This ket without a label is called the "standard ket", and its significance is just as defined in equation (8.18a). Further, another function of the operators ξ operating on this ket vector will lead to a ket vector with the representatives

$$\langle \xi' \mid f(\xi) \mid A \rangle = \langle \xi' \mid f(\xi) \int \psi_A(\xi') \mid \xi' \rangle \, \mathrm{d}\xi'$$

$$= \langle \xi' \Big| f(\xi) \, \psi_A(\xi) \rangle = f(\xi') \, \psi_A(\xi'). \tag{8.19}$$

The equality between the first and third terms in equation (8.19) is merely the distributive law of multiplication while the second term is the justification for the last. With this notation the vertical line on the left is unnecessary and one can write

$$f(\xi) \mid A \rangle = f(\xi) \, \psi_A(\xi) \rangle . \tag{8.19a}$$

This notation brings out the role of the Schroedinger wave function in configuration space. Such a function represents a state, or a ket vector, when the basis kets are eigenkets of cartesian coordinates.

To illustrate this further, the ground state of a simple oscillator may be designated by $(\frac{1}{2}\hbar\omega\rangle$ or by $\mid 0 \rangle$. This is an energy eigenket and its designation is either the energy eigenvalue $\frac{1}{2}\hbar\omega$, or the energy quantum number 0. On the other hand, this ket vector can be expressed in terms of the eigenkets of position on the x-axis.

$$| \tfrac{1}{2}\hbar\omega\rangle = \left(\frac{\alpha}{\pi}\right)^{\frac{1}{4}} \int e^{-\frac{1}{2}\alpha x'^2} \,| x'\rangle \,dx' = \left(\frac{\alpha}{\pi}\right)^{\frac{1}{4}} e^{-\frac{1}{2}\alpha x^2}\rangle. \tag{8.20}$$

A similar notation can be used for the bra vectors, and since the coefficients of a vector which is complex imaginary to $| A\rangle = \psi_A(\xi)\rangle$ must involve the complex conjugate coefficients, it follows that

$$\langle A | = \int \langle \xi' | \langle A | \xi'\rangle \,d\xi' = \int \langle \xi' | \psi^*(\xi') \,d\xi' = \langle \psi^*(\xi). \tag{8.21}$$

The use of the ket and the bra without a label serves merely to emphasize the nature of the "wave" functions $\psi(\xi)$ and $\psi^*(\xi)$. An operator needs to be always placed to the left of $\psi(\xi)$ but to the right of $\psi^*(\xi)$. This is, of course, important only when there may be lack of commutativity between the operator and ξ.

PROBLEM 10. Show that

$$\langle P | \xi | P\rangle = \sum_n W_n \xi^{(n)}$$

when the eigenvalues of ξ are discrete. Write the equivalent expression when the eigenvalues are continuous.

PROBLEM 11. Show that $\sum_n | \xi^{(n)}\rangle \langle^{(n)} |\xi = 1$, and that $\int | \xi'\rangle \langle^{\frac{1}{2}'} | d\xi' = 1$ when the eigenvalues are continuous. These are called identity operators.

PROBLEM 12. Show that the matrix representations of linear operators obey the associative law, in particular that

$$[\alpha\beta] \,| A\rangle = \alpha[\beta \,| A\rangle].$$

Also show that

$$[\langle B | \alpha] \,| A\rangle = \langle B | \,[\alpha \,| A\rangle].$$

PROBLEM 13. Illustrate the nature of the matrix $\langle \xi' | \xi''\rangle$, both for the case where the ξ' are discrete and the case where they form a continuum.

PROBLEM 14. Show that if α is a regular operator and a real linear operator, i.e. if it has only real eigenvalues, it is represented by a matrix whose diagonal terms are real and whose non-diagonal terms, reflected across the diagonal, are complex conjugates.

PROBLEM 15. Show that

$$[AB, C] = [A, C]B + A[B, C]$$

for both Poisson brackets and commutators.

PROBLEM 16. Show that for any function $f(q)$ expansible in a series of positive powers of q, $(pf - fp) = -i\hbar f'(q)$.

6. The Schroedinger representation

In the Schroedinger wave mechanics the cartesian coordinates of a system of point particles constitute a set of commuting variables, and the continuum of eigenkets of these coordinates may be used as a basis for representing other ket vectors. Any ket for a system describable by such coordinates can be written in the standard ket notation as $\psi(q)\rangle$.

In the Schroedinger wave mechanics the momentum is represented by a derivative and so it is necessary to formulate the appropriate definition of a derivative as applied to a ket or a bra vector. Since $\psi(q)$ is a function of commuting variables only, its derivative is defined, when q represents a continuum, and is another function of q. This function also can be used to specify a ket and this ket $\partial\psi(q)/\partial q\rangle$ must be the result of a linear operation on $\psi(q)\rangle$. This operator will be defined to be $\partial/\partial q$. Thus

$$\frac{\partial}{\partial q}\left[\psi(q)\rangle\right] = \frac{\partial\psi(q)}{\partial q}\Big\rangle. \tag{8.22}$$

The partial derivative sign is used because there may be several different q's. There will be at least 3 for a system of a single particle. With the definition in (8.22) it follows from (8.18) and (8.18a) that

$$\frac{\partial\psi(q)}{\partial q}\Big\rangle = \int \frac{\partial\psi(q')}{\partial q'}\,|\,q'\rangle\,\mathrm{d}q'. \tag{8.23}$$

Also it is evident that the derivative operator applied to the standard ket gives zero.

The derivative operator can also be applied to a bra vector written to the left of the operator Let $\phi(q)\rangle$ and $\psi(q)\rangle$ be two arbitrary ket vectors. ($\phi^*(q)$ will be the complex conjugate of $\phi(q)$.) Since the associative law holds

$$\left[\langle\phi^*(q)\frac{\partial}{\partial q}\right]\psi(q) = \langle\phi^*(q)\left[\frac{\partial}{\partial q}\psi(q)\rangle\right]$$

$$= \int \phi^*(q'')\,\langle\,q''\,|\,\mathrm{d}q'' \int \frac{\partial\psi(q')}{\partial q'}\,|\,q'\,\rangle\,\mathrm{d}q'. \tag{8.24}$$

Since $\langle q'' \mid q' \rangle = \delta(q''-q')$, the last expression in (8.24) becomes

$$\int \phi^*(q') \frac{\partial \psi(q')}{\partial q'} \mathrm{d}q' = - \int \frac{\partial \phi^*(q')}{\partial q'} \psi(q') \, \mathrm{d}q' , \qquad (8.24a)$$

when the functions $\phi^*(q')$ and $\psi(q')$ vanish suitably at the limits of integration. From (8.24a) it follows that the first expression in (8.24) should be interpreted as

$$[\langle \phi^*(q)] \frac{\partial}{\partial q} = - \left\langle \frac{\partial \phi^*(q)}{\partial q} \right. = - \int \langle q' \left| \frac{\partial \phi^*(q')}{\partial q'} \right. \mathrm{d}q' . \qquad (8.25)$$

Furthermore, since $\langle \phi^*(q)\partial/\partial q$ is to be the complex imaginary of $(\partial/\partial q)\phi(q)\rangle$, the adjoint of $\partial/\partial q$ must be taken as $-\partial/\partial q$ and $\partial/\partial q$ may be considered as a pure imaginary operator.

It is thus apparent that $(\hbar/\mathrm{i})\partial/\partial q$ is a real linear operator, which satisfies the same commutation relation with q as does the conjugate momentum p. Hence a representation based on eigenkets of the q's, $\mid q_1', q_2', \ldots, q_n' \rangle$, and in which the canonically conjugate momenta are represented by derivatives, is called a Schroedinger representation. Its relationship to the Schroedinger wave mechanics is fairly obvious.

PROBLEM 17. Show that

$$\mid q' \rangle = \int \delta(q''-q') \mid q'' \rangle \, \mathrm{d}q'' = \delta(q-q')\rangle,$$

that

$$\langle q'' \mid \mathsf{q} \mid q' \rangle = q'' \delta(q''-q') \qquad (8.26)$$

and that

$$\left\langle q'' \left| \frac{\partial}{\partial q} \right| q' \right\rangle = \left\langle q'' \left| \frac{\partial}{\partial q} \delta(\mathsf{q}-q') \right. \right\rangle = \frac{\partial}{\partial q''} \delta(q''-q') . \qquad (8.27)$$

The meaning of the difference $(\mathsf{q}-q')$ is made clear by regarding the number q' as multiplied by a unit operator.

The function of two variables, $\langle q'' \mid \mathsf{q} \mid q' \rangle$, may be called a matrix, and it is clear that the matrix of any function of q, on a basis of eigenvectors of q, is a diagonal matrix. It differs from zero only when $q'' = q'$.

PROBLEM 18. Write the expressions for the mean values of q and p when the arbitrary state vector is given in the coordinate representation.

7. The momentum representation

In Chapter 3 it was shown that a wave function of the cartesian momenta serves as well as a function of the coordinates to represent the state of a system of point particles. Hence one may consider using the continuum of eigenvalues of the momentum to define the basic kets and bras.

An eigenket of momentum p with the eigenvalue p' satisfies the equation

$$\mathsf{p} \mid p'\rangle = p' \mid p'\rangle . \tag{8.28}$$

This eigenket can be expressed in terms of the eigenkets of the coordinates as

$$\mid p'\rangle = \int \langle q' \mid p'\rangle \, \mathrm{d}q' \mid q'\rangle . \tag{8.29}$$

From (8.28)

$$\langle q' \mid \mathsf{p} \mid p'\rangle = p' \langle q' \mid p'\rangle \tag{8.30}$$

also

$$\begin{aligned}
\langle q' \mid \mathsf{p} \mid p'\rangle &= \frac{\hbar}{\mathrm{i}} \left\langle q' \left| \frac{\partial}{\partial q} \right| p' \right\rangle = \frac{\hbar}{\mathrm{i}} \left\langle \delta(\mathsf{p}-q') \frac{\partial}{\partial q} \right| p' \right\rangle \\
&= -\frac{\hbar}{\mathrm{i}} \left\langle \frac{\partial}{\partial q} \delta(\mathsf{q}-q') \mid p' \right\rangle = -\frac{\hbar}{\mathrm{i}} \int \mathrm{d}q'' \left\langle q'' \left| \frac{\partial}{\partial q''} \delta(q''-q') \right| p' \right\rangle \\
&= -\frac{\hbar}{\mathrm{i}} \int \frac{\partial}{\partial q''} \delta(q''-q') \langle q'' \mid p'\rangle \, \mathrm{d}q'' \\
&= -\frac{\hbar}{\mathrm{i}} \left[\delta(q''-q') \langle q'' \mid p'\rangle \Big|_{q''=-\infty}^{q''=+\infty} - \int_{-\infty}^{+\infty} \delta(q''-q') \frac{\partial}{\partial q''} \langle q'' \mid p'\rangle \, \mathrm{d}q'' \right].
\end{aligned} \tag{8.31}$$

The integrated term is clearly zero at both limits, and the remaining integral contains a δ function, so one has

$$p' \langle q' \mid p'\rangle = \frac{\hbar}{\mathrm{i}} \frac{\partial}{\partial q'} \langle q' \mid p'\rangle , \tag{8.32}$$

which leads to

$$\langle q' \mid p'\rangle = C \exp\left(\frac{\mathrm{i}}{\hbar} p' q'\right). \tag{8.33}$$

This will be recognized as the wave function for a momentum eigenstate, and one has

$$| p'\rangle = C \int \exp\left(\frac{i}{\hbar} p'q'\right) | q'\rangle \, dq' . \tag{8.34}$$

Correspondingly

$$\langle p' | = C^* \int \exp\left(-\frac{i}{\hbar} p'q'\right) \langle q' | \, dq' . \tag{8.35}$$

There still remains the difficulty, already mentioned in Chapter 3, about the normalization of these ket and bra vectors. As in that case it must be handled by a limiting process:

$$\begin{aligned}\langle p'' | p'\rangle &= | C |^2 \lim_{\varepsilon\to 0} \int_{-\infty}^{+\infty} e^{-\varepsilon q'^2} \exp\left(-\frac{i}{\hbar}(p''-p')q'\right) dq' \\ &= | C |^2 \, 2\pi\hbar\delta(p''-p') .\end{aligned} \tag{8.36}$$

A suitable normalization is then to take

$$C = 1/(2\pi\hbar)^{\frac{1}{2}} . \tag{8.37}$$

In this representation the matrix for p is diagonal

$$\langle p'' | \mathsf{p} | p'\rangle = p'\langle p'' | p'\rangle = p''\delta(p''-p')$$

8. Unitary transformations

If a linear operator U has a reciprocal which is equal to its adjoint, it is called a unitary operator. If any operator α is transformed into a new operator β by a unitary transformation such that

$$\beta = \mathsf{U}\alpha\mathsf{U}^{-1} , \tag{8.38}$$

the operator β has the same eigenvalues as α. If $| A\rangle$ is an eigenket of α with the eigenvalue a, $\mathsf{U} | A\rangle$ is the corresponding eigenket of β with the same eigenvalue. Furthermore, if α is real, or Hermitian, β is also. If a number of different operators are transformed by the same unitary transformation, the algebraic relationships between the operators are preserved.

A unitary transformation (8.38) is analogous to an orthogonal transformation in real space. It may be viewed as a rotation of vectors and operators, or it may be viewed as the opposite rotation of the coordinate system. If the basic ket vectors are $| n'\rangle$ they may be changed to another set by the transformation U. Let the new kets be $| s'\rangle = \mathsf{U} | n'\rangle$ and the bras $\langle s' | = \langle n' | \mathsf{U}^\dagger = \langle n' | \mathsf{U}^{-1}$. Then the transformed operator $\mathsf{U}\alpha\mathsf{U}^{-1}$ combined with the new bras and kets gives the same bracket expressions as the original operator with the original bras and kets:

$$\langle s'' \mid \mathsf{U}\alpha\mathsf{U}^{-1} \mid s' \rangle = \langle n'' \mid \mathsf{U}^{-1}\mathsf{U}\alpha\mathsf{U}^{-1}\mathsf{U} \mid n' \rangle = \langle n'' \mid \alpha \mid n' \rangle . \quad (8.39)$$

Unitary transformations in the ket vector space are also the quantum mechanical analog of the classical contact transformations in the phase space, and infinitesimal unitary transformations can be produced by a generating function. If a unitary operator differs from the identity only by an infinitesimal, whose square and higher powers can be neglected, it can be used to form an infinitesimal transformation. Let $\mathsf{U} = 1+i\varepsilon\mathsf{F}$ where F is a real (Hermitian) operator and ε is a real scalar infinitesimal. Then $\mathsf{U}^{-1} = \mathsf{U}^{\dagger} = 1-i\varepsilon\mathsf{F}$ if terms in ε^2 are neglected, and an operator α is transformed to β as

$$\beta = (1+i\varepsilon\mathsf{F})\alpha(1-i\varepsilon\mathsf{F}) = \alpha+i\varepsilon(\mathsf{F}\alpha-\alpha\mathsf{F}) = \alpha-i\varepsilon[\alpha, \mathsf{F}] . \quad (8.40)$$

This is formally the same as a classical contact transformation expressed in terms of the Poisson brackets.

PROBLEM 19. Show that $(1-i\varepsilon\mathsf{F})$ is not only the complex conjugate of $(1+i\varepsilon\mathsf{F})$ but is also the adjoint.

PROBLEM 20. Show that if a number of different operators are transformed by the same unitary transformation, the algebraic relations between the operators are preserved.

PROBLEM 21. Show that the eigenvalues of an operator are unchanged by a unitary transformation, and also that the transformed eigenbras are the complex imaginaries of the transformed eigenkets.

9. The equation of motion

Just as the wave function in wave mechanics changes continuously with the time, so the representative ket and bra vectors will change continuously as the system moves under the influence of the forces represented by the Hamiltonian function. If at t_0 the state is represented by $\mid P, t_0\rangle$, at a later time t it will be represented by another ket vector $\mid P, t\rangle$ and the two will be related by a unitary operator T:

$$\mid P, t\rangle = \mathsf{T}(t) \mid P, t_0\rangle . \quad (8.41)$$

The requirement that the operator $\mathsf{T}(t)$ be unitary guarantees the normalization of $\mid P, t\rangle$ at all times.

The expectation that the ket vector will change continuously with the time leads to the *assumption* that

$$\frac{\mathrm{d}}{\mathrm{d}t} \mid P, t\rangle = \lim_{t \to t_0} \frac{\mid P, t\rangle - \mid P, t_0\rangle}{t - t_0}$$

exists, and hence may be called the derivative of the ket vector with respect to the time. We may take as a postulate that the limit operator is $-\mathrm{i}/\hbar$ times the Hamiltonian function. The suggestion of this assumption is the analogy with classical mechanics, where the Hamiltonian function generates an infinitesimal transformation in the phase space. The justification, of course, lies in the results to which it leads. Hence we have

POSTULATE B.

$$-\frac{\hbar}{\mathrm{i}} \frac{\mathrm{d}}{\mathrm{d}t} \mid P, t\rangle = \mathsf{H} \mid P, t\rangle. \tag{8.42}$$

Using equation (8.41) this becomes

$$-\frac{\hbar}{\mathrm{i}} \frac{\mathrm{d}\mathsf{T}}{\mathrm{d}t} \mid P, t_0\rangle = \mathsf{HT} \mid P, t_0\rangle$$

and since this is true for any $\mid P, t_0\rangle$,

$$-\frac{\hbar}{\mathrm{i}} \frac{\mathrm{d}\mathsf{T}}{\mathrm{d}t} = \mathsf{HT} \tag{8.43}$$

with the initial condition that $\mathsf{T}(t_0) = 1$.

The Schroedinger equation of Chapter 3 is the Schroedinger representation of equation (8.42), and either (8.42) or (8.43) may be called the Schroedinger form of the equation of motion.

When H is not a function of the time

$$\mathsf{T} = \exp\left\{-\frac{\mathrm{i}}{\hbar} \mathsf{H}(t - t_0)\right\}, \tag{8.44}$$

where the exponential represents a power series in the operator H. If H does contain the time, further attention must be given to the meaning and integration of equation (8.43). Even the solution (8.44) is not easy to use because the exponential represents an infinite series. However, if T is applied to an energy eigenket the result is directly given:

$$\mid H', t\rangle = \mathsf{T}(t) \mid H', t_0\rangle = \exp\{-(\mathrm{i}/\hbar)H'(t - t_0)\} \mid H', t_0\rangle. \tag{8.45}$$

The operator $\mathsf{T}(t)$ applied to any sum of energy eigenkets gives a general solution which is the same as given in Chapter 4.

It is also possible to treat the change with the time in another way, a way more closely analogous to the classical mechanics. In this formulation, often associated with the name of Heisenberg, the operators representing dynamical quantities change with the time, while the state vector remains constant and may be regarded as representing the initial state at $t = t_0$.

From equation (8.41) it follows that

$$| P, t_0\rangle = \mathsf{T}^{-1} | P, t\rangle . \tag{8.46}$$

If each operator is subjected to a unitary transformation with T it becomes a function of the time,

$$\alpha_t = \mathsf{T}^{-1}\alpha_0\mathsf{T}. \tag{8.47}$$

This operator can be applied to the state at t_0 to give the same result as applying α_0 to the state at t:

$$\alpha_t | P, t_0\rangle = \mathsf{T}^{-1}\alpha_0\mathsf{T}\mathsf{T}^{-1} | P, t\rangle = \mathsf{T}^{-1}\alpha_0 | P, t\rangle . \tag{8.48}$$

Physically significant results are obtained only by forming brackets, so the bra vectors must also be considered. The bra vectors at the time t will be related to those at t_0 by

$$\langle P, t | = \langle P, t_0 | \mathsf{T}^{-1} \tag{8.49}$$

since T is a unitary operator. Then

$$\langle Q, t | \alpha_0 | P, t\rangle = \langle Q, t_0 | \mathsf{T}^{-1}\alpha_0\mathsf{T} | P, t_0\rangle = \langle Q, t_0 | \alpha_t | P, t_0\rangle . \tag{8.50}$$

Thus a bracket may be regarded as formed with time dependent bras and kets and a constant operator, or with a time dependent operator operating on constant state vectors. In the latter case the constant state vectors give the initial conditions. From equation (8.47)

$$\frac{\mathrm{d}\mathsf{T}}{\mathrm{d}t}\alpha_t + \mathsf{T}\frac{\mathrm{d}\alpha_t}{\mathrm{d}t} = \alpha_0\frac{\mathrm{d}\mathsf{T}}{\mathrm{d}t} \tag{8.51}$$

so that

$$\begin{aligned} \frac{\mathrm{d}\alpha_t}{\mathrm{d}t} &= \mathsf{T}^{-1}\alpha_0\frac{\mathrm{d}\mathsf{T}}{\mathrm{d}t} - \mathsf{T}^{-1}\frac{\mathrm{d}\mathsf{T}}{\mathrm{d}t}\alpha_t \\ &= (-\mathrm{i}/\hbar)\mathsf{T}^{-1}[\alpha_0\mathsf{T}\mathsf{T}^{-1}\mathsf{H}\mathsf{T} - \mathsf{H}\mathsf{T}\alpha_t] \\ &= (-\mathrm{i}/\hbar)[\alpha_t\mathsf{H}_t - \mathsf{H}_t\alpha_t] . \end{aligned} \tag{8.52}$$

This is formally the same as the classical equation

$$\frac{d\alpha}{dt} = [\alpha, \mathsf{H}]$$

where $[\alpha, \mathsf{H}]$ is the Poisson bracket.

In the analysis leading to equation (8.52) it has been assumed that the operator α is not an explicit function of the time. It has been assumed to be a function of only dynamical variables, such as momenta and coordinates, which do change with the time, and the time dependence of α in equation (8.52) takes account of such changes. If, in addition, there is an explicit time dependence, the proper equation is

$$\frac{d\alpha}{dt} = \frac{\partial\alpha}{\partial t} - (i/\hbar)\,[\alpha\mathsf{H} - \mathsf{H}\alpha]. \tag{8.52a}$$

When H is not a function of the time, the integral of the equation of motion in the Heisenberg form is obtained by using (8.44) in (8.47):

$$\alpha_t = \exp\{(i/\hbar)\mathsf{H}(t-t_0)\}\alpha_0 \exp\{-(i/\hbar)\mathsf{H}(t-t_0)\}. \tag{8.53}$$

Again the formal solution is not easily applicable unless a representation is used in which H is diagonal. Then

$$\begin{aligned}\langle H' \mid \alpha_t \mid H''\rangle &= \exp\{(i/\hbar)H'(t-t_0)\}\langle H' \mid \alpha_0 \mid H''\rangle \exp\{-(i/\hbar)H''(t-t_0)\}\\ &= \exp\{(i/\hbar)\,(H'-H'')\,(t-t_0)\}\langle H' \mid \alpha_0 \mid H''\rangle. \qquad (8.54)\end{aligned}$$

This is the form in which Heisenberg first developed his matrix mechanics. The time dependence of the matrix elements is closely related to the spectroscopic combination principle.

PROBLEM 22. Show that equation (8.45) follows from (8.44).

10. The harmonic oscillator

As an illustration of the use of the algebra of non-commuting quantities, one may treat the simple harmonic oscillator by means different from the wave mechanical means used in Chapter 4. Let the Hamiltonian function be

$$H = \frac{p^2}{2m} + \tfrac{1}{2}m\omega^2 q^2. \tag{8.55}$$

The quantities p and q are the linear momentum and the coordinate respectively, and they satisfy the commutation relation (8.16), so that

$$\mathsf{pq} - \mathsf{qp} = \hbar/i. \tag{8.55a}$$

From the Heisenberg form of the equation of motion (8.52)

$$\dot{q} = -(i/\hbar)(qH-Hq) = p/m, \tag{8.56a}$$

and

$$\dot{p} = -(i/\hbar)(pH-Hp) = -m\omega^2 q. \tag{8.56b}$$

These are identical with the classical equations. If a second derivative is taken the result is still identical with the classical equation of motion. The solution is then obviously

$$q = q_1 e^{i\omega t} + q_2 e^{-i\omega t}. \tag{8.56c}$$

This follows because the time is treated as an ordinary number, but the interpretation of (8.56c) must be carefully considered. The q is an operator, a function of the time; q_1 and q_2 are also operators, but constant. Hence to get physical results one must consider the ket vectors on which the q can operate, and must eventually form brackets.

The form of equations (8.56a) and (8.56b) suggests the use of a transformation to new operators a and $a^\dagger$ defined by

$$a = \frac{p - im\omega q}{(2m\hbar\omega)^{\frac{1}{2}}} \quad \text{and} \quad a^\dagger = \frac{p + im\omega q}{(2m\hbar\omega)^{\frac{1}{2}}}. \tag{8.57}$$

The oscillator can be treated with these "coordinates" as well as with q and p.

From equations (8.56a) and (8.56b) the derivative of a can be written

$$\dot{a} = \frac{1}{(2m\hbar\omega)^{\frac{1}{2}}}(\dot{p} - im\omega\dot{q}) = \frac{1}{(2m\hbar\omega)^{\frac{1}{2}}}(-m\omega^2 q - i\omega p) = -i\omega a. \tag{8.58}$$

The integral is then

$$a = a_0 e^{-i\omega t} \quad \text{and} \quad a^\dagger = a_0^\dagger e^{+i\omega t}. \tag{8.59}$$

Again it must be remembered that a_0 and $a_0^\dagger$ are operators.

From the transformation equations it follows that

$$H = \hbar\omega(a^\dagger a + \tfrac{1}{2}) = \hbar\omega(aa^\dagger - \tfrac{1}{2}) \tag{8.60}$$

and

$$aa^\dagger - a^\dagger a = 1. \tag{8.61}$$

PROBLEM 23. Show from the definitions that $a^\dagger$ is adjoint to a.

PROBLEM 24. Show from equations (8.60) and (8.61) that

$$\mathsf{aH}-\mathsf{Ha} = \hbar\omega\mathsf{a} \tag{8.62a}$$

and

$$\mathsf{a}^\dagger\mathsf{H}-\mathsf{H}\mathsf{a}^\dagger = -\hbar\omega\mathsf{a}^\dagger\,. \tag{8.62b}$$

PROBLEM 25. Show by induction from equation (8.61) that

$$\mathsf{a}^\dagger\mathsf{a}^n-\mathsf{a}^n\mathsf{a}^\dagger = -n\mathsf{a}^{n-1} \tag{8.63a}$$

and

$$\mathsf{a}\mathsf{a}^{\dagger n}-\mathsf{a}^{\dagger n}\mathsf{a} = n\mathsf{a}^{\dagger n-1}\,. \tag{8.63b}$$

To get numerical results from general operator equations it is necessary to deal with a representation, and the most useful representation is based on energy eigenstates. Using such a representation of equation (8.60)

$$\begin{aligned}\hbar\omega\langle H' \mid \mathsf{a}^\dagger\mathsf{a} \mid H'\rangle &= \langle H' \mid \mathsf{H}-\tfrac{1}{2}\hbar\omega \mid H'\rangle \\ &= [H'-\tfrac{1}{2}\hbar\omega]\langle H' \mid H'\rangle\,. \end{aligned} \tag{8.64}$$

The first expression cannot be negative because it is the square of the length of $\mathsf{a} \mid H'\rangle$ multiplied by $\hbar\omega$. In the last expression $\langle H' \mid H'\rangle$ cannot be negative. Hence $H' \geq (\frac{1}{2})\hbar\omega$. If $H' = \frac{1}{2}\hbar\omega$, it follows that $\mathsf{a} \mid H'\rangle = 0$.

From equation (8.62)

$$\mathsf{Ha} \mid H'\rangle = (\mathsf{aH}-\hbar\omega\mathsf{a}) \mid H'\rangle = (H'-\hbar\omega)\mathsf{a} \mid H'\rangle\,. \tag{8.65}$$

This shows that $\mathsf{a} \mid H'\rangle$ is an eigenket of H with the eigenvalue $H'-\hbar\omega$, unless $\mathsf{a} \mid H'\rangle = 0$. This ket is zero if $H' = \frac{1}{2}\hbar\omega$, but for any H' greater than $\frac{1}{2}\hbar\omega$, and for which $(H'-\hbar\omega)$ is positive, $(H'-\hbar\omega)$ is another eigenvalue. The series must end with $H' = \frac{1}{2}\hbar\omega$ so the eigenvalues are $\frac{1}{2}\hbar\omega$, $\frac{3}{2}\hbar\omega$, $\frac{5}{2}\hbar\omega$, ...

If $\mid 0\rangle$ is the eigenket belonging to the eigenvalue $\frac{1}{2}\hbar\omega$,

$$\mathsf{H}\mathsf{a}^\dagger \mid 0\rangle = (\mathsf{a}^\dagger\mathsf{H}+\hbar\omega\mathsf{a}^\dagger) \mid 0\rangle = \tfrac{3}{2}\hbar\omega\mathsf{a}^\dagger \mid 0\rangle \tag{8.66}$$

so that $\mathsf{a}^\dagger \mid 0\rangle$ is an eigenket with the eigenvalue $\frac{3}{2}\hbar\omega$. By continuing this argument it appears that the sequence of eigenkets is

$$\mid 0\rangle,\quad \mathsf{a}^\dagger \mid 0\rangle,\quad \mathsf{a}^{\dagger 2} \mid 0\rangle,\ \ldots. \tag{8.67}$$

If $\langle 0 \mid 0\rangle$ is taken equal to 1, so that $\mid 0\rangle$ is normalized, it follows that

$$\langle 0 \mid \mathsf{a}^n\mathsf{a}^{\dagger n} \mid 0\rangle = n!. \tag{8.68}$$

This fixes the result of operating on one eigenket by $\mathsf{a}^\dagger$ except for a phase factor. If it is adopted as a convention that $\mathsf{a}^\dagger \mid n\rangle = \mathrm{i}\,(n+1)^{\frac{1}{2}} \mid n+1\rangle$ and

$a \mid n\rangle = -i\, n^{\frac{1}{2}} \mid n-1\rangle$, the phases correspond to those used with the energy eigenfunctions of Chapter 4.

PROBLEM 26. Show that $\langle 0 \mid a^n a^{\dagger n} \mid 0\rangle = n!$, and that the phase convention indicated above corresponds to that used in the wave functions of Chapter 4. Show also that consistency requires

$$\langle n \mid a^\dagger = +i n^{\frac{1}{2}} \langle n-1 \mid \quad \text{and} \quad \langle n \mid a = -i(n+1)^{\frac{1}{2}} \langle n+1 \mid .$$

PROBLEM 27. Show that an arbitrary ket can be expressed as a power series in $a^\dagger$ multiplied into $\mid 0\rangle$.

PROBLEM 28. Express the general solution for the harmonic oscillator ket vector as a function of the time.

PROBLEM 29. Work out the Heisenberg matrices for $a^\dagger$ and a as functions of the time, using the energy eigenkets as a basis.

PROBLEM 30. Work out the matrix representation for q^2 in both the Schroedinger and the Heisenberg forms.

PROBLEM 31. Show that

$$\mid Q\rangle = e^{-\frac{1}{2}x_0^2}\, e^{+ix_0 a^\dagger} \mid 0\rangle \tag{8.69}$$

is a normalized ket vector.

PROBLEM 32. If an oscillator is described by the ket $\mid Q\rangle$ of equation (8.69) at time $t = 0$, write the ket vector as a function of the time and show that the expectation value of q is

$$\langle Q, t \mid q \mid Q, t\rangle = -\left(\frac{2\hbar}{m\omega}\right)^{\frac{1}{2}} x_0 \cos \omega t . \tag{8.70}$$

Find also the expectation value of the momentum.

PROBLEM 33. Work out the expectation values of the squares of the coordinate and momentum and so the value of the indetermination $(\Delta q\, \Delta p)$ as a function of the time.

CHAPTER 9

ANGULAR MOMENTUM AND SPIN

It has already been shown that angular momentum is an important quantity in the wave mechanics of simple two and three dimensional problems. Just as in classical mechanics, angular momentum is a constant of the motion for isolated systems, and its eigenvalues occupy a position second in importance only to that of the energy. Much of the development of atomic, molecular, and nuclear spectroscopy is based on the properties of states characteristic of this constant of the motion. In this chapter some of the elementary properties of angular momentum will be studied in terms of non-commutative algebra, which permits some generalization of the previous results.

1. Eigenvalues of angular momentum

In Chapter 6 the orbital angular momentum of a single particle was expressed in terms of the coordinates and the linear momenta, and there were derived the commutation rules:

$$\begin{aligned} J_xJ_y - J_yJ_x &= i\hbar J_z \\ J_yJ_z - J_zJ_y &= i\hbar J_x \\ J_zJ_x - J_xJ_z &= i\hbar J_y. \end{aligned} \tag{9.1}$$

These follow, for the orbital angular momentum of a particle, from the commutation rules for cartesian coordinates and momenta. For our more general treatment they may be taken as the fundamental properties of the angular momentum operators. Although these properties follow from the classical definition of angular momentum, it does not appear possible to invert the argument; it does not appear possible to show from equations (9.1) that the operators J_x, J_y, J_z are necessarily formed from coordinates and momenta. Hence the assumptions involved are somewhat more general than the previously derived conclusions, as will be seen from their use in describing particle spin. In order to emphasize this point the letter J is used instead of L, which is reserved for orbital angular momentum. Hence let there be three operators J_x, J_y, J_z, designated as the three components of angular momentum, and satisfying the commutation relations (9.1).

Let

$$\mathsf{J}^2 = \mathsf{J}_x^2+\mathsf{J}_y^2+\mathsf{J}_z^2. \tag{9.2}$$

From the above commutation relations it can be shown that J^2 commutes with all three of the component operators. Thus J^2 and any one of them, for example J_z, form a set of commuting operators. It is customary to take J^2 and J_z as the physically complete set of commuting operators in terms of which the state vectors are to be described. This set is physically complete, of course, only when no other aspects of the physical system are taken into consideration.

As in the case of the harmonic oscillator it is often convenient to transform to other operators. Let

$$\mathsf{J}_+ = \mathsf{J}_x+i\mathsf{J}_y \quad \text{and} \quad \mathsf{J}_- = \mathsf{J}_x-i\mathsf{J}_y. \tag{9.3}$$

Since both J_x and J_y represent physical quantities they must be Hermitian operators so that J_+ and J_- are mutually adjoint. It follows directly from the assumed commutation relations that

$$\mathsf{J}_+\mathsf{J}_- = \mathsf{J}^2-\mathsf{J}_z^2+\hbar\mathsf{J}_z, \tag{9.4a}$$

$$\mathsf{J}_-\mathsf{J}_+ = \mathsf{J}^2-\mathsf{J}_z^2-\hbar\mathsf{J}_z, \tag{9.4b}$$

$$\mathsf{J}_z\mathsf{J}_+-\mathsf{J}_+\mathsf{J}_z = \hbar\mathsf{J}_+. \tag{9.4c}$$

Let J_z' be an eigenvalue of J_z, and let k be an eigenvalue of J^2. The ket corresponding to these two eigenvalues may be designated by $|\, k, J_z'\rangle$. Note that the vector angular momentum is not precisely specified by these eigenvalues. Although the length of the vector might be thought of as given by the square root of the eigenvalue of J^2, the components are not given by this square root multiplied by direction cosines.

In the state specified by k and J_z'

$$\begin{aligned}\langle k, J_z' \,|\, \mathsf{J}_-\mathsf{J}_+ \,|\, k, J_z'\rangle &= \langle k, J_z' \,|\, \mathsf{J}^2-\mathsf{J}_z^2-\hbar\mathsf{J}_z \,|\, k, J_z'\rangle \\ &= (k-J_z'^2-\hbar J_z')\,\langle k, J_z' \,|\, k, J_z'\rangle.\end{aligned} \tag{9.5}$$

Since we assume that $|\, k, J_z'\rangle \neq 0$, unless $\mathsf{J}_+ \,|\, k, J_z'\rangle$ is zero we must conclude that

$$(k - J_z'^2 - \hbar J_z') > 0.$$

and hence

$$(k + \tfrac{1}{4}\hbar^2) > (J_z' + \tfrac{1}{2}\hbar)^2. \tag{9.6}$$

From a similar consideration of $\langle k, J_z' \mid \mathsf{J}_+ \mathsf{J}_- \mid k, J_z' \rangle$ it follows that

$$(k + \tfrac{1}{4}\hbar^2) > (J_z' - \tfrac{1}{2}\hbar)^2. \tag{9.7}$$

Since both of these inequalities must be satisfied, k can have no negative values.

From equation (9.4c) one has

$$\mathsf{J}_z \mathsf{J}_+ \mid k, J_z' \rangle = (\mathsf{J}_+ \mathsf{J}_z + \hbar \mathsf{J}_+) \mid k, J_z' \rangle = (J_z' + \hbar) \mathsf{J}_+ \mid k, J_z' \rangle, \tag{9.8}$$

which shows that $\mathsf{J}_+ \mid k, J_z' \rangle$, unless it is zero, is another eigenket of J_z with the eigenvalue $(J_z' + \hbar)$. For this reason J_+ is called an "up-stepping" operator. Similarly J_- is a "down-stepping" operator.

Repeated application of J_+ leads to a series of eigenkets with eigenvalues of J_z differing by $\hbar$. Eventually the eigenvalue would become so large that, for a given value of k, equations (9.6) and (9.7) would not be satisfied. Hence the series must stop, and it will stop if

$$\mathsf{J}_+^{s+1} \mid k, J_z' \rangle = \mathsf{J}_+ \mid k, (J_z' + s\hbar) \rangle = 0. \tag{9.9}$$

This makes $\mid k, (J_z' + s\hbar) \rangle$ the last member of the series and makes $(J_z' + s\hbar)$ the largest eigenvalue of J_z.

By a similar argument with the operator J_-, it can be shown that the lowest eigenvalue of J_z is $(J_z' - r\hbar)$. The difference between the highest and the lowest eigenvalues is then $(s+r)\hbar$, where s and r are both integers.

In equation (9.5), if J_z'' is the highest eigenvalue of J_z, $\mathsf{J}_+ \mid k, J_z'' \rangle = 0$. Hence, if $(J_z' + s\hbar)$ is the highest eigenvalue

$$k + \tfrac{1}{4}\hbar^2 = [J_z' + (s + \tfrac{1}{2})\hbar]^2. \tag{9.10}$$

Similarly

$$k + \tfrac{1}{4}\hbar^2 = [J_z' - (r + \tfrac{1}{2})\hbar]^2. \tag{9.10a}$$

From these two equations

$$J_z' = \tfrac{1}{2}(r - s)\hbar, \tag{9.11}$$

and

$$k = \tfrac{1}{2}(s + r)\,[\tfrac{1}{2}(s + r) + 1]\hbar^2 = J(J+1)\hbar^2. \tag{9.12}$$

If $s+r$ is an even number, J is an integer and J_z runs from $J\hbar$ to $-J\hbar$ in steps of $\hbar$. On the other hand, if $s+r$ is an odd number, J is an integer plus a half, but J_z' still runs from $J\hbar$ to $-J\hbar$ in steps of $\hbar$. Note that J is not an eigenvalue. However it may be called a quantum number and used to specify a ket vector.

In Chapter 6 the properties of the orbital angular momentum of a particle moving in a central field were based on the properties of spherical harmonics. The only J values (there called l) permitted were integral values. It can be shown that when the angular momentum operators are formed from particle coordinates and momenta, only integral values are permitted. The above example of analysis by means of non-commutative algebra shows that the assumptions in equations (9.1) are somewhat more general and permit half integral values of J.

PROBLEM 1. Show that equation (9.4c) follows from the definition of J_+ and the commutation rules for the components of angular momentum. Work out the corresponding expression for J_-.

PROBLEM 2. Show that in a state $| k, J_z' \rangle$, the mean value of $\mathrm{J}_x = 0$. This can be done by noting that $\mathrm{J}_x = -(\mathrm{i}/\hbar)(\mathrm{J}_y\mathrm{J}_z - \mathrm{J}_z\mathrm{J}_y)$.

2. Matrix representation of angular momentum operators

On the basis of the algebraic relationships between the various quantities associated with angular momentum there can be constructed matrix representations of the operators. Although these are infinite matrices, they can be formulated so that they consist of blocks along the diagonal such that each block corresponds to one value of J. The matrices for the different values of J can then be treated separately.

A set of basic kets may be taken as $| J, M \rangle$, where $J(J+1)\hbar^2$ is the eigenvalue of J^2 and $M\hbar$ is the eigenvalue of J_z. For each value of J, M runs from $+J$ to $-J$.

From equation (9.4b) one has

$$\begin{aligned}\langle J', M' \mid \mathrm{J}_-\mathrm{J}_+ \mid J, M\rangle &= [J(J+1)-M^2-M]\hbar^2\langle J', M' \mid J, M\rangle \\ &= [J(J+1)-M^2-M]\hbar^2\delta(M, M')\delta(J, J').\end{aligned} \tag{9.13}$$

This is a diagonal matrix, and one can consider separately those elements on the diagonal corresponding to each separate value of J. Similarly

$$\langle J', M' \mid \mathrm{J}_+\mathrm{J}_- \mid J, M\rangle = [J(J+1)-M^2+M]\hbar^2\delta(J, J')\delta(M, M'). \tag{9.14}$$

From equations (9.4a) and (9.4b) as well as from the fact that the basic functions are eigenfunctions of J_z

$$\langle J', M' \mid \mathrm{J}_z \mid J, M\rangle = M\hbar\delta(J, J')\delta(M, M'). \tag{9.15}$$

To evaluate the matrices for J_+ and J_- separately, one may make use of equations (9.8) and (9.9) to show that

$$\langle J', M' \mid \mathrm{J}_+ \mid J, M\rangle = \lambda_M\langle J', M' \mid J', M+1\rangle = \lambda_M\delta(J', J)\delta(M', M+1). \tag{9.16}$$

The introduction of the number λ_M is necessary, since the argument from equations (9.8) and (9.9) is independent of the normalization. Similarly

$$\langle J', M' \mid \mathrm{J}_- \mid J, M\rangle = \mu_M\langle J', M' \mid J', M-1\rangle = \mu_M\delta(J', J)\delta(M', M-1). \tag{9.17}$$

To evaluate the numbers λ_M and μ_M one may multiply the matrices in (9.16) and (9.17) and use equation (9.14)

$$\begin{aligned}\langle J', M' \mid \mathrm{J}_+\mathrm{J}_- \mid J, M\rangle &= \sum_{J'', M''} \langle J', M' \mid \mathrm{J}_+ \mid J'', M''\rangle \langle J'', M'' \mid \mathrm{J}_- \mid J, M\rangle \\ &= \sum_{J'', M''} \lambda_{M''}\delta(J', J'')\delta(M', M''+1)\mu_M\delta(J'', J)\delta(M'', M-1) \\ &= \lambda_{M-1}\, \mu_M\delta(J', J)\delta(M', M) \qquad (9.18) \\ &= [J(J+1)-M^2+M]\hbar^2\delta(J', J)\delta(M', M).\end{aligned}$$

To evaluate the separate numbers two more conditions must be used. The matrix for J_x, which is $\frac{1}{2}(\mathrm{J}_+ + \mathrm{J}_-)$, must be Hermitian. Hence

$$\langle J', M' \mid \mathrm{J}_+ \mid J, M\rangle + \langle J', M' \mid \mathrm{J}_- \mid J, M\rangle = $$
$$= \langle J, M \mid \mathrm{J}_+ \mid J', M'\rangle^* + \langle J, M \mid \mathrm{J}_- \mid J', M'\rangle^*$$

so that

$$\lambda_M\delta(M', M+1) + \mu_M\delta(M', M-1) = \lambda^*_{M-1}\delta(M-1, M') + \mu^*_{M'}\delta(M, M'-1).$$

This requires that

$$\lambda_M = \mu^*_{M+1} \quad \text{or} \quad \mu_M = \lambda^*_{M-1}. \tag{9.19}$$

Then from (9.18)

$$|\lambda_{M-1}|^2 = [J(J+1)-M^2+M]^{\frac{1}{2}}\hbar^2 = |\mu_M|^2. \tag{9.20}$$

This expression leaves undetermined a phase factor which, however, is arbitrary and may be taken to be zero. The result is

$$\lambda_M = [J(J+1)-M^2-M]^{\frac{1}{2}}\hbar. \tag{9.21}$$

Consequently the effect of the operators J_+ and J_- can be expressed in terms of normalized ket vectors:

$$J_+ \mid J, M\rangle = [J(J+1)-M^2-M]^{\frac{1}{2}}\hbar \mid J, M+1\rangle, \tag{9.22a}$$

$$J_- \mid J, M\rangle = [J(J+1)-M^2+M]^{\frac{1}{2}}\hbar \mid J, M-1\rangle. \tag{9.22b}$$

This zero value for the phase factor corresponds to the relationships given in the Appendix, equations (9.20) and (9.21).

PROBLEM 3. Show from the fact that J^2 commutes with J_x, J_y, J_z, J_+ and J_- that the matrix represenstations of these quantities in the (J, M) scheme will have no elements for which $J' \neq J$.

PROBLEM 4. Write out the matrices for J_+, J_-, J_x, J_y and J_z up to and including the terms with $J = 1$.

PROBLEM 5. Write out the results of the operators J_+ and J_- operating to the left on normalized bra vectors.

The ket and bra vectors can be represented as column and row matrices respectively on the basis of the eigenkets and eigenbras of J^2 and J_z. If only the representation for $J = 1$ is considered, an arbitrary ket may be represented by

$$a \mid 1, 1\rangle + b \mid 1, 0\rangle + c \mid 1, -1\rangle \rightarrow \begin{pmatrix} a \\ b \\ c \end{pmatrix}. \tag{9.23}$$

The normalization requires that

$$\mid a \mid^2 + \mid b \mid^2 + \mid c \mid^2 = 1, \tag{9.23a}$$

and the corresponding bra is a row matrix

$$a^*\langle 1, 1 \mid + b^*\langle 1, 0 \mid + c^*\langle 1, -1 \mid \rightarrow (a^*, b^*, c^*). \tag{9.23b}$$

Also one may use the matrix elements to form operators in the dyadic form as in equation (8.20). If only states with one value of J are considered an arbitrary linear operator is

$$Q = \sum_{M', M} Q_{M', M} \mid J, M'\rangle \langle J, M \mid = \sum_{M', M} \langle J, M' \mid Q \mid J, M\rangle \mid J, M'\rangle \langle J, M \mid$$

$$= \sum_{M', M} \mid J, M'\rangle \langle J, M' \mid Q \mid J, M\rangle \langle J, M \mid. \tag{9.24}$$

As was indicated previously, the matrices for the quantities associated with angular momentum have an infinity of rows and columns since J can take all integral and half integral positive values. However, since many quantities commute with J^2, only one value of J need be considered at a time when only these commuting quantities are considered. In more complicated situations, in particular when several interacting particles are involved, it may be necessary to consider larger sections of the matrices.

PROBLEM 6. Write the matrix for the angular momentum about an arbitrary axis whose direction cosines are λ, μ, ν. Write this out specifically for $J = 1$.

PROBLEM 7. Write the matrices for the component angular momentum operators along a system of axes (x', y', z') when the basic kets are eigenkets of the angular momentum along z. Locate (x', y', z') with respect to (x, y, z) by means of the Eulerian angles.

PROBLEM 8. Work out the eigenkets of angular momentum about an arbitrary axis for the case $J = 1$.

3. Electron spin

In the early 1920's it was recognized that a complete description of an electron requires more than the three coordinates of position and the three corresponding momenta. There appeared to be what was called an "irrational duality" in observed atomic and molecular spectra which could be explained only by assigning to an electron at least one other coordinate. In 1925 it was suggested by Uhlenbeck and Goudsmit that this duality could be attributed to an electron "spin", or angular momentum, of magnitude $\frac{1}{2}\hbar$, and an associated magnetic moment. The magnitude $\frac{1}{2}\hbar$ provided just two values of "spin" along any axis, $+\frac{1}{2}\hbar$ and $-\frac{1}{2}\hbar$, so the "duality" could be understood. Later it appeared that protons and neutrons have the same property, and all "fundamental" particles must have a value of "spin" as part of their description.

The letter s may be used to represent spin and the operators s_x, s_y, s_z are the angular momentum operators for the three components of angular momentum when $J = \frac{1}{2}$.

The usual matrix representations of the spin operators follow directly from equations (9.16), (9.17), (9.20) and (9.22):

$$\mathsf{s}_x = \begin{pmatrix} 0 & \frac{1}{2} \\ \frac{1}{2} & 0 \end{pmatrix}\hbar; \quad \mathsf{s}_y = \begin{pmatrix} 0 & -\frac{1}{2} \\ \frac{1}{2} & 0 \end{pmatrix} i\hbar; \quad \mathsf{s}_z = \begin{pmatrix} \frac{1}{2} & 0 \\ 0 & -\frac{1}{2} \end{pmatrix}\hbar. \tag{9.25}$$

In many situations the presence of an electron spin is made evident by the magnetic moment associated with it. Because of this magnetic moment there is an interaction with any magnetic field that may be present. In a crude way an electron is sometimes pictured as a spinning sphere, or spherical shell, of charge, and has been referred to as a spinning electron. A more straightforward procedure seems to be to regard the magnitude of the magnetic moment as given by the experimental observations. The part of the Hamiltonian operator representing the interaction between the spin and the external field is

$$\mathsf{H}_2 = -\mu\, \boldsymbol{s} \cdot \boldsymbol{B}, \tag{9.26}$$

where μ is the magnetic moment per unit of angular momentum. For an electron, $\mu = -|e|/m$. For a proton μ is approximately $5.6\,|e|/M$ and for a neutron approximately $-3.8\,|e|/M$. The quantity $\mu_0 = -|e|\hbar/2m$ is called a Bohr magneton. It represents the magnetic moment of an electron of charge $-|e|$ moving in an orbit of angular momentum $\hbar$. The magnetic moment associated with the electron spin is also $-|e|\hbar/2m$, but since the angular momentum is only $\frac{1}{2}\hbar$, the ratio is μ, or $-|e|/m$.

Although the Hamiltonian H_2 may be regarded as determined by experiment, it can be written by analogy with the classical case of a magnetic dipole in a magnetic field. From this analogy one can infer that the expression (9.26) is valid only in case the motion of the particle is so slow that relativistic effects can be neglected. The presence of a negative instead of a positive sign is paralleled by the corresponding expression for the energy of a moving charge in a magnetic field as derived in Chapter 3. When the result in Chapter 3 is expressed in terms of magnetic moment, the sign is negative as here.

To illustrate the significance of the Hamiltonian function (9.26), and to illustrate the use of the vector operator **s**, it is instructive to compute the energy eigenstates of a free electron in a uniform magnetic field, while neglecting any kinetic energy of motion the electron may have. This involves the function H_2 only, and the equation for the energy eigenstates is

$$H_2\,|\,H_2'\rangle = -\mu(s_x B_x + s_y B_y + s_z B_z)\,|\,H_2'\rangle = H_2'\,|\,H_2'\rangle. \tag{9.27}$$

The eigenket $|\,H_2'\rangle$ may be expressed in terms of the complete set of eigenkets for s_z. Thus let

$$|\,H_2'\rangle = a\,|\,\tfrac{1}{2}\rangle + b\,|\,-\tfrac{1}{2}\rangle. \tag{9.28}$$

Inserting the spin operators and writing out the whole equation leads to

$$\begin{aligned} -a(B_x s_x + B_y s_y + B_z s_z)\,|\,\tfrac{1}{2}\rangle - b(B_x s_x + B_y s_y + B_z s_z)\,|\,-\tfrac{1}{2}\rangle &= \\ = (H_2'/\mu)a\,|\,\tfrac{1}{2}\rangle + (H_2'/\mu)b\,|\,-\tfrac{1}{2}\rangle. \end{aligned} \tag{9.29}$$

If now the equation is multiplied on the left by $\langle \frac{1}{2} |$, and the elements $\langle s_z' | \mathrm{s}_i | s_z'' \rangle$ are taken from (9.25),

$$[B_z + (2H_2'/\mu\hbar)]a + (B_x - \mathrm{i}B_y)b = 0. \tag{9.30a}$$

Similarly, multiplication by $\langle -\frac{1}{2} |$ gives

$$(B_x + \mathrm{i}B_y)a - [B_z - (2H_2'/\mu\hbar)]b = 0. \tag{9.30b}$$

The condition for compatibility of these two equations is

$$H_2' = \pm\tfrac{1}{2}\mu\hbar(B_x^2 + B_y^2 + B_z^2)^{\frac{1}{2}} = \pm\tfrac{1}{2}\mu\hbar B. \tag{9.31}$$

When $H_2' = +\frac{1}{2}\mu\hbar B$

$$a = \sin\tfrac{1}{2}\vartheta\, \mathrm{e}^{-\frac{1}{2}\mathrm{i}\varphi} \quad \text{and} \quad b = -\cos\tfrac{1}{2}\vartheta\, \mathrm{e}^{\frac{1}{2}\mathrm{i}\varphi}. \tag{9.32a}$$

When $H_2' = -\frac{1}{2}\mu\hbar B$

$$a = \cos\tfrac{1}{2}\vartheta\, \mathrm{e}^{-\frac{1}{2}\mathrm{i}\varphi} \quad \text{and} \quad b = \sin\tfrac{1}{2}\vartheta\, \mathrm{e}^{\frac{1}{2}\mathrm{i}\varphi}. \tag{9.32b}$$

When the magnetic field is parallel to the z axis, the two energy states correspond to the two spin states in terms of which the matrices and vectors have been written, and in which the z component of spin is precisely defined. This is clear from equations (9.32a) and (9.32b). When the field is not parallel to the polar axis, the two energy states represent eigenstates of the spin component in the direction of the field.

PROBLEM 9. Show how the solutions in equations (9.31) and (9.32) follow from the Hamiltonian operator.

PROBLEM 10. If the spin of an electron parallel to the z-axis is given, and the magnetic field has an arbitrary direction, what is the distribution of energy? What is the probability that a measurement would find the spin parallel or opposite to the field?

With the Hamiltonian function of equation (9.26), the equation of motion can be written and the behavior of the spin as a function of the time can be computed.

Let $| F, 0 \rangle$ be the state vector at time 0 and $| F, t \rangle$ the vector at time t. Then

$$| F, t \rangle = \mathsf{T} \, | F, 0 \rangle, \tag{9.33}$$

where the operator T satisfies equation (8.43) The solution

$$\mathsf{T} = \mathrm{e}^{-(\mathrm{i}/\hbar)\mathsf{H}t} \tag{9.34}$$

is easily applicable when the basic kets are energy eigenkets and the magnetic field is along the z axis. Thus, if

$$| F, 0\rangle = a \,| \tfrac{1}{2}\rangle + b\, | -\tfrac{1}{2}\rangle,$$

$$\mathsf{T} = \begin{pmatrix} e^{+\frac{1}{2}i\mu Bt} & 0 \\ 0 & e^{-\frac{1}{2}i\mu Bt} \end{pmatrix}$$

and

$$| F, t\rangle = a\, e^{+\frac{1}{2}i\mu Bt}\, | \tfrac{1}{2}\rangle + b\, e^{-\frac{1}{2}i\mu Bt}\, | -\tfrac{1}{2}\rangle. \tag{9.35}$$

From this state vector the distribution of physical quantities at time t can be calculated.

PROBLEM 11. Consider an electron in a magnetic field along the polar axis. At $t = 0$ the component of spin along the direction (ϑ, φ) is given. Find the mean values, as functions of the time, of the components of spin along the x, y and z axes. Show how the result can be interpreted as representing a precession about the magnetic field.

PROBLEM 12. Use equation (9.34) to transform the spin operators to operator functions of the time.

4. Electron in a rotating magnetic field

The case of an electron in a large constant and uniform magnetic field with a small perpendicular field rotating around it is of particular interest in many applications. It may be given an exact solution and does not neeed to be treated by perturbation methods. For this reason, and because electron spin has only two eigenvalues, it is suitable for illustrating the equivalence of various procedures for finding the changes with time of a quantum mechanical system.

Let the constant magnetic field B_0 be along the z-axis, and let there be a smaller field, B_1, in the x–y plane and rotating around B_0. Then the Hamiltonian operator is

$$\mathsf{H} = -\mu\, \boldsymbol{s} \cdot \boldsymbol{B} = -\mu(s_z B_0 + s_x B_1 \cos \omega t + s_y B_1 \sin \omega t) = \mathsf{H}_0 + \mathsf{H}_1. \tag{9.36}$$

This Hamiltonian contains only the spin operators and, as in H_2 of equation (9.27), ignores any kinetic or other potential energy that the electron may have. Again, as in the case of the constant magnetic field

$$| F(t)\rangle = a(t)\, | +\tfrac{1}{2}\rangle + b(t)\, | -\tfrac{1}{2}\rangle \tag{9.37}$$

where $a(t)$ and $b(t)$ must be determined from the Schroedinger equation and the initial conditions. Since the Hamiltonian function contains the time explicitly the integral in equation (8.44) cannot be used directly but the method of Chapter 6 is immediately applicable. Thus

$$-\frac{\hbar}{i}\frac{d}{dt}|F(t)\rangle = -\frac{\hbar}{i}\left[\frac{da}{dt}|+\tfrac{1}{2}\rangle + \frac{db}{dt}|-\tfrac{1}{2}\rangle\right]$$
$$= a(\mathsf{H}_0+\mathsf{H}_1)|+\tfrac{1}{2}\rangle + b(\mathsf{H}_0+\mathsf{H}_1)|-\tfrac{1}{2}\rangle,$$

and this separates into the two equations

$$\begin{aligned} -\frac{\hbar}{i}\frac{da}{dt} &= \langle\tfrac{1}{2}|\mathsf{H}_0+\mathsf{H}_1|\tfrac{1}{2}\rangle a + \langle\tfrac{1}{2}|\mathsf{H}_0+\mathsf{H}_1|-\tfrac{1}{2}\rangle b \\ -\frac{\hbar}{i}\frac{db}{dt} &= \langle-\tfrac{1}{2}|\mathsf{H}_0+\mathsf{H}_1|\tfrac{1}{2}\rangle a + \langle-\tfrac{1}{2}|\mathsf{H}_0+\mathsf{H}_1|-\tfrac{1}{2}\rangle b. \end{aligned} \tag{9.38}$$

Since the basis functions $|\tfrac{1}{2}\rangle$ and $|-\tfrac{1}{2}\rangle$ are eigenfunctions of H_0 with the eigenvalues $-\tfrac{1}{2}\mu\hbar B_0$ and $+\tfrac{1}{2}\mu\hbar B_0$, and H_1 contains only s_x and s_y which have only non-diagonal matrix elements, these equations can be further reduced to

$$\begin{aligned} -\frac{\hbar}{i}\frac{da}{dt} &= -\tfrac{1}{2}\mu\hbar B_0 a + \langle\tfrac{1}{2}|\mathsf{H}_1|-\tfrac{1}{2}\rangle b, \\ -\frac{\hbar}{i}\frac{db}{dt} &= \langle-\tfrac{1}{2}|\mathsf{H}_1|\tfrac{1}{2}\rangle a + \tfrac{1}{2}\mu\hbar B_0 b. \end{aligned} \tag{9.39}$$

The matrix elements of H_1 involve the time, but equations (9.39) can be solved by usual methods of integration. The results, inserted in (9.37) give the ket vector at any time, from which the observable values of the various operators can be determined.

Another approach, which is really another notation for the same thing, is to use the time development operator T whose time rate of change is given by equation (8.43). Using as a basis the spin functions $|m_s\rangle$, where $s_z' = m_s\hbar$, T is represented by a 2×2 matrix

$$\mathsf{T} = \begin{pmatrix} \langle\tfrac{1}{2}|\mathsf{T}|\tfrac{1}{2}\rangle & \langle\tfrac{1}{2}|\mathsf{T}|-\tfrac{1}{2}\rangle \\ \langle-\tfrac{1}{2}|\mathsf{T}|\tfrac{1}{2}\rangle & \langle-\tfrac{1}{2}|\mathsf{T}|-\tfrac{1}{2}\rangle \end{pmatrix} \tag{9.40}$$

where each element is a function of the time. The matrix T must be a unitary matrix which requires that

$$\begin{pmatrix} \langle\tfrac{1}{2}|\mathsf{T}^{-1}|\tfrac{1}{2}\rangle & \langle\tfrac{1}{2}|\mathsf{T}^{-1}|-\tfrac{1}{2}\rangle \\ \langle-\tfrac{1}{2}|\mathsf{T}^{-1}|\tfrac{1}{2}\rangle & \langle-\tfrac{1}{2}|\mathsf{T}^{-1}|-\tfrac{1}{2}\rangle \end{pmatrix} = \begin{pmatrix} \langle\tfrac{1}{2}|\mathsf{T}|\tfrac{1}{2}\rangle^* & \langle-\tfrac{1}{2}|\mathsf{T}|\tfrac{1}{2}\rangle^* \\ \langle\tfrac{1}{2}|\mathsf{T}|-\tfrac{1}{2}\rangle^* & \langle-\tfrac{1}{2}|\mathsf{T}|-\tfrac{1}{2}\rangle^* \end{pmatrix}.$$

At $t = 0$, the non-diagonal elements are zero and the diagonal elements are each equal to one. Using the same basis as in equation (9.37) the Hamiltonian operator has the matrix form

$$\mathsf{H}(t) = -\tfrac{1}{2}\mu\hbar \begin{pmatrix} B_0 & B_1\,\mathrm{e}^{-\mathrm{i}\omega t} \\ B_1\,\mathrm{e}^{+\mathrm{i}\omega t} & -B_0 \end{pmatrix}. \tag{9.41}$$

Equation (8.43) gives a differential equation for each component of T, since the matrix elements for the time derivatives of an operator are the time derivatives of the matrix elements when the basis functions are constant. Thus

$$\begin{aligned} -\frac{\hbar}{\mathrm{i}}\frac{\mathrm{d}}{\mathrm{d}t}\langle\tfrac{1}{2}\,|\,\mathsf{T}\,|\,\tfrac{1}{2}\rangle &= -\tfrac{1}{2}\mu\hbar B_0\langle\tfrac{1}{2}\,|\,\mathsf{T}\,|\,\tfrac{1}{2}\rangle - \tfrac{1}{2}\mu\hbar B_1\,\mathrm{e}^{-\mathrm{i}\omega t}\langle-\tfrac{1}{2}\,|\,\mathsf{T}\,|\,\tfrac{1}{2}\rangle \\ -\frac{\hbar}{\mathrm{i}}\frac{\mathrm{d}}{\mathrm{d}t}\langle-\tfrac{1}{2}\,|\,\mathsf{T}\,|\,\tfrac{1}{2}\rangle &= -\tfrac{1}{2}\mu\hbar B_1\,\mathrm{e}^{\,\mathrm{i}\omega t}\langle\tfrac{1}{2}\,|\,\mathsf{T}\,|\,\tfrac{1}{2}\rangle + \tfrac{1}{2}\mu\hbar B_0\langle-\tfrac{1}{2}\,|\,\mathsf{T}\,|\,\tfrac{1}{2}\rangle \end{aligned} \tag{9.42}$$

with another similar pair for the other two elements. These are identical with equations (9.39) except for a different notation. Equations (9.39) imply arbitrary initial values for a and b. When the matrix notation is used there are four matrix elements to be determined subject to the initial condition that T is defined to be a unit matrix at $t = 0$. The resulting time dependent matrix can then operate on an arbitrary ket vector describing the state at $t = 0$. Similar values would be obtained if equations (9.39) were solved once for $[a(0) = 1, b(0) = 0]$ and again for $[a(0) = 0, b(0) = 1]$.

The equations (9.42) can be made a trifle simpler by defining another operator

$$\mathsf{T}_0 = \exp\{(\mathrm{i}/\hbar)\mathsf{H}_0 t\}\mathsf{T}. \tag{9.43}$$

If then T_0 is evaluated, T can be determined from the inverse operation

$$\mathsf{T} = \exp\{-(\mathrm{i}/\hbar)\mathsf{H}_0 t\}\mathsf{T}_0.$$

The operator T_0 is unitary since T itself is required to be unitary and $\exp\{(\mathrm{i}/\hbar)\mathsf{H}_0 t\}$ is obviously unitary also. Then

$$\begin{aligned} \frac{\mathrm{d}\mathsf{T}_0}{\mathrm{d}t} &= \frac{\mathrm{i}}{\hbar}\mathsf{H}_0\exp\{(\mathrm{i}/\hbar)\mathsf{H}_0 t\}\,\mathsf{T} + \exp\{(\mathrm{i}/\hbar)\mathsf{H}_0 t\}\,\frac{\mathrm{d}\mathsf{T}}{\mathrm{d}t} \\ &= \frac{\mathrm{i}}{\hbar}\exp\{(\mathrm{i}/\hbar)\mathsf{H}_0 t\}\,\mathsf{H}_0\mathsf{T} + \exp\{(\mathrm{i}/\hbar)\mathsf{H}_0 t\}\left[-\frac{\mathrm{i}}{\hbar}\mathsf{H}_0\mathsf{T} - \frac{\mathrm{i}}{\hbar}\mathsf{H}_1\mathsf{T}\right] \\ &= -\frac{\mathrm{i}}{\hbar}\mathsf{H}_1^{(0)}\mathsf{T}_0, \end{aligned} \tag{9.44}$$

where

$$\mathsf{H}_1^{(0)} = \exp\{(i/\hbar)\mathsf{H}_0 t\}\,\mathsf{H}_1 \exp\{-(i/\hbar)\mathsf{H}_0 t\}. \tag{9.44a}$$

When the basis functions of the matrix operators are the eigenfunctions of H_0, the transformation in (9.44a) can be simply expressed:

$$\begin{aligned}\langle m_s'' \mid \mathsf{H}_1^{(0)} \mid m_s' \rangle &= \langle m_s'' \mid \exp\{(i/\hbar)\mathsf{H}_0 t\}\,\mathsf{H}_1 \exp\{-(i/\hbar)\mathsf{H}_0 t\} \mid m_s' \rangle \\ &= \exp\{(i/\hbar)(H_0''-H_0')t\}\,\langle m_s'' \mid \mathsf{H}_1 \mid m_s' \rangle;\end{aligned} \tag{9.45}$$

m_s is a quantum number such that $s_z = m_s\hbar$.

Only the non-diagonal elements of H_1 are different from zero so only the non-diagonal elements of $\mathsf{H}_1^{(0)}$ are different from zero:

$$\begin{aligned}\langle \tfrac{1}{2} \mid \mathsf{H}_1^{(0)} \mid -\tfrac{1}{2} \rangle &= -\tfrac{1}{2}\mu\hbar B_1 \exp\{i(\omega_0-\omega)t\}, \\ \langle -\tfrac{1}{2} \mid \mathsf{H}_1^{(0)} \mid \tfrac{1}{2} \rangle &= -\tfrac{1}{2}\mu\hbar B_1 \exp\{-i(\omega_0-\omega)t\},\end{aligned} \tag{9.46}$$

where $\omega_0 = -\mu B_0$. Because $\mathsf{H}_1^{(0)}$ has only the two non-diagonal terms, each element of its product with T_0 involves only one term. There are then four differential equations for the rates of change of the four elements of the matrix T_0. These divide, as before, into two pairs of simultaneous differential equations:

$$\begin{aligned}\frac{d}{dt}\langle \tfrac{1}{2} \mid \mathsf{T}_0 \mid \tfrac{1}{2} \rangle &= -\frac{i}{\hbar}\langle \tfrac{1}{2} \mid \mathsf{H}_1^{(0)} \mid -\tfrac{1}{2} \rangle \langle -\tfrac{1}{2} \mid \mathsf{T}_0 \mid \tfrac{1}{2} \rangle \\ \frac{d}{dt}\langle -\tfrac{1}{2} \mid \mathsf{T}_0 \mid \tfrac{1}{2} \rangle &= -\frac{i}{\hbar}\langle -\tfrac{1}{2} \mid \mathsf{H}_1^{(0)} \mid \tfrac{1}{2} \rangle \langle \tfrac{1}{2} \mid \mathsf{T}_0 \mid \tfrac{1}{2} \rangle,\end{aligned} \tag{9.47a}$$

$$\begin{aligned}\frac{d}{dt}\langle \tfrac{1}{2} \mid \mathsf{T}_0 \mid -\tfrac{1}{2} \rangle &= -\frac{i}{\hbar}\langle \tfrac{1}{2} \mid \mathsf{H}_1^{(0)} \mid -\tfrac{1}{2} \rangle \langle -\tfrac{1}{2} \mid \mathsf{T}_0 \mid -\tfrac{1}{2} \rangle \\ \frac{d}{dt}\langle -\tfrac{1}{2} \mid \mathsf{T}_0 \mid -\tfrac{1}{2} \rangle &= -\frac{i}{\hbar}\langle -\tfrac{1}{2} \mid \mathsf{H}_1^{(0)} \mid \tfrac{1}{2} \rangle \langle \tfrac{1}{2} \mid \mathsf{T}_0 \mid -\tfrac{1}{2} \rangle.\end{aligned} \tag{9.47b}$$

Each pair can be reduced to a second order equation and solved in the usual way to make T_0 (and hence T) a unit matrix at $t = 0$:

$$\begin{aligned}\langle \tfrac{1}{2} \mid \mathsf{T}_0 \mid \tfrac{1}{2} \rangle &= (1/\Delta)\,[\alpha_1 \exp(-i\alpha_2 t) - \alpha_2 \exp(-i\alpha_1 t)], \\ \langle -\tfrac{1}{2} \mid \mathsf{T}_0 \mid \tfrac{1}{2} \rangle &= (\mu B_1/2\Delta)\,[\exp(+i\alpha_1 t) - \exp(+i\alpha_2 t)],\end{aligned} \tag{9.48a}$$

where

$$\begin{aligned}\alpha_1 &= \tfrac{1}{2}[\omega-\omega_0+\Delta], \\ \alpha_2 &= \tfrac{1}{2}[\omega-\omega_0-\Delta], \\ \Delta &= +\,[(\omega-\omega_0)^2+(\mu B_1)^2]^{\frac{1}{2}}.\end{aligned}$$

Also

$$\langle \tfrac{1}{2} | \mathsf{T}_0 | -\tfrac{1}{2} \rangle = -(\mu B_1/2\Delta)\,[\exp(-\mathrm{i}\alpha_1 t) - \exp(-\mathrm{i}\alpha_2 t)],$$
$$\langle -\tfrac{1}{2} | \mathsf{T}_0 | -\tfrac{1}{2} \rangle = +(1/\Delta)\,[\alpha_1 \exp(\mathrm{i}\alpha_2 t) - \alpha_2 \exp(\mathrm{i}\alpha_1 t)]. \quad (9.48\text{b})$$

This matrix is unitary and the matrix for T will also be unitary:

$$\begin{aligned}
\langle \tfrac{1}{2} | \mathsf{T} | \tfrac{1}{2} \rangle &= \mathrm{e}^{-\frac{1}{2}\mathrm{i}\omega_0 t} \langle \tfrac{1}{2} | \mathsf{T}_0 | \tfrac{1}{2} \rangle = \\
&= \mathrm{e}^{-\frac{1}{2}\mathrm{i}\omega t}\,\{\cos \tfrac{1}{2}\Delta t + \mathrm{i}[(\omega-\omega_0)/\Delta] \sin \tfrac{1}{2}\Delta t\},\\
\langle \tfrac{1}{2} | \mathsf{T} | -\tfrac{1}{2} \rangle &= \mathrm{e}^{-\frac{1}{2}\mathrm{i}\omega_0 t} \langle \tfrac{1}{2} | \mathsf{T}_0 | -\tfrac{1}{2} \rangle = \mathrm{i}(\mu B_1/\Delta)\,\mathrm{e}^{-\frac{1}{2}\mathrm{i}\omega t} \sin \tfrac{1}{2}\Delta t,\\
\langle -\tfrac{1}{2} | \mathsf{T} | \tfrac{1}{2} \rangle &= \mathrm{e}^{+\frac{1}{2}\mathrm{i}\omega_0 t} \langle -\tfrac{1}{2} | \mathsf{T}_0 | \tfrac{1}{2} \rangle = \mathrm{i}(\mu B_1/\Delta)\,\mathrm{e}^{\frac{1}{2}\mathrm{i}\omega t} \sin \tfrac{1}{2}\Delta t, \qquad (9.49)\\
\langle -\tfrac{1}{2} | \mathsf{T} | -\tfrac{1}{2} \rangle &= \mathrm{e}^{+\frac{1}{2}\mathrm{i}\omega_0 t} \langle -\tfrac{1}{2} | \mathsf{T}_0 | -\tfrac{1}{2} \rangle = \\
&= \mathrm{e}^{+\frac{1}{2}\mathrm{i}\omega t}\,\{\cos \tfrac{1}{2}\Delta t - \mathrm{i}[(\omega-\omega_0)/\Delta] \sin \tfrac{1}{2}\Delta t\}.
\end{aligned}$$

The operator T serves to transform the ket vector at the time $t = 0$ to its value at the time t, and from the time dependent ket vector the distribution of any function of the spin variable may be obtained. If $| F(0)\rangle = | \tfrac{1}{2}\rangle$,

$$\begin{aligned}
| F(t)\rangle = \mathrm{e}^{-\frac{1}{2}\mathrm{i}\omega t}\{\cos \tfrac{1}{2}\Delta t + [\mathrm{i}(\omega-\omega_0)/\Delta] \sin \tfrac{1}{2}\Delta t\}\, | \tfrac{1}{2}\rangle + \\
+ \mathrm{i}(\mu B_1/\Delta)\,\mathrm{e}^{+\frac{1}{2}\mathrm{i}\omega t} \sin \tfrac{1}{2}\Delta t\, | -\tfrac{1}{2}\rangle.
\end{aligned} \qquad (9.50)$$

The probability of finding the spin in the $+z$ direction is then

$$P(\tfrac{1}{2}) = | \langle \tfrac{1}{2} | F(t)\rangle |^2 = 1 - (\mu B_1/\Delta)^2 \sin^2 \tfrac{1}{2}\Delta t. \qquad (9.51)$$

PROBLEM 13. Show that the operator T, whose matrix elements are given in equations (9.49) is a unitary operator and is equal to the unit operator when $t = 0$. In the matrix representation of a unitary operator the complex conjugate of the transposed matrix is the reciprocal matrix.

PROBLEM 14. If at $t = 0$ the spin of an electron in a magnetic field described by the Hamiltonian function of equation (9.36) is along the negative $z$axis, find the probability as a function of the time of measuring a spin along the $x$axis equal to $\tfrac{1}{2}\hbar$.

PROBLEM 15. Show that the transformation of an Hermitian matrix by a unitary transformation $\mathsf{H}_\mathrm{H} = \mathsf{U}^{-1}\mathsf{H}\mathsf{U}$ leads to another Hermitian matrix.

Thus far the equation of motion, and the time development operator T, have been used to determine the ket vector as a function of the time. T can also be used to form time dependent operators which operate on the ket vector of the initial state. Such operators will be designated by the subscript H since they are often known as Heisenberg operators, $\alpha_\mathrm{H} = \mathsf{T}^{-1}\alpha\mathsf{T}^{+1}$. Using the matrix for T as given in equations (9.49), the matrices for the three spin components as functions of the time are

$$\langle \tfrac{1}{2} \mid \mathsf{s}_{z\mathrm{H}} \mid \tfrac{1}{2} \rangle = \tfrac{1}{2}\hbar[1-(\mu B_1/\Delta)^2(1 - \cos \Delta t)]$$

$$\langle \tfrac{1}{2} \mid \mathsf{s}_{z\mathrm{H}} \mid -\tfrac{1}{2} \rangle = +\tfrac{1}{2}\hbar(\mu B_1/\Delta) \{[(\omega-\omega_0)/\Delta] (1 - \cos \Delta t)+\mathrm{i} \sin \Delta t\}$$

$$\langle -\tfrac{1}{2} \mid \mathsf{s}_{z\mathrm{H}} \mid \tfrac{1}{2} \rangle = +\tfrac{1}{2}\hbar(\mu B_1/\Delta) \{[(\omega-\omega_0)/\Delta] (1 - \cos \Delta t)-\mathrm{i} \sin \Delta t\}$$

$$\langle -\tfrac{1}{2} \mid \mathsf{s}_{z\mathrm{H}} \mid -\tfrac{1}{2} \rangle = -\tfrac{1}{2}\hbar[1-(\mu B_1/\Delta)^2(1 - \cos \Delta t)]$$

$$\langle \tfrac{1}{2} \mid \mathsf{s}_{x\mathrm{H}} \mid \tfrac{1}{2} \rangle = $$
$$= \tfrac{1}{2}\hbar(\mu B_1/\Delta) \{[(\omega-\omega_0)/\Delta] (1 - \cos \Delta t) \cos \omega t - \sin \Delta t \sin \omega t\}$$

$$\langle \tfrac{1}{2} \mid \mathsf{s}_{x\mathrm{H}} \mid -\tfrac{1}{2} \rangle = $$
$$= \tfrac{1}{2}\hbar(\mu B_1/\Delta)^2\{(1 - \cos \Delta t) \cos \omega t + \mathrm{e}^{\mathrm{i}\omega t}[\cos \Delta t-\mathrm{i}[(\omega-\omega_0)/\Delta] \sin \Delta t]\}$$

$$\langle -\tfrac{1}{2} \mid \mathsf{s}_{x\mathrm{H}} \mid \tfrac{1}{2} \rangle = $$
$$= \tfrac{1}{2}\hbar\{(\mu B_1/\Delta)^2(1 - \cos \Delta t) \cos \omega t + \mathrm{e}^{-\mathrm{i}\omega t} [\cos \Delta t+\mathrm{i}[(\omega-\omega_0)/\Delta] \sin \Delta t]\}$$

$$\langle -\tfrac{1}{2} \mid \mathsf{s}_{x\mathrm{H}} \mid -\tfrac{1}{2} \rangle = $$
$$= -\tfrac{1}{2}\hbar(\mu B_1/\Delta) \{[(\omega-\omega_0)/\Delta] (1 - \cos \Delta t) \cos \omega t - \sin \Delta t \sin \omega t\}$$

$$\langle \tfrac{1}{2} \mid \mathsf{s}_{y\mathrm{H}} \mid \tfrac{1}{2} \rangle = $$
$$= \tfrac{1}{2}\hbar(\mu B_1/\Delta) \{[(\omega-\omega_0)/\Delta] (1 - \cos \Delta t) \sin \omega t + \sin \Delta t \cos \omega t\}$$

$$\langle \tfrac{1}{2} \mid \mathsf{s}_{y\mathrm{H}} \mid -\tfrac{1}{2} \rangle = $$
$$= \tfrac{1}{2}\hbar\{(\mu B_1/\Delta)^2(1 - \cos \Delta t) \sin \omega t - \mathrm{i}\mathrm{e}^{\mathrm{i}\omega t}[\cos \Delta t - \mathrm{i}[(\omega-\omega_0)/\Delta] \sin \Delta t]\}$$

$$\langle -\tfrac{1}{2} \mid \mathsf{s}_{y\mathrm{H}} \mid \tfrac{1}{2} \rangle = $$
$$= \tfrac{1}{2}\hbar\{(\mu B_1/\Delta)^2(1 - \cos \Delta t) \sin \omega t + \mathrm{i}\mathrm{e}^{-\mathrm{i}\omega t}[\cos \Delta t+\mathrm{i}[(\omega-\omega_0)/\Delta] \sin \Delta t]\}$$

$$\langle -\tfrac{1}{2} \mid \mathsf{s}_{y\mathrm{H}} \mid -\tfrac{1}{2} \rangle = $$
$$= -\tfrac{1}{2}\hbar\{ (\mu B_1/\Delta) [(\omega-\omega_0)/\Delta] (1 - \cos \Delta t) \sin \omega t + \sin \Delta t \cos \omega t\} .$$

These spin operators preserve the commutation relations which were valid at $t = 0$. They also satisfy the operator or matrix equation for the time dependence

$$\frac{\mathrm{d}\alpha_\mathrm{H}}{\mathrm{d}t} = -(\mathrm{i}/\hbar) [\alpha_\mathrm{H}\mathsf{H}_\mathrm{H}-\mathsf{H}_\mathrm{H}\alpha_\mathrm{H}] . \tag{9.52}$$

Equation (9.52) is based on the assumption that the operator α is not an explicit function of the time. When an explicit function of the time is under consideration, an additional partial derivative must be used as in equation (8.52a).

PROBLEM 16. Show that equation (9.52) for the spin vector operator $\mathbf{s} = s_x i + s_y j + s_z k$ is equivalent to the classical equation

$$\frac{d\mathbf{s}}{dt} = \mu \mathbf{s} \times \boldsymbol{B}.$$

PROBLEM 17. Show in one case that the time dependent matrices given above satisfy the commutation rules for the spin matrices.

PROBLEM 18. Write the matrix representation of H_H by transforming the representation in (9.41) by the T matrix. Then show that the result so obtained is equal to that obtained by expressing equation (9.36) in terms of the Heisenberg matrices.

PROBLEM 19. Derive equation (9.52) from the definitions of α_H and H_H and the equation for the rate of change of T.

5. An electron with spin in a central field

The case of an electron with spin and magnetic moment moving in an electric field of central symmetry is of particular importance in spectroscopy. It represents roughly the situation of a valence electron of an atom, and is a good approximation for the external electron of monovalent elements such as sodium. In case the velocities are not too great, the operator for the part of the Hamiltonian function representing the interaction of the electron's magnetic moment with the electrical field in which it moves is

$$\mathsf{H}_1 = \frac{\mu}{2mc^2} \frac{1}{r} \frac{d\Phi}{dr} \mathbf{l} \cdot \mathbf{s} \qquad (9.53)$$

where $\mathbf{l}$ is the vector operator for the angular momentum of the electron about the origin, m is the electron mass, and Φ is the potential of the central field. A lower case $\boldsymbol{l}$ is used to designate the orbital angular momentum of a single electron.

The derivation of this form of the operator will not be considered here. The most satisfactory derivation obtains it as an approximation to the exact operator used in relativistic quantum mechanics, but it can also be derived by considering the classical interaction of an electrically charged magnetic dipole moving in an electric field (THOMAS [1926]). A crude way of looking at the problem is to consider the motion of the charge as producing a magnetic field in the direction of the orbital angular momentum $\boldsymbol{l}$. The factor 2 in the denominator comes from an analysis of the field as seen from the moving electron. For a simple Coulomb field $\Phi = -Ze/4\pi\varepsilon_0 r$, and the whole expression is proportional to $1/r^3$. For other fields, however, when account is taken of the shielding of the nucleus by the inner electrons, the form (9.53) can be used.

The energy eigenstates of the system satisfy the equation

$$(H_0+H_1-E') \mid E'\rangle = 0, \qquad (9.54)$$

where H_0 is the Hamiltonian function without spin treated in Chapter 6. The additional part of the Hamiltonian, H_1, makes it necessary to treat the problem by perturbation methods as described in Chapter 7. The treatment given there is easily transcribed into the present notation.

There are now four physically independent quantities, and a state vector can best be designated by four independent eigenvalues of these quantities. For the unperturbed system, when H_1 is neglected, these may be taken as the spin, the two eigenvalues describing the orbital angular momentum, and the energy. The spin has two eigenvalues of its component along the z-axis equal to $\frac{1}{2}\hbar$ and $-\frac{1}{2}\hbar$. These may be designated by two values of s_z' or m_s. $s_z' = m_s\hbar$ with $m_s = \pm\frac{1}{2}$. The orbital angular momentum has eigenvalues of its component along the z-axis equal to $m_l\hbar$ where m_l runs from $-l$ to l with integral values. The square of the total orbital angular momentum has eigenvalues of $l(l+1)\hbar^2$ and these may be designated by the integer l. The energy values may then be designated by another integer n. If the central field is a Coulomb field, the energy is proportional to $1/n^2$, but in other kinds of central field it is a function of n and l.

When the whole Hamiltonian function (9.54) is used, the vector orbital angular momentum and the spin do not separately commute with it and so they are not constants of the motion. Instead one has the total angular momentum $(l+s) = j$, and the component of this total angular momentum about the z axis, j_z. States characteristic of these quantities, as well as of the energy, would be designated as $\mid E, j, m\rangle$. In this and the following expressions $m\hbar = j_z$ and $j(j+1)h^2 = (j^2)'$. In the zero order approximation such states would be linear combinations of the unperturbed states

$$\mid E_0, l, j, m\rangle = \sum_{m_l,\, m_s} a(m_l, m_s) \mid n, l, m_l, m_s\rangle . \qquad (9.55)$$

The index l is included in the specification of the state vector even though the orbital angular momentum it represents is not a constant of the motion. Its inclusion indicates that the zero order function is an eigenfunction of this orbital angular momentum and so is only an approximation to the stationary state. The zero approximation functions are eigenfunctions of l^2 and are good approximations only if the states with other values of l are widely different in energy from the ones considered. The kets $| n, l, m_l, m_s\rangle$ satisfy the equation

$$[H_0 - E_0'(n, l)] \,|\, n, l, m_l, m_s\rangle = 0 \tag{9.56}$$

for all values of m_l and m_s. All of them are represented by the same values of n and l and so by the same unperturbed energy E_0'. Any one member of the sum (9.55) is a state in which the orbital motion and the spin are independent. If there were an external magnetic field, but no spin-orbit interaction, each such term would be an energy eigenstate and the energy would depend also on m_l and m_s. When H_1 is included in the Hamiltonian, the suitable linear combination of different values of m_l and m_s must be taken to give the proper zero order approximate state. Such a sum begins to take account of the interaction between spin and orbital motion and the way they are tied closely together.

The operator H_0 operating on the sum (9.55) multiplies it by the constant $E_0'(n, l)$. If then $E' = E_0'(n, l) + E_1(n, l, j, m)$, the sum must satisfy the equations

$$\langle n, l, m_l', m_s' \,|\, (\mathsf{H}_1 - E_1) \sum_{m_l, m_s} a(m, m_s) \,|\, n, l, m_l, m_s\rangle = 0, \tag{9.57}$$

and these can be used to determine both the first order corrections to the energy, $E_1(n, l, j, m)$, and the coefficients $a(m_l, m_s)$ in the sums.

The application of the operator H_1 is facilitated by expression $\boldsymbol{l} \cdot \boldsymbol{s}$ in the form

$$\begin{aligned} \boldsymbol{l} \cdot \boldsymbol{s} &= \tfrac{1}{2}[(l_x + il_y)(s_x - is_y) + (l_x - il_y)(s_x + is_y) + l_z s_z] \\ &= \tfrac{1}{2}[l_+ s_- + l_- s_+ + 2l_z s_z]. \end{aligned} \tag{9.58}$$

Since H_1 may be written $Q\boldsymbol{l} \cdot \boldsymbol{s}$,

$$\begin{aligned} &\mathsf{H}_1 \sum_{m_l, m_s} a(m_l, m_s) \,|\, n, l, m_l, m_s\rangle = \\ &= \tfrac{1}{2}Q\hbar^2 \sum_{m_l, m_s} \{a(m_l, m_s)(l-m)^{\frac{1}{2}}(l+m+1)^{\frac{1}{2}}(\tfrac{1}{2}+m_s)^{\frac{1}{2}}(\tfrac{3}{2}-m_s)^{\frac{1}{2}} \times \\ &\times |\, m_l + 1, m_s - 1\rangle + (l+m_l)^{\frac{1}{2}}(l-m+1)^{\frac{1}{2}}(\tfrac{1}{2}-m_s)^{\frac{1}{2}}(\tfrac{3}{2}+m_s)^{\frac{1}{2}} \times \\ &\times \,|\, m_l - 1, m_s + 1\rangle + 2m_l, m_s \,|\, m_l, m_s\rangle\}. \end{aligned} \tag{9.59}$$

Forming the bracket with $\langle n, m_l', m_s' \,|$ gives

$$\begin{aligned} &\sum_{m_l, m_s} a(m_l, m_s) \langle n, l, m_l', m_s' \,|\, \mathsf{H}_1 \,|\, n, l, m_l, m_s\rangle = \\ &= \tfrac{1}{2}Q\hbar^2 (l-m'+1)^{\frac{1}{2}}(l+m')^{\frac{1}{2}}(\tfrac{3}{2}+m_s')^{\frac{1}{2}}(\tfrac{1}{2}-m_s')^{\frac{1}{2}} a(m_l'-1, m_s'+1) + \\ &\quad + (l+m'+1)^{\frac{1}{2}}(l-m')^{\frac{1}{2}}(\tfrac{3}{2}-m_s')^{\frac{1}{2}}(\tfrac{1}{2}+m_s')^{\frac{1}{2}} a(m_l'+1, m_s'-1) + \\ &\quad + 2m_l' m_s' a(m_l', m_s') = E_1(n, l, j, m) a(m_l', m_s'). \end{aligned} \tag{9.60}$$

The significance of j and m will be discussed in more detail later. The Q in equation (9.60) is a function of n and l only and can be taken out of the sum.

Although each of the equations indicated in (9.60) appears to contain three of the coefficients $a(m_l, m_s)$ it contains in fact only two. One of the coefficients will always be multiplied by zero. This situation may be illustrated by some examples. For $m_l' = l$ and $m_s' = \frac{1}{2}$ both $(\frac{1}{2}-m_s')$ and $(l-m_l')$ vanish so there is left only

$$\left[\frac{2E_1}{\hbar^2 Q} - l\right] a(l, \tfrac{1}{2}) = 0. \tag{9.61a}$$

For $m' = l$ and $m_s' = -\frac{1}{2}$ there is

$$-(2l)^{\frac{1}{2}} a(l-1, \tfrac{1}{2}) + \left[\frac{2E_1}{\hbar^2 Q} + l\right] a(l, -\tfrac{1}{2}) = 0. \tag{9.61b}$$

For $m' = l-1$ and $m_s' = +\frac{1}{2}$

$$\left[\frac{2E_1}{\hbar^2 Q} - l+1\right] a(l-1, \tfrac{1}{2}) - (2l)^{\frac{1}{2}} a(l, -\tfrac{1}{2}) = 0. \tag{9.61c}$$

If $E_1 = \frac{1}{2}l\hbar^2 Q$, the coefficient $a(l, \frac{1}{2}) \neq 0$, and all the others zero, gives a solution. Equations (9.61b) and (9.61c) contain the same two coefficients, but neither contains the coefficient in (9.61a). The latter two equations are compatible if $E_1 = \frac{1}{2}l\hbar^2 Q$ or $E_1 = -\frac{1}{2}(l+1)\hbar^2 Q$. All of the equations can be paired off in this way except (9.61a) and the corresponding equation for $a(-l, -\frac{1}{2})$. From the whole set of $(4l+2)$ equations there are just two possible values of E_1. These will be designated by two different values of j, and these two values of j will be taken to be $l+\frac{1}{2}$, and $l-\frac{1}{2}$. The solutions of each pair of equations will then be designated by the appropriate value of $m = m_l + m_s$, and by one of the values of j. E_1 is then independent of m, but does depend on j;

$$\begin{aligned} E(n, l, l+\tfrac{1}{2}) &= E_0(n, l) + l(\tfrac{1}{2}\hbar^2 Q) \\ E(n, l, l-\tfrac{1}{2}) &= E_0(n, l) - (l+1)(\tfrac{1}{2}\hbar^2 Q). \end{aligned} \tag{9.62}$$

The solutions of the equations (9.60) have the form

$$\begin{aligned} |\, n, l, l+\tfrac{1}{2}, m_l+\tfrac{1}{2}\rangle &= \left(\frac{l+m_l+1}{2l+1}\right)^{\frac{1}{2}} |\, n, l, m_l, \tfrac{1}{2}\rangle + \left(\frac{l-m_l}{2l+1}\right)^{\frac{1}{2}} |\, n, l, m_l+1, -\tfrac{1}{2}\rangle \\ |\, n, l, l-\tfrac{1}{2}, m_l+\tfrac{1}{2}\rangle &= \left(\frac{l-m}{2l+1}\right)^{\frac{1}{2}} |\, n, l, m_l, \tfrac{1}{2}\rangle - \left(\frac{l+m+1}{2l+1}\right)^{\frac{1}{2}} |\, n, l, m_l+1, -\tfrac{1}{2}\rangle \end{aligned} \tag{9.63}$$

where the last numbers in the kets on the left side are the values of m. The kets on the right side are those of (9.55). Note that in these equations m_l may take on the value $-(l+1)$ since $|\, n, l, m+1, -\frac{1}{2}\rangle$ exists for $m_l+1 = -l$.

For each value of n and l there are these two forms of solution, designated by a value of j and a value of m. The range of m is determined by the form of the coefficients. For $j = l+\frac{1}{2}$, m can be as large as $l+\frac{1}{2}$. When $m_l = l$, the first term in the expression for the function exists but the coefficient of the second term vanishes. Similarly m_l can have the value $-(l+\frac{1}{2})$, for if m_l is taken equal to $-(l+1)$ the coefficient of the first term vanishes but the second term remains. There are then $(2l+2)$, or $(2j+1)$, functions that correspond to one of the energy values. To the other energy value, with $j = l-\frac{1}{2}$, there correspond $2l$ functions, or again $(2j+1)$ functions.

PROBLEM 20. Show that (9.63) is the solution of equation (9.60) for all values of m_l and m_s.

PROBLEM 21. Show that the expressions in (9.63) are eigenkets of the operator $j_z = l_z+s_z$ with the eigenvalues $(m_l+\frac{1}{2})\hbar$, and of the operator $j^2 = (l+s)^2$, with the values $(l+\frac{1}{2})\,(l+\frac{3}{2})\hbar^2$ and $(l-\frac{1}{2})\,(l+\frac{1}{2})\hbar^2$. The importance of this result is further developed in the next section.

6. Vector addition of angular momenta

In the preceding section a spin angular momentum, s, was coupled with an orbital angular momentum l, through a spin-orbit coupling term in the Hamiltonian. This is a special case of the general problem of coupling two or more angular momenta to form a resultant. To treat the general problem consider first a system in which there are two angular momenta, j_1 and j_2, that interact in some manner to form states of total angular momentum J. There are two possible approaches to the description of such states. A state may be such as to be specified by products of kets each representing one of the angular momentum states, or it may be described by a ket specifying the properties of the total angular momentum. Since it is usually only the total angular momentum that commutes with the Hamiltonian and is conserved, the latter description has many advantages. These two types of kets must be linearly related:

$$|\, J, M\rangle = \sum_{m_1, m_2} C^{J, M}_{j_1, j_2, m_1, m_2} \,|\, j_1, m_1\rangle \,|\, j_2, m_2\rangle . \qquad (9.64)$$

The C's are known as the Clebsch-Gordan, or Wigner coefficients. These numbers are also frequently denoted by the bracket symbols to which they are equal:

$$C^{J,M}_{j_1,j_2,m_1,m_2} = \langle j_1, m_1, j_2, m_2 \mid J, M\rangle . \tag{9.64a}$$

One may define a vector linear operator for the total angular momentum by

$$\mathbf{J} = \mathbf{j}_1 + \mathbf{j}_2 \tag{9.65}$$

where $J_z = j_{1,z} + j_{2,z}$, etc.

The notation in (9.64) implies that

$$J_z \mid J, M\rangle = M\hbar \mid J, M\rangle = (m_1+m_2)\hbar \mid J, M\rangle ,$$

so that (9.64) may be rewritten

$$\mid J, M\rangle = \sum_{m_1} C^{J,M}_{j_1,j_2,m_1,M-m_1} \mid j_1, m_1\rangle \mid j_2, M-m_1\rangle . \tag{9.66}$$

The designation of the ket vector $\mid J, M\rangle$ also implies that the state is an eigenvector of the operator $\mathbf{J}^2$ which by (9.65) is given by

$$\begin{aligned} \mathbf{J}^2 &= \mathbf{J}\cdot\mathbf{J} = (\mathbf{j}_1+\mathbf{j}_2)\cdot(\mathbf{j}_1+\mathbf{j}_2) = \mathbf{j}_1^2+\mathbf{j}_2^2+2\mathbf{j}_1\cdot\mathbf{j}_2 \\ &= \mathbf{j}_1^2+\mathbf{j}_2^2+2\mathrm{j}_{1,z}\mathrm{j}_{2,z}+(\mathrm{j}_{1+}\mathrm{j}_{2-})+(\mathrm{j}_{1-}\mathrm{j}_{2+}), \end{aligned} \tag{9.67}$$

where the commutation of j_1 and j_2 has been assumed and the stepping operators have been defined for the angular momenta j_1 and j_2.

Employing (9.66) and (9.67)

$$\begin{aligned} \mathbf{J}^2 \mid J, M\rangle = {} & \{\mathbf{j}_1^2+\mathbf{j}_2^2+2\mathrm{j}_{1,z}\mathrm{j}_{2,z}+(\mathrm{j}_{1+}\mathrm{j}_{2-})+(\mathrm{j}_{1-}\mathrm{j}_{2+})\} \times \\ & \times \sum_{m_1} C^{J,M}_{j_1,j_2,m_2,M-m_1} \mid j_1, m_1\rangle \mid j_2, M-m_1\rangle , \end{aligned}$$

which gives

$$\begin{aligned} & \sum_{m_1} \{J(J+1)\hbar^2 - j_1(j_1+1)\hbar^2 - j_2(j_2+1)\hbar^2 - 2m_1(M-m_1)\hbar^2\} \times \\ & \quad \times C^{J,M}_{j_1,j_2,m_1,M-m_1} \mid j_1, m_1\rangle \mid j_2, M-m_1\rangle = \\ & \quad = \sum_{m_1} C^{J,M}_{j_1,j_2,m_1,M-m_1} \{(j_{1+}j_{2-})+(j_{1-}j_{2+})\} \mid j_1, m_1\rangle \mid j_2, M-m_1\rangle . \end{aligned} \tag{9.68}$$

This equation may be multiplied in turn by the product bras $\langle j_1, m_1' \mid \langle j_2, M-m_1' \mid$ and the resulting set of simultaneous equations solved for the Clebsch-Gordan coefficients.

As an example of the use of equation (9.68) consider the case in which $j_1 = j_2 = \frac{1}{2}$. Then for $M = 1$

$$[J(J+1) - \tfrac{3}{2} - \tfrac{1}{2}]\hbar^2 C^{J,1}_{\frac{1}{2},\frac{1}{2},\frac{1}{2},\frac{1}{2}} = 0 .$$

Hence, if $J(J+1) = 2$, $J = 1$, and the coefficient $C^{J,1}_{\frac{1}{2},\frac{1}{2},\frac{1}{2},\frac{1}{2}}$ can be taken equal to unity. In case $M = 0$ there are two equations

$$[J(J+1)-1]C^{J,0}_{\frac{1}{2},\frac{1}{2},\frac{1}{2},-\frac{1}{2}} - C^{J,0}_{\frac{1}{2},\frac{1}{2},-\frac{1}{2},\frac{1}{2}} = 0,$$

$$-C^{J,0}_{\frac{1}{2},\frac{1}{2},\frac{1}{2},-\frac{1}{2}} + [J(J+1)-1]C^{J,0}_{\frac{1}{2},\frac{1}{2},-\frac{1}{2},\frac{1}{2}} = 0.$$

The compatibility of these two equations requires that $J = 0$, or $J = 1$.

Hence for $J = 0$ there is the one state with $M = 0$, and for $J = 1$, M may be $+1$, 0 or -1.

The coefficients for $j_1 = l$ and $j_2 = \frac{1}{2}$ were treated when treating an electron in a central field. Some other coefficients are given in the Appendix.

PROBLEM 22. Calculate the Clebsch-Gordan coefficients for the vector addition of two angular momenta of magnitude $\hbar$.

PROBLEM 23. Consider two electrons whose spins act on each other to give a Hamiltonian equal to $(\boldsymbol{s}_1 \cdot \boldsymbol{s}_2)$. Let them also be in a uniform magnetic field. Compute the energy eigenvalues and the corresponding states.

PROBLEM 24. Show that

$$s_x s_y + s_y s_x = 0 \tag{9.69}$$

and that

$$s_x s_y = \tfrac{1}{2} i\hbar s_z . \tag{9.70}$$

CHAPTER 10

SYSTEMS OF IDENTICAL PARTICLES

In case a system consists of N particles the wave function is a function in $3N$ dimensions, in addition to the spin. Such a function, for a pure state, must be an eigenfunction of $3N$ commuting operators, or of $4N$ commuting operators if the particles have a spin. Thus the "wave" representing one particle does not "interfere" with the wave representing another particle because they are waves in different portions of the $3N$ or $4N$ dimensional space.

In the application of quantum mechanics to a system of electrons, special attention must be paid to the fact that electrons are indistinguishable. The same is true for a system of protons, or of neutrons. Not only do they all behave in the same way, but no experiment can distinguish one from another. As a consequence the Hamiltonian function describing a system of electrons, or other identical particles, will be symmetrical in the coordinates of the particles. Such symmetry is merely the mathematical formulation of the statement that the particles are identical. One may then introduce interchange operators P_{ij} which interchange, in any function on which they operate, the coordinates of particles i and j. Thus

$$\mathsf{P}_{ij}\psi(r_1, r_2, \ldots, r_i, \ldots, r_j, \ldots, r_N) = \psi(r_1, r_2, \ldots, r_j, \ldots, r_i, \ldots, r_N). \tag{10.1}$$

These operators commute with the Hamiltonian function and hence represent constants of the motion. However they do not commute with each other, and so cannot all be given precise values at the same time. The simplest situation is that in which only two identical particles are involved.

1. Systems of two identical particles

When there are only two particles there is only one permutation operator in addition to the identity. It is the interchange operator P_{12}. Since applying this operator twice returns any function to its original form, $\mathsf{P}_{12}^2 = 1$, and the possible eigenvalues of P_{12} are ± 1. Thus we have symmetric and anti-symmetric states only. This property is conserved as the system changes with time and so the eigenvalue of the operator P_{12} can serve as part of the description of the state.

If $U(q_1, q_2, t)$ is a solution of the Schroedinger equation it is not necessarily an eigenfunction of P_{12}. However, since P_{12} commutes with H, $\mathsf{P}_{12}U(q_1, q_2, t)$ is also a solution, and eigenstates of P_{12} may be formed by combining these two. Thus

$$U_{(+)} = \frac{1}{\sqrt{2}}\{U(q_1, q_2, t)+U(q_2, q_1, t)\} \tag{10.2}$$

is a symmetric function of (q_1, q_2) and an eigenfunction of P_{12} with the eigenvalue $+1$. Similarly

$$U_{(-)} = \frac{1}{\sqrt{2}}\{U(q_1, q_2 t)-U(q_2, q_1, t)\}\,. \tag{10.3}$$

$U(q_1, q_2, t)$ is not necessarily orthogonal to $U(q_2, q_1, t)$ but $U_{(+)}$ is orthogonal to $U_{(-)}$.

The above considerations are quite general but frequently they must be applied to approximate rather than to exact solutions. If the Hamiltonian represents two identical particles moving in an external field of force and subject to an interaction, the interaction may often be neglected as a first approximation. An approximate solution is then a product of functions for the two particles. An approximate energy eigenfunction could then be written

$$U_{\mathrm{a}} = U_n(\boldsymbol{r}_1)U_m(\boldsymbol{r}_2)\,. \tag{10.4a}$$

But

$$U_{\mathrm{b}} = U_n(\boldsymbol{r}_2)U_m(\boldsymbol{r}_1) \tag{10.4b}$$

is also an approximate eigenfunction with the same eigenvalue for the approximate energy. The use of such functions is called the independent particle approximation. Neither of these product functions, however, is an eigenfunction of the operator P_{12}. Such eigenfunctions can be formed as in (10.2) and (10.3) so that

$$\psi(E', +) = \frac{1}{\sqrt{2}}\{U_n(\boldsymbol{r}_1)U_m(\boldsymbol{r}_2)+U_n(\boldsymbol{r}_2)U_m(\boldsymbol{r}_1)\}\,, \tag{10.5a}$$

$$\psi(E'', -) = \frac{1}{\sqrt{2}}\{U_n(\;_1)U_m(\boldsymbol{r}_2)-U_n(\;_2)U_m(\boldsymbol{r}_1)\}. \tag{10.5b}$$

In this case U_a and U_b are orthogonal as well as $\psi(E',+)$ and $\psi(E'',-)$. The functions are designated by the two energy values E' and E'' since they may represent different energies when the interaction is included. If the interaction is applied as a perturbation, the correction terms will all have the appropriate symmetry.

2. The Pauli exclusion principle

It has been found experimentally that electrons, protons and neutrons exist only in antisymmetric states, and that alpha particles exist only in symmetric states. Direct experimental evidence on this point comes from the analysis of atomic and molecular spectra. Also the scattering of protons by protons and of alpha particles by alpha particles shows the behavior required by this exclusion principle.

For two particle systems this observation means that only half of the solutions of the Schroedinger equation really represent the behavior of physical systems. The others are "excluded". For systems of more than two particles the states representing electrons, protons and neutrons must be antisymmetric to the interchange of the coordinates, including the spin coordinates, of any pair of identical particles. For alpha particles the states must remain unchanged for any exchange of particle coordinates. Thus only a small fraction of the possible states are allowed. Most of them must be excluded.

Those particles that must be represented by antisymmetric states are particles with spin, and the spin coordinates must be exchanged along with the others in applying the Pauli principle.

To form symmetric states in the independent particle approximation it is only necessary to take a product function $U_1(\boldsymbol{r}_1)U_2(\boldsymbol{r}_2)\dots U_n(\boldsymbol{r}_n)$ and perform on it all possible permutations of the coordinates. The sum of these product functions, divided by $\sqrt{n!}$ is then a normalized function invariant to all permutations, and hence to all exchanges.

Slater has suggested a simple way to form an antisymmetric function in the independent particle approximation:

$$\psi(\boldsymbol{r}_1, \dots, \boldsymbol{r}_n) = \frac{1}{\sqrt{n!}} \begin{vmatrix} U_1(\boldsymbol{r}_1) & U_2(\boldsymbol{r}_1) & U_3(\boldsymbol{r}_1)\cdots U_n(\boldsymbol{r}_1) \\ U_1(\boldsymbol{r}_2) & U_2(\boldsymbol{r}_2) & U_3(\boldsymbol{r}_2)\cdots U_n(\boldsymbol{r}_2) \\ U_1(\boldsymbol{r}_3) & U_2(\boldsymbol{r}_3) & U_3(\boldsymbol{r}_3)\cdots U_n(\boldsymbol{r}_3) \\ \cdot & \cdot & \cdot \quad\quad \cdot \\ \cdot & \cdot & \cdot \quad\quad \cdot \\ \cdot & \cdot & \cdot \quad\quad \cdot \\ U_1(\boldsymbol{r}_n) & U_2(\boldsymbol{r}_n) & U_3(\boldsymbol{r}_n)\cdots U_n(\boldsymbol{r}_n) \end{vmatrix} . \tag{10.6}$$

This "Slater determinant" clearly changes sign if $\boldsymbol{r}_i$ and $\boldsymbol{r}_j$ are exchanged. Furthermore it vanishes if any two functions U_k and U_l are identical. This latter fact is the basis for one statement of Pauli's principle, which is that no two electrons can be in the same state. Such a statement applies, of course, only to one-electron states and to approximate functions expressed as products of such states. An exact solution of a problem must consist of a sum of determinants of the form (10.6).

One interesting property of antisymmetric states is that the probability of two particles with parallel spins coinciding in space is zero. For since $\psi_a(q_2, q_1) = -\psi_a(q_1, q_2)$,

$$\psi_a(q_1 = q_2 = q_0) = -\psi_a(q_2 = q_1 = q_0) = 0 .$$

If this property of mutual exclusion in ordinary space were independent of spin it would provide a very satisfactory interpretation of the Pauli principle. Even as it is, the property is often helpful in understanding the consequences of the antisymmetry.

It is a consequence of the exclusion principle that matter does not shrink toward zero volume as the temperature is lowered toward the absolute zero. Also except for this principle, the atoms of heavy elements would be much smaller than those of the light elements, instead of about the same size.

3. Separation of spin and orbital motion

When the Hamiltonian function contains no terms representing an interaction between the spin and the orbital motion of the electrons, it is often convenient to separate the wave functions, or the state vectors, into a part depending on the orbital motion only and another part depending on the spin. For a single particle this is trivial.

For two particles an antisymmetric function can be written as a product of a symmetric function of the position coordinates multiplied by an antisymmetric function of the spin. But such a function can also be written as a product of an antisymmetric function of the position and a symmetric function of the spins. Thus

$$\begin{aligned}\psi_{a1} = \tfrac{1}{2}[U_1(\boldsymbol{r}_1)U_2(\boldsymbol{r}_2)+U_1(\boldsymbol{r}_2)U_2(\boldsymbol{r}_1)] \times \\ \times [|m_s'(1)\rangle\, |m_s''(2)\rangle - |m_s'(2)\rangle\, |m_s''(1)\rangle]\end{aligned} \tag{10.7a}$$

$$\begin{aligned}\psi_{a2} = \tfrac{1}{2}[U_1(\boldsymbol{r}_1)U_2(\boldsymbol{r}_2)-U_1(\boldsymbol{r}_2)U_2(\boldsymbol{r}_1)] \times \\ \times [|m_s'(1)\rangle\, |m_s''(2)\rangle + |m_s'(2)\rangle\, |m_s''(1)\rangle] .\end{aligned} \tag{10.7b}$$

Also

$$\psi_a = \frac{1}{\sqrt{2}}[\psi_{a_1}+\psi_{a_2}] =$$
$$= \frac{1}{\sqrt{2}}[U_1(\boldsymbol{r}_1)U_2(\boldsymbol{r}_2)\mid m_s'(1)\rangle\mid m_s''(2)\rangle - U_1(\boldsymbol{r}_2)U_2(\boldsymbol{r}_1)\mid m_s'(2)\rangle\mid m_s''(1)\rangle]. \tag{10.8}$$

As long as functions U_1 and U_2 are different, and m_s' and m_s'' are different, antisymmetric functions can always be written as above. But if $U_1 = U_2$, ψ_{a2} vanishes. This puts some limits on the kinds of functions that can be formed, but generally the functions U_n form an infinite set and the identity of U_1 and U_2 is a special case.

With the spin functions the situation is different. There are only two spin functions and so the number of possible combinations is very much limited. If the total spin about the z-axis is $\hbar$, m_s' and m_s'' are both $\frac{1}{2}$. In this case ψ_{a1} vanishes and there is only

$$\psi_a\begin{pmatrix} S=1 \\ M_s=1 \end{pmatrix} = \frac{1}{\sqrt{2}}[U_1(\boldsymbol{r}_1)U_2(\boldsymbol{r}_2) - U_1(\boldsymbol{r}_2)U_2(\boldsymbol{r}_1)]\mid\tfrac{1}{2}(1)\rangle\mid\tfrac{1}{2}(2)\rangle. \tag{10.9a}$$

Note that S and M_s are used to specify the sum of the spins. For $S_z' = 0$ there are two possibilities

$$\psi_a\begin{pmatrix} S=1 \\ M_s=0 \end{pmatrix} = \tfrac{1}{2}[U_1(\boldsymbol{r}_1)U_2(\boldsymbol{r}_2) - U_1(\boldsymbol{r}_2)U_2(\boldsymbol{r}_1)] \times$$
$$\times [\mid\tfrac{1}{2}(1)\rangle\mid-\tfrac{1}{2}(2)\rangle + \mid\tfrac{1}{2}(2)\rangle\mid-\tfrac{1}{2}(1)\rangle], \tag{10.9b}$$

$$\psi_a\begin{pmatrix} S=0 \\ M_s=0 \end{pmatrix} = \tfrac{1}{2}[U_1(\boldsymbol{r}_1)U_2(\boldsymbol{r}_2) + U_1(\boldsymbol{r}_2)U_2(\boldsymbol{r}_1)] \times$$
$$\times [\mid\tfrac{1}{2}(1)\rangle\mid-\tfrac{1}{2}(2)\rangle - \mid\tfrac{1}{2}(2)\rangle\mid-\tfrac{1}{2}(1)\rangle]. \tag{10.9b}$$

For $M_s = -1$ there is again only one:

$$\psi_a\begin{pmatrix} S=1 \\ M_s=-1 \end{pmatrix} = \frac{1}{\sqrt{2}}[U_1(\boldsymbol{r}_1)U_2(\boldsymbol{r}_2) - U_1(\boldsymbol{r}_2)U_2(\boldsymbol{r}_1)]\mid-\tfrac{1}{2}(1)\rangle\mid-\tfrac{1}{2}(2)\rangle. \tag{10.9c}$$

The spin factors of three of these functions, those labelled $S = 1$, are the three combinations corresponding to $(S^2)' = S(S+1)\hbar^2 = 2\hbar^2$. The other corresponds to $(S^2)' = 0$. The orbital functions associated with these two types of spin functions are quite different and hence the energy of the states with $S = 1$ may be quite different from that of those states with $S = 0$, *even though the Hamiltonian function does not contain the spin at all.* It is in this way that the electron spin comes to be of such importance in the behavior of systems of electrons.

The situation is easily illustrated in this case of two electrons. Let the interaction energy be $e^2/4\pi\kappa_0 r_{12}$ and let it be treated as a perturbation. The first order energy correction is then the mean value of the perturbation taken over the unperturbed wave functions. For the states with $S = 1$, the energy correction is

$$\Delta E(S = 1) = J - K \tag{10.10}$$

where

$$J = \frac{1}{4\pi\kappa_0} \int U_1^*(\boldsymbol{r}_1) U_1(\boldsymbol{r}_1) \frac{e^2}{r_{12}} U_2^*(\boldsymbol{r}_2) U_2(\boldsymbol{r}_2)\, \mathrm{d}v_1\, \mathrm{d}v_2, \tag{10.11a}$$

and

$$K = \frac{1}{4\pi\kappa_0} \int U_1^*(\boldsymbol{r}_1) U_2(\boldsymbol{r}_1) \frac{e^2}{r_{12}} U_2^*(\boldsymbol{r}_2) U_1(\boldsymbol{r}_2)\, \mathrm{d}v_1\, \mathrm{d}v_2. \tag{10.11b}$$

For the state with $S = 0$

$$\Delta E(S = 0) = J + K. \tag{10.12}$$

It is clear from the form of the expressions for J and K that both are invariant to an interchange of $\boldsymbol{r}_1$ and $\boldsymbol{r}_2$. The integral K is often called the exchange integral.

4. Systems of three electrons

The extension of the above considerations to systems containing large numbers of electrons involves the application of the theory of the symmetric group, the group of permutations. The results can be formulated elegantly in terms of group theory, but the development of that theory is beyond the scope of this treatment. Here the properties of the permutations of the coordinates of three identical particles will be treated as a special case, illustrative of the more general methods.

With three objects there are six possible permutations which may be designated as follows:

$$\mathsf{P}_1 = \begin{pmatrix} 1 & 2 & 3 \\ 1 & 2 & 3 \end{pmatrix} \quad \mathsf{P}_2 = \begin{pmatrix} 1 & 2 & 3 \\ 2 & 3 & 1 \end{pmatrix} \quad \mathsf{P}_3 = \begin{pmatrix} 1 & 2 & 3 \\ 3 & 1 & 2 \end{pmatrix}$$

$$\mathsf{P}_4 = \begin{pmatrix} 1 & 2 & 3 \\ 1 & 3 & 2 \end{pmatrix} \quad \mathsf{P}_5 = \begin{pmatrix} 1 & 2 & 3 \\ 3 & 2 & 1 \end{pmatrix} \quad \mathsf{P}_6 = \begin{pmatrix} 1 & 2 & 3 \\ 2 & 1 & 3 \end{pmatrix}.$$

P_1 is the identity, P_2 indicates that object 1 is to be replaced by 2, 2 by 3, and 3 by 1. These permutations can be performed one after the other and the result is equivalent to the application of some other permutation. The successive application of permutations may be called a product as is done with operators. Thus $\mathsf{P}_2\mathsf{P}_3 = \mathsf{P}_1$ means that if permutation P_2 is applied after P_3 the result is that the "operand" is returned to its original state as if the only permutation applied had been P_1. These six permutations on three objects can be described by a multiplication table.

	P_1	P_2	P_3	P_4	P_5	P_6
P_1	P_1	2	P_3	P_4	P_5	P_6
P_2	P_2	P_3	P_1	P_6	P_4	P_5
P_3	P_3	P_1	P_2	P_5	P_6	P_4
P_4	P_4	P_5	P_6	P_1	P_2	P_3
P_5	P_5	P_6	P_4	P_3	P_1	P_2
P_6	P_6	P_4	P_5	P_2	P_3	P_1

Each of the above permutations (operators) commutes with the Hamiltonian operator for a system of three identical particles, and so represents a constant of the motion. However they do not commute with each other, and a state vector cannot be an eigenvector of all of them at the same time. The situation is similar to that with angular momentum where all three components commute with the Hamiltonian but not with each other. As in the case of angular momentum, it is necessary to find such functions if the permutation operators as do commute with each other.

Such commuting functions can be formed by taking the sum of the permutations of each class. The above six permutations comprise 3 classes. The identity P_1 is a class by itself. P_2 and P_3 form a class, and P_4, P_5 and P_6 constitute the third class. The defining property of a class is that any member of it, P_j, when transformed by any member of the group, P_k, according to the prescription

$$\mathsf{P}_i = \mathsf{P}_k\mathsf{P}_j\mathsf{P}_k^{-1}$$

is unchanged or becomes another member of the same class. Thus

$$\begin{aligned}
\mathsf{P}_2\mathsf{P}_3\mathsf{P}_2^{-1} &= \mathsf{P}_2\mathsf{P}_3\mathsf{P}_3 = \mathsf{P}_3\\
\mathsf{P}_3\mathsf{P}_3\mathsf{P}_3^{-1} &= \mathsf{P}_3\mathsf{P}_3\mathsf{P}_2 = \mathsf{P}_3\\
\mathsf{P}_4\mathsf{P}_3\mathsf{P}_4^{-1} &= \mathsf{P}_4\mathsf{P}_3\mathsf{P}_4 = \mathsf{P}_2\\
\mathsf{P}_5\mathsf{P}_3\mathsf{P}_5^{-1} &= \mathsf{P}_5\mathsf{P}_3\mathsf{P}_5 = \mathsf{P}_2\,.
\end{aligned}$$

There are then three operators

$$\mathsf{C}_0 = \mathsf{P}_1\,, \qquad \mathsf{C}_1 = \tfrac{1}{2}(\mathsf{P}_2+\mathsf{P}_3)\,, \qquad \mathsf{C}_2 = \tfrac{1}{3}(\mathsf{P}_4+\mathsf{P}_5+\mathsf{P}_6)\,,$$

that commute with each other as well as with the Hamiltonian operator and so can be used simultaneously to designate symmetry states of the system.

The operator C_0 is trivial in this respect for its only eigenvalue is 1. There are three possible pairs of eigenvalues for C_1 and C_2 as follows:

Case	C_1'	C_2'
A	1	1
B	1	-1
C	$-\frac{1}{2}$	0

The difference between the permutations in C_1 and those in C_2 can be seen by inspection. Those in C_1, P_2 and P_3, are cyclic permutations of all three quantities. Those in C_2 are exchanges. It is the latter class with which the Pauli principle is concerned. Only case A eigenfunctions can be used for alpha particles but only case B states for electrons. Antisymmetric states can always be formed as Slater determinants, but further analysis on the basis of separated spin and orbital functions can give additional information about the possibilities.

5. Orbital functions of three electrons

If all reference to the electron spin is omitted from the Hamiltonian function, an energy eigenfunction may be designed by $U(\boldsymbol{r}_1, \boldsymbol{r}_2, \boldsymbol{r}_3)$. If the particles are independent this may be written as a product

$$U(\boldsymbol{r}_1, \boldsymbol{r}_2, \boldsymbol{r}_3) = U_1(\boldsymbol{r}_1)U_2(\boldsymbol{r}_2)U_3(\boldsymbol{r}_3)\,. \tag{10.13}$$

Many of the following considerations apply to either form. To the function (10.13) the operators P_1–P_6 may be applied to give a total of 6 energy eigenfunctions, all having the same energy eigenvalue:

$$\begin{aligned}
\mathsf{P}_1 U(\boldsymbol{r}_1, \boldsymbol{r}_2, \boldsymbol{r}_3) &= U(\boldsymbol{r}_1, \boldsymbol{r}_2, \boldsymbol{r}_3) = |\,\alpha\rangle \\
\mathsf{P}_2 U(\boldsymbol{r}_1, \boldsymbol{r}_2, \boldsymbol{r}_3) &= U(\boldsymbol{r}_2, \boldsymbol{r}_3, \boldsymbol{r}_1) = |\,\beta\rangle \\
\mathsf{P}_3 U(\boldsymbol{r}_1, \boldsymbol{r}_2, \boldsymbol{r}_3) &= U(\boldsymbol{r}_3, \boldsymbol{r}_1, \boldsymbol{r}_2) = |\,\gamma\rangle \\
\mathsf{P}_4 U(\boldsymbol{r}_1, \boldsymbol{r}_2, \boldsymbol{r}_3) &= U(\boldsymbol{r}_1, \boldsymbol{r}_3, \boldsymbol{r}_2) = |\,\delta\rangle \\
\mathsf{P}_5 U(\boldsymbol{r}_1, \boldsymbol{r}_2, \boldsymbol{r}_3) &= U(\boldsymbol{r}_3, \boldsymbol{r}_2, \boldsymbol{r}_1) = |\,\varepsilon\rangle \\
\mathsf{P}_6 U(\boldsymbol{r}_1, \boldsymbol{r}_2, \boldsymbol{r}_3) &= U(\boldsymbol{r}_2, \boldsymbol{r}_1, \boldsymbol{r}_3) = |\,\varphi\rangle .
\end{aligned} \tag{10.14}$$

It is now possible to form six orthogonal combinations of these functions which transform under the various permutations in simpler ways. Let

$$\begin{aligned}
|\,R_1\rangle &= \frac{1}{\sqrt{6}}\{|\,\alpha\rangle + |\,\beta\rangle + |\,\gamma\rangle + |\,\delta\rangle + |\,\varepsilon\rangle + |\,\varphi\rangle\} \\
|\,R_2\rangle &= \frac{1}{\sqrt{6}}\{|\,\alpha\rangle + |\,\beta\rangle + |\,\gamma\rangle - |\,\delta\rangle - |\,\varepsilon\rangle - |\,\varphi\rangle\} \\
|\,R_3\rangle &= \frac{1}{\sqrt{3}}\{|\,\alpha\rangle - \tfrac{1}{2}|\,\beta\rangle - \tfrac{1}{2}|\,\gamma\rangle - \tfrac{1}{2}|\,\delta\rangle + |\,\varepsilon\rangle - \tfrac{1}{2}|\,\varphi\rangle\} \\
|\,R_4\rangle &= \tfrac{1}{2}\{|\,\beta\rangle - |\,\gamma\rangle + |\,\delta\rangle - |\,\varphi\rangle\} \\
|\,R_5\rangle &= \tfrac{1}{2}\{-|\,\beta\rangle + |\,\gamma\rangle + |\,\delta\rangle - |\,\varphi\rangle\} \\
|\,R_6\rangle &= \frac{1}{\sqrt{3}}\{|\,\alpha\rangle - \tfrac{1}{2}|\,\beta\rangle - \tfrac{1}{2}|\,\gamma\rangle + \tfrac{1}{2}|\,\delta\rangle - |\,\varepsilon\rangle + \tfrac{1}{2}|\,\varphi\rangle\} .
\end{aligned} \tag{10.15}$$

Consideration of the effects of the permutations on these functions shows that R_1 is unchanged no matter what permutation is applied. R_2 is unchanged by permutations P_1, P_2, P_3, but has its sign changed by P_4, P_5 and P_6. When a permutation is applied to R_3 or R_4 it is transformed into a linear combination of these two, and similarly for R_5 and R_6.

If the six functions $|\,\alpha\,\rangle$–$|\,\phi\rangle$ have the second form of (10.13) as zero approximation functions, the application of an electron interaction term

$$V = V(\boldsymbol{r}_1, \boldsymbol{r}_2) + V(\boldsymbol{r}_1, \boldsymbol{r}_3) + V(\boldsymbol{r}_2, \boldsymbol{r}_3) \tag{10.16}$$

will give different values of the mean interaction energy, $\langle R_i\,|\,V\,|\,R_i\rangle$ for the different states $|\,R_i\rangle$. For example

$$\langle R_1\,|\,V\,|\,R_1\rangle = \sum_{i \neq j} \tfrac{1}{2}\{J(i,j) + K(i,j)\}$$

where

$$J(i,j) = \int |\,U_i(1)\,|^2\, V(\boldsymbol{r}_1, \boldsymbol{r}_2)\, |\,U_j(2)\,|^2\, \mathrm{d}v_1\, \mathrm{d}v_2$$

and

$$K(i,j) = \int U_i(r_1)U_j^*(r_1)V(r_1, r_2)U_i^*(r_2)U_j(r_2)\, dv_1\, dv_2\,.$$

On the other hand

$$\langle R_2 \mid V \mid R_2 \rangle = \sum_{i \neq j} \tfrac{1}{2}\{J(i,j) - K(i,j)\}\,.$$

Thus the functions $\mid R_1 \rangle$ and $\mid R_2 \rangle$ have different energies as soon as the interaction energy is included. Similarly the other states will have their own interaction energies. Functions $\mid R_3 \rangle$ and $\mid R_4 \rangle$ lead to the same mean value of the interaction energy,

$$\langle R_3 \mid V \mid R_3 \rangle = \tfrac{1}{2} \sum_{i \neq j} J(i,j) + \{K(1,3) - \tfrac{1}{2}K(1,2) - \tfrac{1}{2}K(2,3)\}\,,$$

and functions $\mid R_5 \rangle$ and $\mid R_6 \rangle$ to

$$\langle R_5 \mid V \mid R_5 \rangle = \tfrac{1}{2} \sum_{i \neq j} J(i,j) - \{K(1,3) - \tfrac{1}{2}K(1,2) - \tfrac{1}{2}K(2,3)\}.$$

In case $U_1 = U_2 = U_3$ only $\mid R_1 \rangle$ is different from zero. In case only two functions are the same there are only three different functions of the coordinates. If $U_1 = U_2$ and U_3 is a different function, $\mid R_2 \rangle = 0$, $\mid R_3 \rangle = -(1/\sqrt{3}) \times \mid R_4 \rangle$, and $\mid R_6 \rangle = -\sqrt{3} \mid R_5 \rangle$. Similarly, if $U_2 = U_3$, $\mid R_2 \rangle = 0$, and in that case $\mid R_3 \rangle = (1/\sqrt{3}) \mid R_4 \rangle$, and $R_5 = (1/\sqrt{3}) \mid R_6 \rangle$. In using such special cases the normalization coefficients must be appropriately changed.

Not all of these states and not all of these energies occur. To get the functions that can be used to describe a system of three identical particles with spin, the functions (10.15) must be combined with suitable functions of the spin.

6. Spin functions of three electrons

To treat the spin functions one can proceed in much the same way as above, except that there are only two independent eigenfunctions available for each electron. Since the component of the spin along the z-axis is a constant of the motion it is convenient to form combinations of spin functions that are eigenfunctions of $\mathsf{S}_z = \Sigma_{i=1}^{3} \mathsf{S}_z(i)$.

For $S_z' = \frac{3}{2}\hbar$ there is but one function:

$$\mid S_z' = \tfrac{3}{2}\hbar \rangle = \mid +\tfrac{1}{2}(1) \rangle \mid +\tfrac{1}{2}(2) \rangle \mid +\tfrac{1}{2}(3) \rangle\,. \qquad (10.17)$$

This is also an eigenfunction of S^2 with the eigenvalue $\frac{15}{4}\hbar^2$ as well as of the permutation operators C_1 and C_2 with the eigenvalues $+1$. It can be multiplied by $| R_2\rangle$ to give a function which is antisymmetric to any exchange of all four of the coordinates of any two electrons. No other orbital function can be so used. Also there is no other function with $S_z' = \frac{3}{2}\hbar$ and so no other function of the position coordinates which can be used to give a function satisfying the Pauli principle.

For $S_z' = \frac{1}{2}\hbar$ there are three possible functions. Starting with one product function, a total of 6 can be produced by applying permutations, but they are identical in pairs. Using a notation similar to that used for the spatial functions, let

$$\begin{aligned} | a\rangle &= \mathsf{P}_1 \,| a\rangle = | +\tfrac{1}{2}(1)\rangle \,| +\tfrac{1}{2}(2)\rangle \,| -\tfrac{1}{2}(3)\rangle = \mathsf{P}_6 \,| a\rangle = | f\rangle \\ | b\rangle &= \mathsf{P}_2 \,| a\rangle = | +\tfrac{1}{2}(2)\rangle \,| +\tfrac{1}{2}(3)\rangle \,| -\tfrac{1}{2}(1)\rangle = \mathsf{P}_5 \,| a\rangle = | e\rangle \\ | c\rangle &= \mathsf{P}_3 \,| a\rangle = | +\tfrac{1}{2}(3)\rangle \,| +\tfrac{1}{2}(1)\rangle \,| -\tfrac{1}{2}(2)\rangle = \mathsf{P}_4 \,| a\rangle = | d\rangle . \end{aligned} \tag{10.18}$$

The linear combinations corresponding to $R_1, \ldots, R_6$ are

$$\begin{aligned} | S_1\rangle &= \frac{1}{\sqrt{3}} \{| a\rangle + | b\rangle + | c\rangle\} \\ | S_2\rangle &= 0 \\ | S_3\rangle &= \frac{1}{\sqrt{6}} \{| a\rangle + | b\rangle - 2\,| c\rangle\} = -\,| S_5\rangle \\ | S_4\rangle &= \frac{1}{\sqrt{2}} \{| b\rangle - | a\rangle\} = -\,| S_6\rangle . \end{aligned} \tag{10.19}$$

There are just these three independent spin states possible for $S_z' = \frac{1}{2}\hbar$. The state $| S_1\rangle$ is an eigenstate of both C_1 and C_2 with the eigenvalue 1 for each. It represents case A. It is also an eigenstate of S^2 with the eigenvalue $\frac{15}{4}\hbar^2$. This state $| S_1\rangle$ together with the state in equation (10.17) and the corresponing state for the negative value of S_z' provides three independent spin states with $\mathsf{S}^{2\prime} = \frac{15}{4}\hbar_2$ and values of 1 for C_1' and C_2'.

The states $| S_3\rangle$ and $| S_4\rangle$ are both eigenstates of S^2 with the eigenvalue $\frac{3}{4}\hbar^2$. They are also eigenstates of C_1 with the eigenvalue $-\frac{1}{2}$ and of C_2 with the eigenvalue 0.

7. Combined spin and orbital functions

To satisfy the Pauli exclusion principle the spin and orbital functions must be combined in such ways that the product will change sign when any one of the permutations P_4, P_5 or P_6 is applied to the whole function. For $S'_z = \frac{3}{2}\hbar$ there is just one such combination as has been shown. It is the spin function of equation (10.17) multiplied by the orbital function $| R_2\rangle$. There is the corresponding function for $S'_z = -\frac{3}{2}\hbar$ which also involves the orbital function $| R_2\rangle$.

For $S'_z = \frac{1}{2}\hbar$ there are three possibilities. If the spin function $| S_1\rangle$ is combined with $| R_2\rangle$ the product is antisymmetric. Thus there are three functions

$$| S'_z = \tfrac{3}{2}\hbar\rangle\, | R_2\rangle, \quad | S_1\rangle\, | R_2\rangle, \quad | S'_z = -\tfrac{3}{2}\hbar\rangle\, | R_2\rangle$$

that satisfy the Pauli exclusion principle, represent eigenstates of S^2, and have the same orbital function. With a Hamiltonian that includes no reference to the spin they will all represent the same zero approximation energy.

The function $| S_2\rangle$ is an attempt to form an antisymmetric spin function and it vanishes. No state satisfying the Pauli principle can be formed with the symmetric orbital function $| R_1\rangle$.

It is possible, however, to form two combinations of the other functions that satisfy the condition of antisymmetry. Without going into the proof of why this is so, or the method of finding the functions, they will be given:

$$\begin{aligned} | S = \tfrac{1}{2}\hbar, S'_z = \tfrac{1}{2}\hbar, 1\rangle &= \frac{1}{\sqrt{2}}\{| R_3\rangle\, | S_4\rangle - | R_4\rangle\, | S_3\rangle\} = \\ &= \frac{1}{\sqrt{12}}\{(-\,| \alpha\rangle + | \gamma\rangle - | \varepsilon\rangle + |\varphi\rangle)\, | a\rangle + \\ &\quad + (| \alpha\rangle - | \beta\rangle - | \delta\rangle + | \varepsilon\rangle)\, | b\rangle + \\ &\quad + (| \beta\rangle - | \gamma\rangle + | \delta\rangle - | \varphi\rangle)\, | c\rangle\}, \qquad (10.20) \end{aligned}$$

$$\begin{aligned} | S = \tfrac{1}{2}\hbar, S'_z = \tfrac{1}{2}\hbar, 2\rangle &= \frac{1}{\sqrt{2}}\{| R_6\rangle\, | S_3\rangle - | R_5\rangle\, | S_4\rangle\} = \\ &= \tfrac{1}{6}\{(-\,| \alpha\rangle + 2\,| \beta\rangle - | \gamma\rangle - 2\,| \delta\rangle + | \varepsilon\rangle + | \varphi\rangle)\, | a\rangle + \\ &\quad + (-\,| \alpha\rangle - | \beta\rangle + 2\,| \gamma\rangle + | \delta\rangle + | \varepsilon\rangle - 2\,| \varphi\rangle)\, | b\rangle + \\ &\quad + (2\,| \alpha\rangle - | \beta\rangle - | \gamma\rangle + | \delta\rangle - 2\,| \varepsilon\rangle + | \varphi\rangle)\, | c\rangle\}\,. \qquad (10.21) \end{aligned}$$

These two functions represent the same energy in this zero approximation and with this Hamiltonian.

The above considerations, and their generalization in the language of group theory to larger numbers of electrons, serve to provide the detailed basis for the series of rules developed in the analysis of atomic spectra. The case of three electrons appears in lithium, or for practical purposes in any element with three electrons outside of a closed shell, such as boron.

In the case of lithium the lowest energy state before the inter-electron repulsion is taken into account can be formed from two 1s states and one other state. This can be represented by letting U_2 and U_3 in equation (10.13) be the 1s states. When this is done

$$|\,\alpha\rangle = |\,\delta\rangle, \qquad |\,\beta\rangle = |\,\varphi\rangle, \qquad |\,\gamma\rangle = |\,\varepsilon\rangle,$$

and the permitted state, equation (10.20) takes the form

$$\begin{aligned} |\,S = \tfrac{1}{2}, S_z = \tfrac{1}{2}\hbar\rangle = \frac{1}{\sqrt{6}}\{(|\,\beta\rangle - |\,\alpha\rangle)\,|\,a\rangle + \\ + (|\,\gamma\rangle - |\,\beta\rangle)\,|\,b\rangle + (|\,\alpha\rangle - |\,\gamma\rangle)\,|\,c\rangle\}. \end{aligned} \tag{10.22}$$

The form in equation (10.21) gives the same function for this case so that there is only one $S = \frac{1}{2}$ state when $U_2 = U_3$ instead of two as in the general case.

The three parts of this state are identical except for the permutation of three electrons, and the interaction energy in the first approximation is

$$E_1 = 2J(1, 2) + J(2, 2) - K(1, 2)$$

where the $J(1, 2)$ is the interaction energy between an electron in a 1s state, designated by U_2 and some other state designated by U_1. $K(1, 2)$ is the corresponding exchange integral. Hence the result, in this approximation, is that of a single electron moving in a central field composed of the nuclear field and a field due to two electronic charges distributed as $|\,U_2\,|^2$ but corrected by the exchange integral. Such states are the only ones appearing in the optical spectrum of neutral lithium, since the excitation of an electron from the 1s state to a 2s state would require more than enough energy to remove the U_1 electron altogether.

In atoms such as boron, the three electrons outside of the closed shell will have, as a zero approximation, states corresponding to a single electron moving in a central but non-Coulomb field. Hence the 2s and 2p states will have different energies and although the ground state will be a doublet state as for lithium, excited states of higher multiplicity will be available. Since S can take on the values $\frac{1}{2}$ and $\frac{3}{2}$ for a three electron system, the values of C_3' are 0 and 1.

PROBLEM 1. By evaluating the first order correction to the wave functions, show that the functions (10.5a) and (10.5b) retain their symmetries when a symmetric perturbation is added to the Hamiltonian.

PROBLEM 2. Show that

$$\psi_a\begin{pmatrix} S = 1 \\ M_s = 0 \end{pmatrix} \quad \text{and} \quad \psi_a\begin{pmatrix} S = 0 \\ M_s = 0 \end{pmatrix}$$

of (10.9b) are eigenfunctions of $[\sigma(1)+\sigma(2)]^2$ with eigenvalues $2\hbar^2$ and 0 respectively.

PROBLEM 3. Show that the function $| R_2\rangle | S_1\rangle$ is an eigenfunction of S^2 and of C_1 and C_2.

PROBLEM 4. Show that the functions in equations (10.19) are eigenfunctions of S^2 and of C_1 and C_2.

PROBLEM 5. Show that the indicated properties of the functions (10.15) follow from the properties of the permutation operators.

PROBLEM 6. Show that the functions (10.20) and (10.21) satisfy the Pauli exclusion principle.

CHAPTER 11

QUANTIZATION OF ELECTROMAGNETIC RADIATION IN EMPTY SPACE

The necessity for some departure from classical physics in the direction of quantum mechanics was first recognized in the study of the interaction of radiation and matter. In Chapter 1 some of the difficulties in understanding the properties of black-body radiation were described, and it was shown how Planck's postulate resolved one dilemma. This postulate, although successful in its specific application, was quite irrational and did not lead to a completely satisfying theory. Similarly, the phenomena of the photoelectric effect and the Compton effect required that one ascribe to electromagnetic radiation many of the properties more easily associated with material particles. It is the task of quantum mechanics to include these properties under the general scheme developed for the treatment of all physical phenomena; and in the next two chapters it will be shown that the general scheme can be employed to give a description of the electromagnetic field which, although not satisfactory in every respect, presents only difficulties which seem to be connected with the nature of electrons.

The previous chapters have been devoted to the wave-like properties of electrons, atomic nuclei, and other entities that one often likes to picture as essentially particle-like. In this chapter we begin to deal with the particle like properties of phenomena at first thought to be essentially wave-like.

1. Classical mechanics of a vibrating string

As an introduction to the treatment of the electromagnetic field, it is instructive to consider a problem which can be treated by the same methods but is less complicated. A stretched string vibrating in one plane displays many of the characteristics of a field of electromagnetic radiation, but its treatment avoids the excessive accumulation of subscripts associated with the two components each of electric and magnetic vectors, and the necessity for specifying the direction of propagation of a plane wave.

The first problem in treating the string is the selection of the appropriate coordinates. The procedure of treating the displacement x as a function of the position z along the string and the time t, $x(z, t)$, leads to partial differential equations of motion whose solutions can be written down in the usual manner. It is more convenient here to introduce a new set of coordinates by expanding the displacement in a Fourier series. Hence let

$$x(z, t) = \sum_{s=0}^{\infty} \left\{ q_{s,1}(t) \cos \frac{2\pi s z}{L} + q_{s,2}(t) \sin \frac{2\pi s z}{L} \right\}. \tag{11.1}$$

The portion of the string to be considered lies between $z = 0$ and $z = L$, and equation (11.1) implies a periodic boundary condition. It implies that, at any time, the displacement at $z = L$ is the same as that at $z = 0$. Such a boundary condition is less restrictive than fixing the ends, but it does imply that reflections from the ends are not important. It suggests that L is large enough to make L/v large compared with the duration of any phenomena to be studied. v is the velocity with which a disturbance travels along the string. Such a boundary condition is customarily applied in radiation problems when it is desired to use discrete coordinates such as $q_{s,\mu}(t)$. The infinity of discrete coordinates $q_{s,\mu}(t)$ serves to describe the configuration of the string at any time t.

The energy of the string can be transformed from its expression in terms of $x(z, t)$ and $\dot{x}(z, t)$ to the corresponding expression in terms of the $q_{s,\mu}(t)$ and $\dot{q}_{s,\mu}(t)$. The kinetic energy is

$$T = \tfrac{1}{2}\rho \int_0^L [\dot{x}(z, t)]^2 \, dz = \tfrac{1}{4}M \sum_{s=0}^{\infty} (\dot{q}_{s,1}^2 + \dot{q}_{s,2}^2), \tag{11.2}$$

where ρ is the mass of the string per unit length and M is the mass in the length L. This is clearly a transformation to normal coordinates, since the result is a sum of squares with no cross products. In the same way the potential energy is

$$V = \tfrac{1}{2}\tau \int_0^L \left(\frac{\partial x}{\partial z}\right)^2 dz = \sum_{s=0}^{\infty} s \, \frac{\pi^2 \tau}{L} (q_{s,1}^2 + q_{s,2}^2), \tag{11.3}$$

where τ is the tension. Then, following the usual procedure in forming a Hamiltonian function, $p_{s,\mu} = \tfrac{1}{2}M\dot{q}_{s,\mu}$, so that

$$H = \sum_{s=0}^{\infty} \sum_{\mu=1}^{2} \left[\frac{1}{M} p_{s,\mu}^2 + \frac{s^2 \pi^2 \tau}{L} q_{s,\mu}^2 \right]. \tag{11.4}$$

This is the sum of the Hamiltonian functions of a set of two-dimensional isotropic harmonic oscillators. The angular frequency of the s'th oscillator is $\omega_s = 2\pi s(\tau/ML)^{\frac{1}{2}}$. Note that the positive root is always taken. Some treatments use the negative root also, but here its place is taken by using a second value of μ. The case $s = 0$, as is seen in equation (11.1), represents a uniform displacement of the whole string in the x direction and with it there is, of course, no restoring force, $\omega_0 = 0$.

If the two coordinates $q_{s,1}$ and $q_{s,2}$, that have the same frequency of vibration, are pictured as orthogonal cartesian coordinates in a plane, the general solution of the system described by one value of s in (11.4) can be described as a sum of two opposite circular motions:

$$\begin{aligned} q_{s,1} &= R_{+s}\cos(\omega_s t - \varphi_{+s}) + R_{-s}\cos(\omega_s t - \varphi_{-s}) \\ q_{s,2} &= R_{+s}\sin(\omega_s t - \varphi_{+s}) - R_{-s}\sin(\omega_s t - \varphi_{-s}). \end{aligned} \tag{11.5}$$

When these solutions are used in equation (11.1) the expression for $x(z, t)$ becomes a series of travelling waves:

$$x(z, t) = \sum_{s=0}^{\infty} \left\{ R_{+s}\cos\left[\omega_s t - \frac{2\pi s}{L} z - \varphi_{+s}\right] + R_{-s}\cos\left[\omega_s t + \frac{2\pi s}{L} z - \varphi_{-s}\right]\right\}. \tag{11.6}$$

Each value of s occurs in a wave travelling in each direction. The φ_{+s} and φ_{-s} are phase constants. Expressed in terms of the constants in equation (11.6) the energy of the string is

$$E = \tfrac{1}{2}M \sum_{s=-\infty}^{\infty} \omega_s^2 R_s^2, \tag{11.7}$$

where the separate occurrence of R_{+s} and R_{-s} is replaced by taking the sum from $-\infty$ to $+\infty$. In equation (11.7) $\omega_s = \omega_{-s}$ but $R_s \neq R_{-s}$.

When considering the coordinates $q_{s,1}$ and $q_{s,2}$ as cartesian coordinates in a plane it is obvious that there is another constant of the motion,

$$L_s = q_{s,1}p_{s,2} - q_{s,2}p_{s,1}. \tag{11.8}$$

This has the appearance of angular momentum but it clearly does not represent any angular momentum of the string. The motion of the string is strictly perpendicular to its length. The waves travelling on it do not carry any momentum, and certainly no angular momentum. If this quanty L is expressed in terms of the constants in equation (11.5), (this L need not be confused with the length of the string)

$$L = \sum_{s=0}^{\infty} L_s = \tfrac{1}{2}M \sum_{s=0}^{\infty} \omega_s(R_{+s}^2 - R_{-s}^2). \tag{11.9}$$

From a knowledge of L as well as of E it is possible to determine how much of the energy is associated with waves travelling toward positive z and how much with waves in the opposite direction.

PROBLEM 1. Show how equations (11.2), (11.3) and (11.4) follow from the definition of the q's in equation (11.1).

PROBLEM 2. Show that equation (11.5) is a general solution of the canonical equations associated with the Hamiltonian function (11.4).

PROBLEM 3. Show that (11.9) follows from the definition (11.8) and the solutions (11.5).

2. Quantum mechanics of a vibrating string

To treat the problem of a vibrating string quantum mechanically, one could use the wave mechanical treatment of Chapter 4 for a two dimensional oscillator and apply it to each pair of coordinates, $q_{s,1}$ and $q_{s,2}$. It is often more convenient to use the methods of noncommutative algebra and to formulate creation and destruction operators for the various quanta of vibration, which, in the case of a string, may be called phonons. To do this, commutation rules for the operators involved must be assumed.

Let

$$\mathsf{p}_{s,\mu}\mathsf{q}_{s',\mu'} - \mathsf{q}_{s',\mu'}\mathsf{p}_{s,\mu} = \frac{\hbar}{i}\,\delta_{\mu,\mu'}\delta_{s,s'}. \tag{11.10}$$

These commutation rules express the fact that the different coordinates, $q_{s,\mu}$, are to be treated as independent cartesian coordinates. For any one s, however, the $q_{s,1}$ and $q_{s,2}$ can be taken together and treated as representing a plane simple harmonic oscillator. For each such oscillator

$$\mathsf{H}_s = \frac{1}{M}(\mathsf{p}_{s,1}^2 + \mathsf{p}_{s,2}^2) + \tfrac{1}{4}M\omega_s^2(\mathsf{q}_{s,1}^2 + \mathsf{q}_{s,2}^2) \tag{11.11}$$

and

$$\mathsf{L}_s = \mathsf{q}_{s,1}\mathsf{p}_{s,2} - \mathsf{q}_{s,2}\mathsf{p}_{s,1} \tag{11.12}$$

with

$$\tfrac{1}{4}M\omega_s^2 = s^2\pi^2\tau/L.$$

It follows from the commutation rules and the definition of L_s that

$$\mathsf{L}_s\mathsf{H}_s - \mathsf{H}_s\mathsf{L}_s = 0, \tag{11.13}$$

so it is possible to describe the system by simultaneous eigenstates of H_s and L_s. These two quantities form a physically complete set so the oscillator is completely described if they are given. The H and the L are of course specified at the same time, since they are merely sums over s of H_s and L_s. A ket vector for the s oscillator may then be written as $|\, H'_s, L'_s\rangle$. A ket vector for the whole string would be designated by values of H'_s and L'_s for all values of s. Now let

$$\mathsf{R}_s^\dagger = \frac{1}{(2M\hbar\omega_s)^{\frac{1}{2}}}(\mathsf{p}_{s,1}+\mathrm{i}\mathsf{p}_{s,2})+\mathrm{i}\left(\frac{M\omega_s}{8\hbar}\right)^{\frac{1}{2}}(\mathsf{q}_{s,1}+\mathrm{i}\mathsf{q}_{s,2}), \tag{11.14a}$$

$$\mathsf{R}_s = \frac{1}{(2M\hbar\omega_s)^{\frac{1}{2}}}(\mathsf{p}_{s,1}-\mathrm{i}\mathsf{p}_{s,2})-\mathrm{i}\left(\frac{M\omega_s}{8\hbar}\right)^{\frac{1}{2}}(\mathsf{q}_{s,1}-\mathrm{i}\mathsf{q}_{s,2}), \tag{11.14b}$$

$$\mathsf{R}_{-s}^\dagger = \frac{1}{(2M\hbar\omega_s)^{\frac{1}{2}}}(\mathsf{p}_{s,1}-\mathrm{i}\mathsf{p}_{s,2})+\mathrm{i}\left(\frac{M\omega_s}{8\hbar}\right)^{\frac{1}{2}}(\mathsf{q}_{s,1}-\mathrm{i}\mathsf{q}_{s,2}), \tag{11.14c}$$

$$\mathsf{R}_{-s} = \frac{1}{(2M\hbar\omega_s)^{\frac{1}{2}}}(\mathsf{p}_{s,1}+\mathrm{i}\mathsf{p}_{s,2})-\mathrm{i}\left(\frac{M\omega_s}{8\hbar}\right)^{\frac{1}{2}}(\mathsf{q}_{s,1}+\mathrm{i}\mathsf{q}_{s,2}). \tag{11.14d}$$

These are four independent linear combinations of the four operators in H_s and L_s and can be used in place of them. The commutation relationships between these quantities follow from the definitions and the assumed relationships in (11.10). Among them are

$$\begin{aligned} &\mathsf{R}_s\mathsf{R}_s^\dagger - \mathsf{R}_s^\dagger\mathsf{R}_s = 1 \qquad &&\mathsf{R}_{-s}\mathsf{R}_{-s}^\dagger - \mathsf{R}_{-s}^\dagger\mathsf{R}_{-s} = 1 \\ &\mathsf{R}_s\mathsf{R}_{-s} - \mathsf{R}_{-s}\mathsf{R}_s = 0 \qquad &&\mathsf{R}_s^\dagger\mathsf{R}_{-s} - \mathsf{R}_{-s}\mathsf{R}_s^\dagger = 0. \end{aligned} \tag{11.15}$$

With these new variables

$$\begin{aligned} \mathsf{H}_s &= (\mathsf{R}_s^\dagger\mathsf{R}_s + \mathsf{R}_{-s}^\dagger\mathsf{R}_{-s} + 1)\hbar\omega_s \\ \mathsf{L}_s &= (\mathsf{R}_s^\dagger\mathsf{R}_s - \mathsf{R}_{-s}^\dagger\mathsf{R}_{-s})\hbar\,. \end{aligned} \tag{11.16}$$

There are also among the commutation relationships

$$\mathsf{R}_s^\dagger\mathsf{H}_s - \mathsf{H}_s\mathsf{R}_s^\dagger = -\hbar\omega_s\mathsf{R}_s^\dagger \qquad \mathsf{R}_s\mathsf{H}_s - \mathsf{H}_s\mathsf{R}_s = \hbar\omega_s\mathsf{R}_s \tag{11.17a}$$

$$\mathsf{R}_s^\dagger\mathsf{L}_s - \mathsf{L}_s\mathsf{R}_s^\dagger = -\hbar\mathsf{R}_s^\dagger \qquad \mathsf{R}_s\mathsf{L}_s - \mathsf{L}_s\mathsf{R}_s = \hbar\mathsf{R}_s\,. \tag{11.17b}$$

There are similar relationships for $\mathsf{R}_{-s}^\dagger$ and R_{-s}. It should be noted that H_s and L_s occur with positive s only. H_s represents the energy associated with the normal vibration of the string which has the frequency ω_s. This vibration can always be resolved into a wave travelling in the positive direction and one in the negative direction. The value of L_s indicates the division of energy between these two directions of wave travel.

From the definitions and the commutation relationships it follows that when the normalization is ignored

$$\mathsf{R}_s^\dagger \mid H_s', L_s'\rangle = \mid H_s'+\hbar\omega_s, L_s'+\hbar\rangle \tag{11.18a}$$

$$\mathsf{R}_s \mid H_s', L_s'\rangle = \mid H_s'-\hbar\omega_s, L_s'-\hbar\rangle \tag{11.18b}$$

$$\mathsf{R}_{-s}^\dagger \mid H_s', L_s'\rangle = \mid H_s'+\hbar\omega_s, L_s'-\hbar\rangle \tag{11.18c}$$

$$\mathsf{R}_{-s} \mid H_s', L_s'\rangle = \mid H_s'-\hbar\omega_s, L_s'+\hbar\rangle . \tag{11.18d}$$

By the methods used in Chapter 8 it can be shown that the lowest value of the energy is $\hbar\omega_s$, and that for a given value of H_s', denoted by $(k_s+1)\hbar\omega_s$, the eigenvalues of L_s run from $L_s' = k_s\hbar$ in steps of $2\hbar$ down to $-k_s\hbar$. The possible states are indicated in the figure along with the effects of the various operators. These states are just those derived from the wave mechanics of the two dimensional oscillator in Chapter 4. The two treatments are entirely equivalent.

Vibrations travelling through a crystal are often referred to as phonons, and the same term may be used here. Thus a state with energy $(k_s+1)\hbar\omega_s$ may be thought of as representing k_s phonons. If at the same time $L_s' = m_s\hbar$, $\frac{1}{2}(k_s+m_s)$ phonons are travelling in the positive z direction and $\frac{1}{2}(k_s-m_s)$ are moving in the opposite direction. As can be seen in the figure, and as was shown in Chapter 4 both $\frac{1}{2}(k_s+m_s)$ and $\frac{1}{2}(k_s-m_s)$ are always integers.

With the above ideas of phonons $\mathsf{R}_s^\dagger$ may be called a creation operator. It "creates" a phonon in the postive direction with the energy $\hbar\omega_s$. Similarly $\mathsf{R}_{-s}^\dagger$ creates a phonon of the same energy in the opposite direction. The adjoint operators are destruction operators.

Since the energy H_s always appears in units of $\hbar\omega_s$, a state can be designated by the integer

$$k_s = (H_s'/\hbar\omega_s)-1 \tag{11.19}$$

together with the integer

$$m_s = L_s'/\hbar . \tag{11.20}$$

On the other hand the state could equally well be designated by $n_s = \frac{1}{2}(k_s+m_s)$ and $n_{-s} = \frac{1}{2}(k_s-m_s)$. n_s and n_{-s} then represent the numbers of phonons travelling in the $+z$ and $-z$ directions. These numbers, n_s and n_{-s}, will usually be used to designate the states of the string.

Associated with the idea of "phonons" is the idea of localizability. A phonon can be localized only in the same sense an electron can. An electron can be localized only when its momentum and its energy are not precisely determined. The same is true of a phonon. A localized phonon may best be pictured as a wave packet built up of a sum over a number of values of s. The creation operator for such a wave packet is a sum over a number of $\mathsf{R}_s^\dagger$ (including $\mathsf{R}_{-s}^\dagger$), and one can build up such linear combinations to represent various positions and energies.

PROBLEM 4. Show that equation (11.13) follows from the commutation rule.

PROBLEM 5. Show that equations (11.15) follow from the commutation rule and the definitions.

PROBLEM 6. Show that equations (11.16) follows from the commutation rule and the definitions.

PROBLEM 7. Show that equations (11.17) follow from the commutation rule and the definitions.

PROBLEM 8. Show that

$$2\mathsf{R}_s^\dagger \mathsf{R}_s = (\mathsf{H}_s/\hbar\omega_s) - 1 + (\mathsf{L}_s/\hbar) = 2\mathsf{n}_s$$

and that $\mathsf{R}_s^\dagger \mathsf{R}_s$ has eigenvalues of $\frac{1}{2}(n_s+m_s) = n_s$.

PROBLEM 9. Show that

$$2\mathsf{R}_{-s}^\dagger \mathsf{R}_{-s} = (\mathsf{H}_s/\hbar\omega_s) - 1 - (\mathsf{L}_s/\hbar) = 2\mathsf{n}_{-s}$$

and that $\mathsf{R}_{-s}^\dagger \mathsf{R}_{-s}$ has the eigenvalues $\frac{1}{2}(n_s-m_s) = n_{-s}$.

PROBLEM 10. Show that equations (11.18a) and (11.18b) follow from the commutation rules, and that

$$\mathsf{R}_s^\dagger \mid n_s, n_{-s}\rangle = \mathrm{i}(n_s+1)^{\frac{1}{2}} \mid n_s+1, n_{-s}\rangle \tag{11.21}$$

and

$$\mathsf{R}_s \mid n_s, n_{-s}\rangle = -\mathrm{i}\, n_s^{\frac{1}{2}} \mid n_s-1, n_{-s}\rangle$$

when the kets are understood to be normalized.

PROBLEM 11. Show that R_{-s} gives zero when applied to one of the states on the right-hand edge of the scheme in fig. 11.1.

The total Hamiltonian for the string can be written in the form

$$\mathsf{H} = \sum_{s=-\infty}^{\infty} (\mathsf{R}_s^\dagger \mathsf{R}_s + \tfrac{1}{2})\hbar\omega_s = \sum_{s=-\infty}^{\infty} (\mathsf{n}_s + \tfrac{1}{2})\hbar\omega_s, \tag{11.22}$$

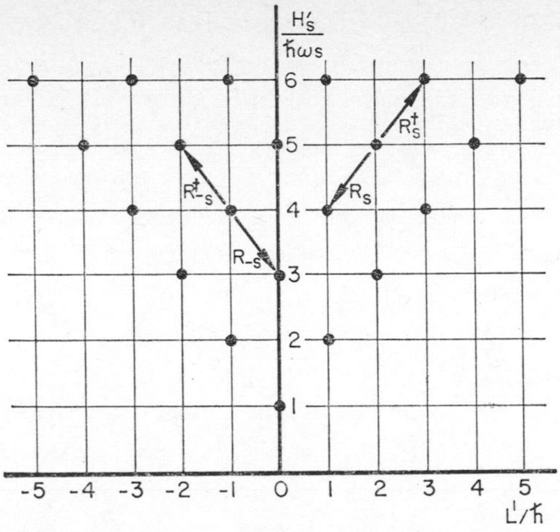

Fig. 11.1. A graphical representation of the effects of the operators R and R† in transforming eigenfunctions of the two operators H and L.

where the n_s are operators with integral eigenvalues and where the negative values of s represent waves moving in the negative z direction. It must be remembered that $\omega_s = \omega_{-s} > 0$.

At this point there comes to light in the problem of the vibrating string a difficulty which plagues all treatments of the electromagnetic field. Equation (11.22) shows that the minimum value of the energy is infinite. Even when all of the n_s' are zero, the sum of $\frac{1}{2}\hbar\omega_s$, over the infinity of values of s, does not converge. Even when the string is not vibrating, when it has its lowest possible energy, that energy is infinite.

There are two things which can be said about the difficulty in this simple problem. In the first place the blame can be placed on the assumption of an infinity of coordinates for a material string. It can be claimed that a string consists, at most, of only a finite number of particles and so the number of coordinates must really be finite. According to this point of view, the expansion in equation (11.1) is valid only for purposes in which the values of s above a certain large value N play no part. This would reduce the minimum value of E in equation (11.22) to a finite, although a large, value, and might be conceptually less objectionable. Nevertheless this somewhat evades the problem and does not give a satisfactory treatment of a continuum. If it is adopted, one must hold that there is no real continuum in nature, and that apparent continua merely appear so because of the small scale of their constituent parts. The application of this idea to the electromagnetic field has not been satisfactorily carried out.

On the other hand, it can be maintained that the zero point of energy is entirely arbitrary and that the minimum value of E in equation (11.22) merely represents an arbitrary additive constant that follows from the method of analysis used. The term $\frac{1}{2}\hbar\omega_s$ in each member of the sum is then not really a significant part of the energy and might well be omitted. Such a treatment might seem satisfactory for the energy, but the same difficulty crops up in other places as will be shown later.

The states represented by the ket vectors in (11.18) are eigenstates of the energy and the quantity L. In such states the quantities $q_{s,\mu}$ are not specified. However, it is possible to consider other kinds of states, for example, those in which the displacement of the string at t_0 is known and each of the $q_{s,\mu}$ has a precise value. The situation would be represented by δ functions in the coordinate space of each two dimensional harmonic oscillators. These values will then not only change, but will also become indeterminate.

The behavior of a simple oscillator is unusual in that the wave function oscillates, and expands and contracts, with the classical frequency. A half period after t_0, when each normal coordinate has the value $q^0_{s,\mu}$, each $q_{s,\mu}$ will have a precisely defined value equal to $-q^0_{s,\mu}$. Then since all frequencies are multiples of the fundamental, it follows that at intervals of half the fundamental period the string will have a precisely defined configuration, first on one side of its rest position and then on the other. That this is true in quantum mechanics can be seen from the fact that half of the energy eigenfunctions are odd and the other half are even. However since the length of the string is taken long enough so that reflections from the ends are unimportant, the fundamental period is, in effect, infinitely long.

The fact that the operators for the p's do not commute with those for the corresponding q's indicates that if the configuration of the string is precisely known, its velocity cannot be known at all. Such a statement is a conclusion from the theory. To test the validity of the conclusion in the light of experimental results, it is necessary to consider the methods that might be used to determine the configuration. An analysis along the lines of the gamma-ray microscope considerations in Chapter 2 would seem to be appropriate. If light of sufficiently short wavelength is used to illuminate the string, and if it is of sufficient intensity that one quantum will be scattered from each of a sufficiently large number of points on the string, the momentum of recoil will invalidate for future use any previously obtained knowledge of the velocity. Although no such experiments have been tried for a material string where the expected effects are far below the limits of practical observation, a corresponding analysis has been given by Bohr and Rosenfeld for the electromagnetic field.

The theoretically predicted indetermination relationships between the positions and the velocities of various points on the string can be determined by examining the commutation relationships between the corresponding operators. It was shown in Chapter 3 that if two operators do not commute, an indetermination relationship holds between the two quantities they represent.

The operator for the displacement x of the string at the point z is connected with the operators $\mathsf{q}_{s,\mu}$ by the equation

$$\mathsf{x}(z_1) = \sum_{s=0}^{\infty} \left[\mathsf{q}_{s,1} \cos \frac{2\pi s z_1}{L} + \mathsf{q}_{s,2} \sin \frac{2\pi s z_1}{L} \right]. \tag{11.23}$$

Similarly the operator for the velocity at z_2 is

$$\dot{\mathsf{x}}(z_2) = \sum_{s=0}^{\infty} \frac{2}{M} \left[\mathsf{p}_{s,1} \cos \frac{2\pi s z_2}{L} + \mathsf{p}_{s,2} \sin \frac{2\pi s z_2}{L} \right]. \tag{11.24}$$

Since all of the operators p and q cummute except each $\mathsf{p}_{s,\mu}$ with its conjugate $\mathsf{q}_{s,\mu}$ it follows that

$$\begin{aligned} &\mathsf{x}(z_1)\dot{\mathsf{x}}(z_2) - \dot{\mathsf{x}}(z_2)\mathsf{x}(z_1) = \\ &\quad = -\frac{\hbar}{\mathrm{i}} \frac{2}{M} \sum_{s=0}^{\infty} \left[\cos \frac{2\pi s z_1}{L} \cos \frac{2\pi s z_2}{L} + \sin \frac{2\pi s z_1}{L} \sin \frac{2\pi s z_2}{L} \right] \\ &\quad = -\frac{\hbar}{\mathrm{i}} \frac{4}{\rho} \delta(z_1 - z_2). \end{aligned} \tag{11.25}$$

That the infinite series in equation (11.25) is equivalent to a δ function can be seen by considering its properties.

$$\sum_{s=0}^{\infty} \left[\cos \frac{2\pi s z_1}{L} \cos \frac{2\pi s z_2}{L} + \sin \frac{2\pi s z_1}{L} \sin \frac{2\pi s z_2}{L} \right] = \sum_{s=0}^{\infty} \cos \frac{2\pi s}{L} (z_1 - z_2).$$

If $z_1 - z_2 \neq 0$, the sum over s will not converge but will oscillate, and the oscillations will be large for small $(z_1 - z_2)$ but smaller if $(z_1 - z_2)$ is large. However, for $(z_1 - z_2) = 0$, each term in the sum will be unity and the sum will increase without limit. On the other hand, each member of the sum gives zero when integrated with respect to z_1 from 0 to L, except the term with $s = 0$, which integrates to L. Hence the sum for the point $z_1 = z_2$ is infinitely larger than at any other pair of points and yet the integral of the sum has a finite value.

The conclusion is that the displacement at one point z_1 and the momentum, or velocity, at another point z_2 are not connected by an indetermination relationship. However, the displacement and the velocity at one given point are so related.

It was shown in Chapter 4 that an oscillator in an energy eigenstate has zero for the mean values of both the coordinate and the momentum and has an indetermination in the coordinate and momentum which increases with the energy. These conclusions apply directly to the problem of the stretched string.

From equations (11.14a)–(11.14d) the quantities $q_{s,1}$ and $q_{s,2}$ can be expressed in terms of the R operators. Then it follows that

$$\mathsf{x}(z) = -\mathrm{i} \sum_{s=-\infty}^{\infty} (\hbar/2M\omega_s)^{\frac{1}{2}} [\mathsf{R}_s^\dagger \, \mathrm{e}^{-(2\pi \mathrm{i} sz/L)} - \mathsf{R}_s \, \mathrm{e}^{(2\pi \mathrm{i} sz/L)}]. \quad (11.26)$$

It follows immediately from the fact that $\langle n_s' \mid \mathsf{R}_s \mid n_s' \rangle = 0$ that in any state in which the n_s' are given, $\langle x \rangle = 0$. A corresponding conclusion follows for $\langle \dot{x} \rangle$.

However, the expectation value of x^2 is not zero. From (11.26) it follows that

$$\mathsf{x}^2 = -\sum_{s,r=-\infty}^{\infty} \frac{\hbar}{2M} (\omega_s \omega_r)^{-\frac{1}{2}} [\mathsf{R}_s^\dagger \, \mathrm{e}^{-(2\pi \mathrm{i} sz/L)} - \mathsf{R}_s \, \mathrm{e}^{(2\pi \mathrm{i} sz/L)}] \times$$
$$\times [\mathsf{R}_r^\dagger \, \mathrm{e}^{-(2\pi \mathrm{i} rz/L)} - \mathsf{R}_r \, \mathrm{e}^{(2\pi \mathrm{i} rz/L)}]. \quad (11.27)$$

Since the cross products will not contribute to the expectation value we may write

$$\langle Q \mid \mathsf{x}^2 \mid Q \rangle = -\sum_{s=-\infty}^{\infty} \frac{\hbar}{2M\omega_s} [\mathsf{R}_s^\dagger \, \mathrm{e}^{-(2\pi \mathrm{i} sz/L)} - \mathsf{R}_s \, \mathrm{e}^{(2\pi \mathrm{i} sz/L)}]^2$$
$$= -\sum_s \frac{\hbar}{2M\omega_s} \langle Q \mid \mathsf{R}_s^{\dagger 2} \, \mathrm{e}^{-(4\pi \mathrm{i} sz/L)} - \mathsf{R}_s^\dagger \mathsf{R}_s - \mathsf{R}_s \mathsf{R}_s^\dagger + \mathsf{R}_s^2 \, \mathrm{e}^{(4\pi \mathrm{i} sz/L)} \mid Q \rangle . \quad (11.28)$$

The symbol $\mid Q \rangle$ is used to designate a ket vector in which each n_s is given a value. The terms in $\mathsf{R}_s^{\dagger 2}$ and R_s^2 give zero for the expectation value so that

$$\langle Q \mid \mathsf{x}^2(z) \mid Q \rangle = \sum_s \frac{\hbar}{2M\omega_s} \langle Q \mid \mathsf{R}_s^\dagger \mathsf{R}_s + \mathsf{R}_s \mathsf{R}_s^\dagger \mid Q \rangle$$
$$= \sum_s \frac{\hbar}{2M\omega_s} (2n_s + 1) . \quad (11.29)$$

Here again the sum over the infinity of values of s leads to a divergent value for $\langle Q \mid \mathsf{x}^2 \mid Q\rangle$.

To represent a state of the string in which the displacement has a significant value it is necessary to use state vectors that are not eigenvectors of the operators n_s.

PROBLEM 12. Let $\mid Q\rangle = N(\mathsf{R}_s^{\dagger 3} + \mathsf{R}_s^{\dagger 4} + \mathsf{R}_s^{\dagger 5}) \mid 0\rangle$ where $\mid 0\rangle$ represents the state in which all n_s' are zero. Find the expection values of x and of x^2 as functions of z and t. N is a normalization constant.

3. The electromagnetic field in empty space

In Chapter 1 the Maxwell equations for an electromagnetic field, in the absence of charges and currents, were solved in terms of a vector potential. In other words, they were reduced to a pair of partial differential equations involving a single vector function of the coordinates and the time. The equations are

$$\nabla^2 \boldsymbol{A} = \mu_0 \varepsilon_0 \, (\partial^2 \boldsymbol{A}/\partial t^2) \tag{11.30a}$$

and

$$\operatorname{div} \boldsymbol{A} = 0. \tag{11.30b}$$

If a vector function $\boldsymbol{A}$ can be found which satisfied these conditions, and if the field vectors are determined from $\boldsymbol{A}$ by the relations

$$\boldsymbol{E} = -(\partial \boldsymbol{A}/\partial t), \qquad \boldsymbol{B} = \operatorname{curl} \boldsymbol{A}, \tag{11.30c}$$

the vectors $\boldsymbol{E}$ and $\boldsymbol{B}$ will satisfy Maxwell's equations.

Canonical equations of motion based on a Hamiltonian function can be introduced by means of a Fourier expansion as in the case of the vibrating string. Here again the boundary conditions need some consideration. If it is imagined that the radiation is enclosed by perfectly conducting walls the boundary conditions can be strictly formulated, but they are complicated. It seems also that the complication is not essential to the problem in hand, since the box may be made as large as desired and the walls removed as far as desired from any process of interest. Hence it is generally believed that the essential features of the problem are retained when periodic boundary conditions are applied. The field quantities, in particular the vector potential $\boldsymbol{A}$, are then taken to be periodic with the period L along the three coordinate axes. It is understood also that L is to be taken so large that times of the order of L/c are long compared with times of importance in the problems to be considered. The trigonometric expansion then has the form

$$A = \frac{1}{V^{\frac{1}{2}}} \sum_{\boldsymbol{K},\lambda} \boldsymbol{\varepsilon}_{\boldsymbol{K},\lambda}[Q_{\boldsymbol{K},\lambda,1} \cos(\boldsymbol{K}\cdot\boldsymbol{r}) + Q_{\boldsymbol{K},\lambda,2} \sin(\boldsymbol{K}\cdot\boldsymbol{r})] \qquad (11.31)$$

where the vectors $\boldsymbol{a}_{\boldsymbol{K},\lambda,\mu}$ of Chapter 1 are written in terms of magnitudes $Q_{\boldsymbol{K},\lambda,\mu}$ and unit vectors $\boldsymbol{\varepsilon}_{\boldsymbol{K},\lambda}$. The time dependent vectors are replaced by fixed unit vectors multiplied by time dependent amplitudes. $V = L^3$, and the quantities $Q_{\boldsymbol{K},\lambda,\mu}$ are the discrete coordinates in terms of which the field is to be described. The boundary conditions are satisfied by using only values of $\boldsymbol{K}$ given by

$$\boldsymbol{K} = (2\pi/L)(k_1\boldsymbol{i} + k_2\boldsymbol{j} + k_3\boldsymbol{k}), \qquad (11.32)$$

where the k_i are integers. In fact only half of the values of $\boldsymbol{K}$ given by (11.32) can be used without duplicating terms in (11.31), so only those values of K contained in a solid angle of 2π will be used. To satisfy equation (11.30b) and to take advantage of the convenience of orthogonal unit vectors, let

$$\boldsymbol{\varepsilon}_{\boldsymbol{K},\lambda}\cdot\boldsymbol{K} = 0 \quad \text{and} \quad \boldsymbol{\varepsilon}_{\boldsymbol{K},1}\cdot\boldsymbol{\varepsilon}_{\boldsymbol{K},2} = 0. \qquad (11.33)$$

The first of equations (11.33) guarantees the satisfaction of (11.30b), but to satisfy (11.30a) it is necessary that

$$\frac{\mathrm{d}^2 Q_{\boldsymbol{K},\lambda,\mu}}{\mathrm{d}t^2} + \omega_{\boldsymbol{K}}^2 Q_{\boldsymbol{K},\lambda,\mu} = 0, \qquad (11.34)$$

where $\omega_{\boldsymbol{K}}^2 = K^2c^2 = K^2/\mu_0\varepsilon_0$. This is of the same form as the equations satisfied by the coordinate $q_{s,\mu}$ of the vibrating string. In this case there are four coordinates $Q_{\boldsymbol{K},\lambda,\mu}$ for each value of $\boldsymbol{K}$ instead of only two as for the string.

In terms of the coordinates $Q_{\boldsymbol{K},\lambda,\mu}$ the electromagnetic energy of the field is

$$W = \frac{1}{2}\int\left[\varepsilon_0 E^2 + \frac{1}{\mu_0}B^2\right]\mathrm{d}v = \tfrac{1}{4}\varepsilon_0 \sum_{\boldsymbol{K},\lambda,\mu}[\dot{Q}_{\boldsymbol{K},\lambda,\mu}^2 + \omega_{\boldsymbol{K}}^2 Q_{\boldsymbol{K},\lambda,\mu}^2]. \quad (11.35)$$

The sum is over the four pairs of values of λ and μ as well as over those values of $\boldsymbol{K}$ which satisfy (11.32) and also lie on one side of a plane through the origin. The Hamiltonian function of the field can then be written by expressing the variables $\dot{Q}_{\boldsymbol{K},\lambda,\mu}$ in terms of the momenta $P_{\boldsymbol{K},\lambda,\mu}$:

$$H = \frac{1}{\varepsilon_0}\sum_{\boldsymbol{K},\lambda,\mu}[P_{\boldsymbol{K},\lambda,\mu}^2 + \tfrac{1}{4}\varepsilon_0^2\omega_{\boldsymbol{K}}^2 Q_{\boldsymbol{K},\lambda,\mu}^2]. \qquad (11.36)$$

The resulting canonical equations are just the defining equations of the $P_{\boldsymbol{K},\lambda,\mu}$ and the equations (11.34), so that together with the definitions of the coordinates, this Hamiltonian function is equivalent to the Maxwell equations for the field in the absence of charges and currents.

This expansion of the electromagnetic field in terms of plane-polarized plane waves is not at all the only useful expansion. It is in some respects the simplest, but the radiation emitted from an atom is such that a description in terms of circularly or elliptically polarized waves is often better suited to the physical situation. Furthermore a plane wave is not entirely satisfactory for representing the emission of radiation from atoms; for the radiation emitted is usually such as to have an angular momentum about the location of the atom so that the angular momentum of the field and the atom together is conserved. However the plane wave expansion is conceptually simple and useful for understanding the spectral distribution of black-body radiation.

To treat the electromagnetic field quantum mechanically, the coordinates and momenta in the Hamiltonian function (11.36) are treated as operators with the commutation rules

$$\mathsf{P}_{\boldsymbol{K},\lambda,\mu}\mathsf{Q}_{\boldsymbol{K}',\lambda',\mu'}-\mathsf{Q}_{\boldsymbol{K}',\lambda',\mu'}\mathsf{P}_{\boldsymbol{K},\lambda,\mu} = \frac{\hbar}{\mathrm{i}}\,\delta_{\boldsymbol{K},\boldsymbol{K}'}\delta_{\lambda,\lambda'}\delta_{\mu,\mu'}. \tag{11.37}$$

This differs from the vibrating string in that there are four sets of coordinates and momenta for each vector $\boldsymbol{K}$. However, these may be taken as two independent sets of two pairs each when one is dealing only with plane polarized radiation. For each $\boldsymbol{K}$, λ there can be defined the operators

$$\mathsf{R}^{\dagger}_{\boldsymbol{K},\lambda} = (2\varepsilon_0\hbar\omega_{\boldsymbol{K}})^{-\frac{1}{2}}(\mathsf{P}_{\boldsymbol{K},\lambda,1}+\mathrm{i}\mathsf{P}_{\boldsymbol{K},\lambda,2})+\mathrm{i}\left(\frac{\varepsilon_0\omega_{\boldsymbol{K}}}{8\hbar}\right)^{\frac{1}{2}}(\mathsf{Q}_{\boldsymbol{K},\lambda,1}+\mathrm{i}\mathsf{Q}_{\boldsymbol{K},\lambda,2}) \tag{11.38a}$$

$$\mathsf{R}_{\boldsymbol{K},\lambda} = (2\varepsilon_0\hbar\omega_{\boldsymbol{K}})^{-\frac{1}{2}}(\mathsf{P}_{\boldsymbol{K},\lambda,1}-\mathrm{i}\mathsf{P}_{\boldsymbol{K},\lambda,2})-\mathrm{i}\left(\frac{\varepsilon_0\omega_{\boldsymbol{K}}}{8\hbar}\right)^{\frac{1}{2}}(\mathsf{Q}_{\boldsymbol{K},\lambda,1}-\mathrm{i}\mathsf{Q}_{\boldsymbol{K},\lambda,2}) \tag{11.38b}$$

$$\mathsf{R}_{-\boldsymbol{K},\lambda} = (2\varepsilon_0\hbar\omega_{\boldsymbol{K}})^{-\frac{1}{2}}(\mathsf{P}_{\boldsymbol{K},\lambda,1}-\mathrm{i}\mathsf{P}_{\boldsymbol{K},\lambda,2})+\mathrm{i}\left(\frac{\varepsilon_0\omega_{\boldsymbol{K}}}{8\hbar}\right)^{\frac{1}{2}}(\mathsf{Q}_{\boldsymbol{K},\lambda,1}-\mathrm{i}\mathsf{Q}_{\boldsymbol{K},\lambda,2}) \tag{11.38c}$$

$$\mathsf{R}^{\dagger}_{-\boldsymbol{K},\lambda} = (2\varepsilon_0\hbar\omega_{\boldsymbol{K}})^{-\frac{1}{2}}(\mathsf{P}_{\boldsymbol{K},\lambda,1}+\mathrm{i}\mathsf{P}_{\boldsymbol{K},\lambda,2})-\mathrm{i}\left(\frac{\varepsilon_0\omega_{\boldsymbol{K}}}{8\hbar}\right)^{\frac{1}{2}}(\mathsf{Q}_{\boldsymbol{K},\lambda,1}+\mathrm{i}\mathsf{Q}_{\boldsymbol{K},\lambda,2}). \tag{11.38d}$$

These satisfy the same commutation rules as the corresponding operators in equations (11.14a)–(11.14d). The use of $-\boldsymbol{K}$ as well as $+\boldsymbol{K}$ corresponds to the use of $-s$ and $+s$ and shows that these operators represent creation and destruction of "photons" in all directions. Furthermore the operator

$$\mathsf{L}_{\boldsymbol{K},\lambda} = [\mathsf{Q}_{\boldsymbol{K},\lambda,1}\mathsf{P}_{\boldsymbol{K},\lambda,2} - \mathsf{Q}_{\boldsymbol{K},\lambda,2}\mathsf{P}_{\boldsymbol{K},\lambda,1}]\boldsymbol{K} \tag{11.39}$$

represents the linear momentum of the field due to the $\boldsymbol{K}, \lambda$ component of the vector potential. But this can take on both positive and negative values, indicating that the momentum can be in the direction of $\boldsymbol{K}$ or opposite to it. However the operator L can also be written in terms of the $\mathsf{R}_{\boldsymbol{K},\lambda}$ as

$$\mathsf{L}_{\boldsymbol{K},\lambda} = [\mathsf{R}_{\boldsymbol{K},\lambda}\mathsf{R}_{\boldsymbol{K},\lambda} - \mathsf{R}_{-\boldsymbol{K},\lambda}\mathsf{R}_{\boldsymbol{K},\lambda}]\hbar\boldsymbol{K} \tag{11.40}$$

where the different directions of $\boldsymbol{K}$ over which a sum can be taken to get the total momentum lie on only one side of a plane through the origin. If, however, all possible directions of $\boldsymbol{K}$ are considered, the operator for the total field momentum may be written

$$\mathsf{G} = \sum_{\boldsymbol{K},\lambda} \mathsf{R}_{\boldsymbol{K},\lambda}\mathsf{R}_{\boldsymbol{K},\lambda}h\boldsymbol{K} \tag{11.41}$$

where the sum is now over all directions.

From a knowledge of the energy in the $\boldsymbol{K}, \lambda$ component of the field, and also of the momentum in the same component, it is possible to formulate an operator for the number of quanta in each direction:

$$\mathsf{H} = \sum_{\boldsymbol{K},\lambda} (\mathsf{R}^{\dagger}_{\boldsymbol{K},\lambda}\mathsf{R}_{\boldsymbol{K},\lambda} + \tfrac{1}{2})\hbar\omega_{\boldsymbol{K}} = \sum_{\boldsymbol{K},\lambda} (\mathsf{n}_{\boldsymbol{K},\lambda} + \tfrac{1}{2})\hbar\omega\,. \tag{11.42}$$

Here again the sum over $\boldsymbol{K}$ includes the whole solid angle of 4π, just as Problems 8 and 9 gave operators for the two directions along the string.

PROBLEM 13. Show that $\mathsf{L}_{\boldsymbol{K},\lambda}$ represents the integral of the Poynting vector associated with the $\boldsymbol{K}, \lambda$ component of the vector potential, and that G in equation (11.41) represents the integral of the Poynting vector over the field.

PROBLEM 14. Write the operator for the electric field at the point (x, y, z) in terms of the creation and destruction operators.

PROBLEM 15. Write the operator for the magnetic field in terms of the creation and destruction operators.

PROBLEM 16. Show that in states in which the number of phonons is precisely given, the average values of the electric and magnetic fields are zero.

It is brought out by problem 16 that the quantum mechanical treatment of the electromagnetic field which emphasizes eigenvalues of the energy and momentum, and hence of the number of quanta in the field, is complementary to a treatment in which the values of the fields are specified. To specify a field it is necessary that the coordinates $Q_{\boldsymbol{K},\lambda,\mu}$ have precise values. When this is the case the number of quanta is entirely indeterminate. Of course there are intermediate possibilities. Each pair of coordinates $Q_{\boldsymbol{K},\lambda,1}$ and $Q_{\boldsymbol{K},\lambda,2}$ represented by δ functions will represent an electric field, but the energy and the number of quanta will be indeterminate.

It was shown in Problem 12 that it is possible to formulate a state vector in which the displacement of the string has a quantum mechanical mean value which is that of a wave moving along the string. Similar formulations are possible for the electromagnetic field. As was illustrated in Problem 12, such a state is not an energy eigenstate but must contain two or more such states.

The idea of a "phonon" on a string or a "photon" in the electromagnetic field often carries with it the connotation of localization in space. However it is clear that an energy eigenstate in either case extends over the whole system. In the case of the string it extends over the whole string. In the case of the electromagnetic field it extends over the whole volume considered. To get any degree of localization it is necessary to use a combination of operators representing different frequencies and so to build up a wave packet. The following problems illustrate some of these situations.

PROBLEM 17. Let $\boldsymbol{K}$ be a vector in the direction of the z axis and let $\varepsilon_{\boldsymbol{K},1}$ be a unit vector in the direction of the x axis. Then consider the state of the electromagnetic field represented by

$$|\,Q\rangle = N[(\mathsf{R}^{\dagger}_{\boldsymbol{K},1})^3+(\mathsf{R}^{\dagger}_{\boldsymbol{K},1})^4+(\mathsf{R}^{\dagger}_{\boldsymbol{K},1})^5+(\mathsf{R}^{\dagger}_{\boldsymbol{K},1})^6]\,|\,0\rangle$$

where $|\,0\rangle$ represents the state in which all the $n_{\boldsymbol{K},\lambda}$ are zero.

a. Find the mean value of the energy in the field (over and above the zero point energy) and find the distribution of the energy over its various eigenvalues.
b. Find the quantum mechanical mean values of the electric and magnetic fields as functions of position and time.

PROBLEM 18. Let $\boldsymbol{K}_1$, $\boldsymbol{K}_2$, $\boldsymbol{K}_3$ be vectors along the z axis of "successive" magnitudes,

$$K_1 = 2\pi(k-1)/L, \qquad K_2 = 2\pi k/L, \qquad K_3 = 2\pi(k+1)/L.$$

Let

$$\tfrac{1}{2}\left[\left(\mathsf{R}^{\dagger}_{\mathbf{K}_{1},1} + \frac{1}{\sqrt{2}}\,\mathsf{R}^{\dagger}_{\mathbf{K}_{2},1} + \mathsf{R}^{\dagger}_{\mathbf{K}_{3},1}\right)\right] = \mathsf{S}^{\dagger}.$$

The state $\mathsf{S}^{\dagger} \mid 0\rangle$ will represent a certain amount of localization of a wave packet, but very little. The packet will have a maximum but will be spread over the whole distance L.

a. Find the distribution of the energy (above the zero point energy) in the states

$$\mid S_1\rangle = \mathsf{S}^{\dagger} \mid 0\rangle \quad \text{and} \quad \mid S_2\rangle = N[\mathsf{S}^{\dagger} + (\mathsf{S}^{\dagger})^2] \mid 0\rangle .$$

b. Find the field momentum in the above states.

c. Evalutate the quantum mechanical mean values of the electric and magnetic fields as functions of the time and position for the above states.

CHAPTER 12

INTERACTION OF RADIATION AND MATTER

To treat the interaction of matter and radiation, one considers the material particles and the radiation field as composing a single system. The Hamiltonian function of this system can be written down and the usual quantum-mechanical treatment applied to it. As an illustration of the process it is convenient again to consider an analogous but less complicated problem.

1. Vibrating string coupled to a vibrating particle

Consider the vibrating string treated in the previous chapter, and suppose that at the point $z = z_0$ there is a particle of mass m attached to the point $(z = z_0, x = 0)$ by a spring of force constant k. Let the particle be constrained to move in the x direction only. The string and the particle can be regarded as one composite system, but as long as they are not physically connected, nothing follows from so regarding them. Now let the particle and the string be connected by a weak spring of force constant γ, so that the particle is pulled toward the string with a force γ times the difference of their displacements. If the coordinate and the momentum of the particle are designated by q and p, and the notation of the previous chapter is used to describe the string, the Hamiltonian function is

$$H = \sum_{s=0}^{\infty} \sum_{\mu=1}^{2} \left\{ \frac{1}{2M} p_{s,\mu}^2 + (s^2\pi^2\tau/L) q_{s,\mu}^2 \right\} + \frac{p^2}{2m} + \tfrac{1}{2}kq^2 + \\ + \tfrac{1}{2}\left\{ q - \sum_{s=0}^{\infty} \left[q_{s,1} \cos \frac{2\pi s z_0}{L} + q_{s,2} \sin \frac{2\pi s z_0}{L} \right] \right\}^2 . \qquad (12.1)$$

The canonical equations of motion lead to the following expressions

$$\frac{d^2 q_{s,1}}{dt^2} + \frac{4\pi^2 s^2 \tau}{ML} q_{s,1} =$$
$$= \frac{2\gamma}{M} \cos \frac{2\pi s z_0}{L} \left\{ q - \sum_s \left[q_{s,1} \cos \frac{2\pi s z_0}{L} + q_{s,2} \sin \frac{2\pi s z_0}{L} \right] \right\}, \quad (12.2a)$$

$$\frac{d^2 q_{s,2}}{dt^2} + \frac{4\pi^2 s^2 \tau}{ML} q_{s,2} =$$
$$= \frac{2\gamma}{M} \sin \frac{2\pi s z_0}{L} \left\{ q - \sum_s \left[q_{s,1} \cos \frac{2\pi s z_0}{L} + q_{s,2} \sin \frac{2\pi s z_0}{L} \right] \right\}, \quad (12.2b)$$

$$\frac{d^2 q}{dt^2} + \frac{k}{m} q = \frac{-\gamma}{m} \left\{ q - \sum_s \left[q_{s,1} \cos \frac{2\pi s z_0}{L} + q_{s,2} \sin \frac{2\pi s z_0}{L} \right] \right\}. \quad (12.2c)$$

Equations (12.2a)–(12.2c) are linear. In principle, a further transformation to normal coordinates of the combined system could be made and a general solution written down explicitly. However, the infinity of coordinates $q_{s,\mu}$ makes this difficult. There is a set, of course, those $q_{s,1}$ for which $\cos (2\pi s z_0\ L) = 0$ and those $q_{s,2}$ for which $\sin (2\pi s z_0\ L) = 0$, which are already normal coordinates of the whole system. But they are of little interest because they do not affect the coordinate q.

The motion of the other coordinates can be studied by approximate means if the coupling is weak. Weak coupling means that γ is small compared with k and compared with $(s^2\pi^2\tau/L)$ for those s in which one is interested. These approximate methods are even more appropriate here than would be the exact solution, since they more nearly illustrate the methods to be used with the electromagnetic field.

If at $t = 0$ the string is stationary in its rest position, and the particle is oscillating, so that $q = A \sin (k/m)^{\frac{1}{2}} t$, this special value of q may be inserted on the right-hand side of equation (12.2a). The remainder of the right-hand side may be neglected. Every $q_{s,\mu} = 0$ at $t = 0$, and since γ is small, the initial growth of each $q_{s,\mu}$ is slow. All of them will be unimportant for a while. During this short time (12.2a) becomes

$$\frac{d^2 q_{s,1}}{dt^2} + \frac{4\pi^2 s^2 \tau}{ML} q_{s,1} = \frac{2\gamma A}{M} \cos \frac{2\pi s z_0}{L} \sin \left(\frac{k}{m}\right)^{\frac{1}{2}} t. \quad (12.3a)$$

This is the usual equation for an oscillator with a sinusoidal forcing term and its solution can be written down at once:

$$q_{s,1} = \frac{2\gamma A}{M(\omega_s^2 - \omega^2)} \cos \frac{2\pi s z_0}{L} \left[\sin \omega t - \frac{\omega}{\omega_s} \sin \omega_s t \right] \quad (12.3b)$$

where $\omega^2 = k/m$ and $\omega_s^2 = 4\pi^2 s^2 \tau/ML$. This shows that $q_{s,1}$ will increase slowly from zero, that all of the modes of vibration of the string will be excited except those for which $\cos(2\pi s z_0/L) = 0$, but that the most strongly excited will be those for which $(4s^2\pi^2\tau/ML)$ is very close to (k/m). As time goes on the effect of this resonance becomes more and more pronounced and finally the only oscillations of importance are those for which $(4s^2\pi^2\tau/ML) = (k/m)$. In the meantime, the vibration of the string reacts on the motion of the particle according to equation (12.2c) and the modified motion of the particle can be inserted in equations (12.2a) and (12.2b) for a better approximation. It is difficult, however, to follow the motion very far by this procedure.

A similar treatment can be carried through if at $t = 0$ the mass is at rest and one or more normal modes of the string are excited. The particle will be set in motion according to equation (12.2c), and this motion will again react on the string.

The central point in the above treatment of a string and a vibrating particle is that they are treated as separate systems in the zero approximation. The interaction between them is then introduced as a perturbation which causes changes in the "constants" of the unperturbed motion. Such a treatment is valid only when the perturbation is weak enough to permit the unperturbed motion to give a fair description of the situation. The treatment could be exact if the equations could be given an exact solution since the unperturbed states form a complete set. However, the usual treatment is an approximate one and the nature of the approximation must always be kept in mind.

When the interaction term in the Hamiltonian (12.1) is squared out, the term $\frac{1}{2}\gamma q^2$ could be added to the $\frac{1}{2}kq^2$ to give a slightly modified frequency for the motion of the oscillator. A similar disposition could be made of the square of each $q_{s,\mu}$. The remaining cross products then give a perturbation energy which leads to the interaction between the string and the particle and to transitions. Such corrections to the basic frequencies are not usually made since they affect only the higher orders of approximation. They are important, however, in the more sophisticated electrodymanics.

To treat the problem quantum mechanically the operators $\mathsf{R}_s^\dagger$ and R_s of the previous chapter can be used. The coordinate q and momentum p may be expressed in terms of the quantities $\mathsf{a}^\dagger$ and a of Chapter 8. The cross products, representing the interaction between the vibrations of the string and the motion of the particle and also the interactions between the various modes of vibration of the string, then constitute the perturbing part of the Hamiltonian. The whole function is

$$H = \sum_{s=0}^{\infty} \sum_{\mu=1}^{2} \frac{1}{2M} [p_{s,\mu}^2 + M^2 \omega_s^2 q_{s,\mu}^2] + \frac{1}{2m} [p^2 + m^2\omega^2 q^2] + H_1 \quad (12.4a)$$

with

$$H_1 = -\gamma q \sum_s \left[q_{s,1} \cos \frac{2\pi s z_0}{L} + q_{s,2} \sin \frac{2\pi s z_0}{L} \right]$$

$$+\tfrac{1}{2}\gamma \sum_{s,r} q_{s,1} q_{r,1} \cos \frac{2\pi s z_0}{L} \cos \frac{2\pi r z_0}{L}$$

$$+\tfrac{1}{2}\gamma \sum_{s,r} q_{s,2} q_{r,2} \sin \frac{2\pi s z_0}{L} \sin \frac{2\pi r z_0}{L}$$

$$+\gamma \sum_{s,r} q_{s,1} q_{r,2} \cos \frac{2\pi s z_0}{L} \sin \frac{2\pi r z_0}{L}. \quad (12.4b)$$

In this form for H_1 the term $\frac{1}{2}\gamma q^2$ is combined with $\frac{1}{2}kq^2$ to give $\frac{1}{2}m\omega^2 q^2$. The terms containing $q_{s,\mu}^2$, however, are left in H_1 so that ω_s can be proportional to s. These terms are not used in what follows. Now, as in Chapter 8, let

$$\mathsf{a}^\dagger = (2m\hbar\omega)^{-\frac{1}{2}}(\mathsf{p} + im\omega\mathsf{q}) \quad \text{and} \quad \mathsf{a} = (2m\hbar\omega)^{-\frac{1}{2}}(\mathsf{p} - im\omega\mathsf{q}). \quad (12.5)$$

Using these operators, and the R operators of the previous chapter

$$H_0 = \sum_{s=-\infty}^{\infty} (\mathsf{R}_s^\dagger \mathsf{R}_s + \tfrac{1}{2})\hbar\omega_s + (\mathsf{a}^\dagger \mathsf{a} + \tfrac{1}{2})\hbar\omega. \quad (12.6a)$$

Also

$$\mathsf{q} = i(\hbar/2m\omega)^{\frac{1}{2}}(\mathsf{a} - \mathsf{a}^\dagger) \quad (12.6b)$$

$$q_{s,1} = i(\hbar/2M\omega_s)^{\frac{1}{2}}(\mathsf{R}_s - \mathsf{R}_s^\dagger + \mathsf{R}_{-s} - \mathsf{R}_{-s}^\dagger) \quad (12.6c)$$

$$q_{s,2} = -(\hbar/2M\omega_s)^{\frac{1}{2}}(\mathsf{R}_s + \mathsf{R}_s^\dagger - \mathsf{R}_{-s} - \mathsf{R}_{-s}^\dagger). \quad (12.6d)$$

Using these expressions and a considerable amount of algebraic manipulation one can write the perturbation part of the Hamiltonian

$$H_1 = \gamma \left(\frac{\hbar}{2m\omega}\right)^{\frac{1}{2}} (\mathsf{a} - \mathsf{a}^\dagger) \sum_{s=-\infty}^{\infty} \left(\frac{\hbar}{2M\omega_s}\right)^{\frac{1}{2}} \times$$

$$\times \left[\mathsf{R}_s \exp\left(\frac{2\pi i s z_0}{L}\right) - \mathsf{R}_s^\dagger \exp\left(-\frac{2\pi i s z_0}{L}\right) \right] +$$

$$+\tfrac{1}{2}\gamma \sum_{s,r=-\infty}^{\infty} \left(\frac{\hbar}{2M\omega_s}\right)^{\frac{1}{2}} \left(\frac{\hbar}{2M\omega_r}\right)^{\frac{1}{2}} \times$$

$$\times \left\{(\mathsf{R}_s\mathsf{R}_r^\dagger + \mathsf{R}_s^\dagger\mathsf{R}_r) \cos \frac{2\pi(s-r)z_0}{L} - (\mathsf{R}_s\mathsf{R}_r + \mathsf{R}_s^\dagger\mathsf{R}_r^\dagger) \cos \frac{2\pi(s+r)z_0}{L}\right\} +$$

$$-\frac{\mathrm{i}\hbar\gamma}{2M} \sum_{s,r=-\infty}^{\infty} (\omega_s\omega_r)^{-\frac{1}{2}}[\mathsf{R}_s\mathsf{R}_r + \mathsf{R}_s\mathsf{R}_r^\dagger - \mathsf{R}_s^\dagger\mathsf{R}_r - \mathsf{R}_s^\dagger\mathsf{R}_r^\dagger] \times$$

$$\times \left[\sin 2\pi(s+r)\frac{z_0}{L} - \sin 2\pi(s-r)\frac{z_0}{L}\right]. \tag{12.7}$$

The first part of H_1 represents the direct interaction between the string and the particle. The remainder represents the interaction between the different modes of vibration of the string due to the presence of the particle.

An energy eigenstate of the whole system without H_1 may be designated by the number n for the energy of the vibrating particle and the infinite set of numbers n_s for the energy of the string. The energy of such a state is

$$H_0' = (n+\tfrac{1}{2})\hbar\omega + \sum_{s=-\infty}^{\infty} (n_s+\tfrac{1}{2})\hbar\omega_s. \tag{12.8}$$

The ket vector may be designated by

$$|\, n, \ldots, n_s, \ldots\rangle. \tag{12.8a}$$

Although (12.8a) represents an energy eigenket of the system with $\gamma = 0$, or without H_1, it is not an energy eigenket when this interaction is included. It will change with the time and the operator T will describe the change.

According to Chapter 8

$$-\frac{\hbar}{\mathrm{i}}\frac{\mathrm{d}\mathsf{T}}{\mathrm{d}t} = \mathsf{HT} = \mathsf{H}_0\mathsf{T} + \mathsf{H}_1\mathsf{T}. \tag{12.9}$$

To treat H_1 as a perturbation define

$$\mathsf{T}_0 = \exp\left(\frac{\mathrm{i}}{\hbar} H_0 t\right)\mathsf{T}. \tag{12.10}$$

This change of operator is helpful because $\exp\{(\mathrm{i}/\hbar)H_0 t\}$ can be used in a convenient way with the eigenkets of H_n, such as (12.8).
Then

$$\frac{\mathrm{d}\mathsf{T}_0}{\mathrm{d}t} = -\frac{\mathrm{i}}{\hbar}\mathsf{H}_1^{(0)}\mathsf{T}_0 \tag{12.10a}$$

where $\mathsf{H}_1^{(0)} = \exp\{(\mathrm{i}/\hbar)H_0 t\}\mathsf{H}_1 \exp\{-(\mathrm{i}/\hbar)H_0 t\}$. Now using the eigenkets (12.8) as the basic kets

$$\frac{\mathrm{d}}{\mathrm{d}t}\langle n'', n_s'' \,|\, \mathsf{T}_0 \,|\, n', n_s'\rangle = -\frac{\mathrm{i}}{\hbar}\langle n'', n_s'' \,|\, \mathsf{H}_1^{(0)} \,|\, n''', n_s'''\rangle \langle n''', n_s''' \,|\, \mathsf{T}_0 \,|\, n', n_s'\rangle =$$

$$= -\frac{\mathrm{i}}{\hbar}\exp\left\{\frac{\mathrm{i}}{\hbar}\left[(n''-n''')\hbar\omega + \sum_s (n''-n_s''')\hbar\omega\right]t\right\} \times$$

$$\times \langle n''n_s'' \,|\, \mathsf{H}_1 \,|\, n'''n_s'''\rangle \langle n'''n_s''' \,|\, \mathsf{T}_0 \,|\, n'n_s'\rangle \qquad (12.11)$$

where a sum over (n''', n'''_s) is implied, and n_s' represents the infinity of numbers n_s' for all values of s.

At $t = 0$, $\mathsf{T}_0 = \mathsf{T}$ and is a unit matrix, so the matrix product on the right-hand side of (12.11) contains only one term. Hence the equation can be integrated and leads to

$$\langle n'', n_s'' \,|\, \mathsf{T}_0 \,|\, n', n_s'\rangle = \delta_{n'', n'}\delta_{ns'', n's} +$$

$$+ \frac{1 - \exp\{(\mathrm{i}/\hbar)\left[(n''-n')\hbar\omega + \sum_s (n_s''-n_s')\hbar\omega_s\right]t\}}{\left[(n''-n')\hbar\omega + \sum_s (n_s''-n_s')\hbar\omega_s\right]}\langle n'', n_s'' \,|\, \mathsf{H}_1 \,|\, n', n_s'\rangle. \qquad (12.12)$$

Of course this is an approximation valid for only a short time, since the initial value of T_0 has been used in equation (12.11).

With equation (12.12) the problem is reduced to that of evaluating the matrix elements of H_1. For that purpose it is necessary to known the effects of the various operators when operating on ket vectors. By the methods of the previous chapters it can be shown that the following relations hold:

$$\mathsf{a}^\dagger \,|\, n'\rangle = \mathrm{i}(n'+1)^{\frac{1}{2}} \,|\, n'+1\rangle \qquad (12.13\mathrm{a})$$

$$\mathsf{a} \,|\, n'\rangle = -\mathrm{i}n'^{\frac{1}{2}} \,|\, n'-1\rangle \qquad (12.13\mathrm{b})$$

$$\mathsf{R}_s^\dagger \,|\, n_s'\rangle = \mathrm{i}(n_s'+1)^{\frac{1}{2}} \,|\, n_s'+1\rangle \qquad (12.13\mathrm{c})$$

$$\mathsf{R}_s \,|\, n_s'\rangle = -\mathrm{i}n_s'^{\frac{1}{2}} \,|\, n_s'-1\rangle. \qquad (12.13\mathrm{d})$$

The kets on which the operators act may also be eigenkets of other operators. Only the eigenvalues affected are indicated in the above expressions.

As an example of the use of equation (12.12) consider only the first sum in the interaction Hamiltonian (12.7). Let this first part be called $\mathsf{H}_{1,1}$. Then

$$\langle n'', n_r'' \,|\, \mathsf{H}_{1,1} \,|\, n', n_r'\rangle = \tfrac{1}{2}\gamma h(m\omega M\omega_r)^{\frac{1}{2}}[\langle n'', n_r'' \,|\, (\mathsf{a}-\mathsf{a}^\dagger)\mathsf{R}_r \,|\, n', n_r'\rangle\, \mathrm{e}^{2\pi \mathrm{i} r z_0/L}$$

$$- \langle n'', n_r'' \,|\, (\mathsf{a}-\mathsf{a}^\dagger)\mathsf{R}_r^\dagger \,|\, n', n_r'\rangle\, \mathrm{e}^{-2\pi \mathrm{i} r z_0/L}] =$$

$$= \tfrac{1}{2}\gamma\hbar(m\omega M\omega_r)^{\frac{1}{2}}[n'^{\frac{1}{2}}\delta(n''-n'+1)+(n'+1)^{\frac{1}{2}}\delta(n''-n'-1)] \times$$
$$\times\ [e^{2\pi irz_0/L}n_r'^{\frac{1}{2}}\delta(n_r''-n_r'+1)+e^{-2\pi irz_0/L}(n_r'+1)^{\frac{1}{2}}\delta(n_r''-n_r'-1)]\,. \quad (12.14)$$

In this expression the designation of the matrix element implies that all of the $n_s'' = n_s'$ except for the particular case in which $s = r$. Since $n_r'' \neq n_r'$ the term in $\mathsf{H}_{1,1}$ for which $s = r$ will give the results of equation (12.14). All other terms in the sum will give zero.

The various δ functions show that for each value of r there are four cases:

$$\langle n'-1, n_r'-1 \mid \mathsf{H}_{1,1} \mid n', n_r'\rangle = -\gamma[\tfrac{1}{2}\hbar(mM\omega\omega_r)^{\frac{1}{2}}]n'^{\frac{1}{2}}n_r'^{\frac{1}{2}}\, e^{2\pi irz_0/L}, \quad (12.14a)$$

$$\langle n'+1, n_r'-1 \mid \mathsf{H}_{1,1} \mid n', n_r'\rangle = -\gamma[\tfrac{1}{2}\hbar(mM\omega\omega_r)^{\frac{1}{2}}]\,(n'+1)^{\frac{1}{2}}n_r'^{\frac{1}{2}}\, e^{2\pi irz_0/L}, \quad (12.14b)$$

$$\langle n'-1, n_r'+1 \mid \mathsf{H}_{1,1} \mid n', n_r'\rangle = \gamma[\tfrac{1}{2}\hbar(mM\omega\omega_r)^{\frac{1}{2}}]n'^{\frac{1}{2}}(n_r'+1)^{\frac{1}{2}}\, e^{-2\pi irz_0/L}, \quad (12.14c)$$

$$\langle n'+1, n_r'+1 \mid \mathsf{H}_{1,1} \mid n', n_r'\rangle = \gamma[\tfrac{1}{2}\hbar(mM\omega\omega_r)^{\frac{1}{2}}]\,(n'+1)^{\frac{1}{2}}(n_r'+1)^{\frac{1}{2}}\, e^{-2\pi irz_0/L}. \quad (12.14d)$$

One can now consider a number of particular problems. The simplest is the case in which the initial state is represented by all the $n_s' = 0$ and $n' \neq 0$. This implies that the string is at rest as far as is permitted by the quantum conditions. Each mode of vibration is in its state of lowest energy, $n_s' = 0$. On the other hand, the particle is oscillating in an energy eigenstate, $n' \neq 0$. The matrix elements of $\mathsf{H}_{1,1}$ which may differ from zero are those given by equations (12.14c) and (12.14d) for all values of r. The matrix elements given by (12.14a) and (12.14b) cannot appear since each n_r' is already zero and the state $\mid n_r'-1\rangle$ does not exist. Consider first then the matrix elements of T_0 that come from equation (12.14c)

$$\langle n'+1, n_r'' = 1 \mid \mathsf{T}_0 \mid n', n_r' = 0\rangle =$$
$$= \gamma(4m\omega M\omega_r)^{-\frac{1}{2}}(n'+1)^{\frac{1}{2}}\, e^{-2\pi irz_0/L}[1-e^{i(\omega+\omega_r)t}]/(\omega+\omega_r)\,. \quad (12.15)$$

Since all the ω_r are positive, this function will oscillate rapidly with t and the denominator will keep the maximum value small. In this way the probability that both the string and the particle show an increase in energy is shown to be negligible, since, when applied to the ket vector of the initial state, this element of T_0 will not produce for the time t a ket containing a significant amount of $\mid n'+1, n_r'' = 1\rangle$.

On the other hand

$$\langle n'-1, n_r'' = 1 \mid \mathsf{T}_0 \mid n', n_r' = 0\rangle =$$
$$= \gamma(4m\omega M\omega_r)^{-\frac{1}{2}}n'^{\frac{1}{2}}\, e^{-2\pi irz_0/L}[1-e^{-i(\omega-\omega_r)t}]/(\omega-\omega_r)\,. \quad (12.16)$$

In this case there is an ω_r such that the denominator is very small and the function oscillates only with a very long period. The matrix element may then attain a significant size.

It is through the resonance denominator of equation (12.16) that the conservation of energy makes itself apparent. When $\omega = \omega_r$ the energy lost by the vibrating mass is $\hbar\omega$ and is just equal to the energy gained by the string, $\hbar\omega_r$. The total energy of the whole system is a constant of the motion. The changes induced by the interaction H_1 are redistributions of the energy between the particle and the string.

However the fact that some vibrations of the string for which ω_r is not exactly equal to ω are excited, shows that the energy ascribed to the individual parts of the system is not exactly conserved. The energy of the zero approximation is *not* the whole energy of the system. The interaction energy H_1 must be included also. When the whole energy, $H_0 + H_1$, is considered the initial state is not an energy eigenstate but represents a distribution of energy over the possible eigenenergies of the whole system. This distribution over the eigenenergies does not change with the time although the distribution among the components of the system does change. Some idea of the energy distribution could be obtained by evaluating the mean of $(H_0 + H_1)^2$:

$$
\begin{aligned}
| S, t\rangle &= \mathsf{T} \,| S, 0\rangle = \mathrm{e}^{-(\mathrm{i}/\hbar)H_0 t}\mathsf{T}_0 \,| n', n_r' = 0\rangle = \\
&= \mathrm{e}^{-(\mathrm{i}/\hbar)E_0 t} \,| n', n_r' = 0\rangle + \gamma \sum_{r=-\infty}^{\infty} (4m\omega M\omega_r)^{-\frac{1}{2}} \,\mathrm{e}^{-2\pi \mathrm{i} r z_0/L} n'^{\frac{1}{2}} \,\mathrm{e}^{-(\mathrm{i}/\hbar)E_r t} \times \\
&\quad \times [(1-\mathrm{e}^{-\mathrm{i}(\omega-\omega_r)t})/(\omega-\omega_r)] \,| n'-1, n_r' = 1\rangle . \qquad (12.17)
\end{aligned}
$$

In this expression E_0 represents the H_0 part of the energy of the initial state. It includes $(n' + \frac{1}{2})\hbar\omega$ and the zero point energy of the string, $\frac{1}{2} \sum_r \hbar\omega_r$. E_r is the H_0 part of the energy of the state with $n_r' = 1$, all the remaining $n_s' = 0$, and with the motion of the particle described by $n' - 1$:

$$E_r = E_0 + \hbar\omega_r - \hbar\omega . \qquad (12.18)$$

The ket vector in (12.17) may be considered as composed of two parts. The first part $\exp\{-(\mathrm{i}/\hbar)E_0 t\} \,| n', n_r' = 0\rangle$ shows no effect of the interaction term in the Hamiltonian. It is just the time dependent ket vector of the particle and the string as though they were unconnected. The remainder is small since it is multiplied by γ. In fact it is correct only to the extent that it is small since it represents only the first time derivative of the initial state. Since the initial state is assumed to be normalized the whole state of equation (12.17) is not normalized. Higher order approximations would introduce a reduction in this first part and the exact solution would preserve the normalization.

In spite of the approximate nature of (12.17) it can be used to find the probability that a quantum of energy has left the particle and gone into the string by the time t. This involves only the evaluation of $\langle \mathsf{n} \rangle$ as a function of the time:

$$\langle s, t \mid \mathsf{n} \mid s, t\rangle = \langle s, t \mid \mathsf{a}^\dagger \mathsf{a} \mid s, t\rangle + \\ + \langle s, 0 \mid \mathsf{a}^\dagger \mathsf{a} \mid s, 0\rangle + \langle s, p \mid \mathsf{a}^\dagger \mathsf{a} \mid s, 0\rangle + \langle s, 0 \mid \mathsf{a}^\dagger \mathsf{a} \mid s, p\rangle + \langle s, p \mid \mathsf{a}^\dagger \mathsf{a} \mid s, p\rangle \tag{12.19}$$

where $\mid s, p\rangle$ represents the second part of the ket vector of (12.17). The first term on the right-hand side of (12.19) gives n', the initial value of n which is unchanged in this approximation. The second and third terms vanish since all the $n_r' = 0$ in $\mid s, 0\rangle$. Only the last term gives an approximate result:

$$\langle s, p \mid \mathsf{a}^\dagger \mathsf{a} \mid s, p\rangle = \\ = \gamma^2 \sum_{r=-\infty}^{\infty} (2m\omega M\omega_r)^{-\frac{1}{2}} n'(n'-1)\,[1 - \cos{(\omega-\omega_r)t}]/(\omega-\omega_r)^2 . \tag{12.20}$$

The ket vector $\mid s, p\rangle$ is an eigenket of n with the eigenvalue $(n'-1)$. The probability that an observation of the oscillator would give $(n'-1)$ for the value of n is then just given by the coefficient of $(n'-1)$ in (12.20). Thus

$$P(t) = \gamma^2 n' \sum_{r=-\infty}^{\infty} (2m\omega M\omega_r)^{-1} [1 - \cos{(\omega-\omega_r)t}]/(\omega-\omega_r)^2 . \tag{12.21}$$

$P(t)$ is the probability that by the time t a quantum of energy has moved from the particle to a mode of vibration of the string. Equation (12.21) shows that this probability is a sum of probabilities for the different modes of vibration, but it also shows that the probability is large for only those modes whose ω_r is close to ω.

When the string is long enough, the sum over the values of r can be replaced by an integral over ω_r. Thus

$$P(t) = \gamma^2 n' \frac{1}{4\pi m\omega} \left(\frac{L}{M}\right)^{\frac{1}{2}} \int_0^\infty \frac{1 - \cos{(\omega-\omega_r)t}}{\omega_r(\omega-\omega_r)^2}\, d\omega_r . \tag{12.22}$$

The integrand has a sharp peak at $\omega_r = \omega$, and unless $\omega < 0$ the integral can be evaluated to give

$$P(t) = n'\gamma^2 \frac{1}{4m\omega^2} \left(\frac{L}{M\tau}\right)^{\frac{1}{2}} t . \tag{12.23}$$

Because of the proportionality to the time, it is appropiate to call the coefficient of t a transition probability. In this case it is a spontaneous transition probability proportional to n', the energy at first in the oscillator.

If instead of considering an initial state in which all of the $n_r' = 0$, a more general situation is envisaged, all of the matrix elements in (12.14a),..., (12.14d) may be different from zero. Then

$$
\begin{aligned}
| Q, t\rangle = \mathrm{e}^{-(\mathrm{i}/\hbar)E_0 t} | Q, 0\rangle - \mathrm{e}^{-(\mathrm{i}/\hbar)E_0 t}\gamma \left(\frac{1}{2m\omega}\right)^{\frac{1}{2}} \sum_{r=-\infty}^{\infty} \left(\frac{1}{2M\omega_r}\right)^{\frac{1}{2}} \times \\
\times \Bigg[\mathrm{e}^{2\pi \mathrm{i} r z_0/L} \frac{1-\mathrm{e}^{\mathrm{i}(\omega+\omega_r)t}}{(\omega+\omega_r)} (n' n_r')^{\frac{1}{2}} \, | n'-1, n_r'-1\rangle + \\
+ \mathrm{e}^{2\pi \mathrm{i} r z_0/L} \frac{1-\mathrm{e}^{\mathrm{i}(\omega-\omega_r)t}}{(\omega-\omega_r)} [n'(n_r'+1)]^{\frac{1}{2}} \, | n'-1, n_r'+1\rangle + \\
+ \mathrm{e}^{-2\pi \mathrm{i} r z_0/L} \frac{1-\mathrm{e}^{-\mathrm{i}(\omega+\omega_r)t}}{(\omega+\omega_r)} [(n'+1)(n_r'+1)]^{\frac{1}{2}} \, | n'+1, n_r'+1\rangle + \\
+ \mathrm{e}^{-2\pi \mathrm{i} r z_0/L} \frac{1-\mathrm{e}^{-\mathrm{i}(\omega-\omega_r)t}}{(\omega-\omega_r)} [(n'+1)n_r']^{\frac{1}{2}} \, | n'+1, n_r'-1\rangle\Bigg]. \qquad (12.24)
\end{aligned}
$$

The probability that the oscillator has given up a quantum of energy to the string is the sum of the squares of the absolute values of those ket vectors with $(n'-1)$. Since all the ω's are positive those terms with $(\omega+\omega_r)$ in the denominator can be ignored so that

$$
\begin{aligned}
W(t) &= \frac{4\gamma^2}{2m\omega} \sum_{r=-\infty}^{\infty} \frac{1}{2M\omega_r} \frac{\sin^2[\frac{1}{2}(\omega-\omega_r)t]}{(\omega-\omega_r)^2} n'(n_r'+1)t \\
&= n'(\overline{n_r'}+1) \frac{\gamma^2}{4m\omega^2} \left(\frac{L}{M}\right)^{\frac{1}{2}} t. \qquad (12.25)
\end{aligned}
$$

The use of a mean value of n, written $\overline{n_r'}$, is based on the assumption that the initial state ket vector is an eigenvector of the various operators n_r with eigenvalues which vary with r in such a way that a mean value $\overline{n_r'}$ can be taken outside the integral.

The transition probability in equation (12.25) differs from that in (12.24) in containing the factor $(\overline{n_r'}+1)$. This is the slightly surprising result that the probability that the oscillator will transfer energy to the string is greater when the string already has energy than when it has not. However a similar situation occurs in a classical treatment, since the work done per unit time on a moving string is proportional to the force multiplied by the velocity of the string. When the force and the velocity have the proper phase relationship, energy is transferred to the string. The total transition probability thus consists of a "spontaneous" transition probability and a "stimulated" or "forced" transition probability which is proportional to $\overline{n'_r}$ in the neighborhood of $\omega_r = \omega$.

The inverse problem in which the energy is transferred from the string to the oscillator can be treated in a similar manner.

PROBLEM 1. Compute the probability of absorption of energy by the oscillator when one of the modes of vibration of the string is the only one initially excited.

PROBLEM 2. In the general situation treated above compute the probability that the oscillator will gain energy from the string and compare it with the probability that the oscillator will lose energy.

2. The Hamiltonian treatment of a system composed of charged particles and the radiation field

To describe the combined system it is necessary to use the coordinates and momenta of the field, $Q_{\boldsymbol{K},\lambda,\mu}$ and $P_{\boldsymbol{K},\lambda,\mu}$, defined in the preceding chapter, as well as the ordinary coordinates and Hamiltonian momenta of the charged particles. The Hamiltonian function will consist of terms giving the energy of the field, terms giving the energy of the particles, and terms giving the interaction between the two. The Hamiltonian function is in the nature of a postulate, but it can be built up from a knowledge of classical electromagnetic theory.

For the treatment of problems in which the velocity of the particles is much less than the velocity of light, the Hamiltonian function may be taken to be

$$H = \sum_{\boldsymbol{K},\lambda,\mu} \frac{1}{\varepsilon_0} [P^2_{\boldsymbol{K},\lambda,\mu} + \tfrac{1}{4}\varepsilon_0^2 \omega_{\boldsymbol{K}}^2 Q^2_{\boldsymbol{K},\lambda,\mu}] + \frac{1}{4\pi\varepsilon_0} \sum_{j,j'} \frac{e_j e_{j'}}{r_{j,j'}} +$$

$$+ \sum_{j,\alpha} \frac{1}{2m_j} \left\{ p_{j,\alpha} - \frac{e_j}{V^{\frac{1}{2}}} \sum_{\boldsymbol{K},\lambda} [Q_{\boldsymbol{K},\lambda,1} \cos(\boldsymbol{K}\cdot\boldsymbol{r}_j) + Q_{\boldsymbol{K},\lambda,2} \sin(\boldsymbol{K}\cdot\boldsymbol{r}_j)] \varepsilon_{\boldsymbol{K},\lambda,\alpha} \right\}^2 . \tag{12.26}$$

The different particles are designated by j, and α takes on three values corresponding to the axes of rectangular coordinates. $\boldsymbol{\varepsilon}_{\boldsymbol{K},\lambda,\alpha}$ is the α component of the unit vector $\boldsymbol{\varepsilon}_{\boldsymbol{K},\lambda}$.

The first sum in equation (12.26) is the energy of that part of the electromagnetic field which can be called radiation. It represents a field derived from a vector potential whose divergence is zero. The second sum represents the mutual potential energy of the charges when they are at rest, or moving with low velocities. There is omitted from this term the electrostatic field energy due to each particle alone. The third term includes the kinetic energy of the particles and the interaction between the particles and the field. As indicated above, this Hamiltonian is incomplete. It is not suitable for the treatment of very high energy phenomena, but it does give the basis for the emission, absorption, and scattering of light, X-rays, and gamma rays of moderate energy.

To show that (12.26) is a suitable Hamiltonian one must show that it leads to the proper equations of motion. The first set of canonical equations leads to

$$\frac{\mathrm{d}Q_{\boldsymbol{K},\lambda,\mu}}{\mathrm{d}t} = \frac{\partial H}{\partial P_{\boldsymbol{K},\lambda,\mu}} = \frac{2}{\varepsilon_0} P_{\boldsymbol{K},\lambda,\mu} \tag{12.27a}$$

$$\frac{\mathrm{d}r_{j,\alpha}}{\mathrm{d}t} = \frac{\partial H}{\partial p_{j,\alpha}} =$$

$$= \frac{1}{m_j}\left\{p_{j,\alpha} - \frac{e_j}{V^{\frac{1}{2}}} \sum_{\boldsymbol{K},\lambda} [Q_{\boldsymbol{K},\lambda,1} \cos(\boldsymbol{K}\cdot\boldsymbol{r}_j) + Q_{\boldsymbol{K},\lambda,2} \sin(\boldsymbol{K}\cdot\boldsymbol{r}_j)]\varepsilon_{\boldsymbol{K},\lambda,\alpha}\right\}. \tag{12.27b}$$

These equations serve to express $P_{\boldsymbol{K},\lambda,\mu}$ and $P_{j,\alpha}$ in terms of the coordinates and their time derivatives. The other canonical equations lead to

$$\frac{\mathrm{d}P_{\boldsymbol{K},\lambda,\mu}}{\mathrm{d}t} = -\frac{\partial H}{\partial Q_{\boldsymbol{K},\lambda}\begin{pmatrix}1\\2\end{pmatrix}} = -\tfrac{1}{2}\varepsilon_0\omega_K^2 Q_{\boldsymbol{K},\lambda}\begin{pmatrix}1\\2\end{pmatrix} + \sum_j \frac{e_j}{V^{\frac{1}{2}}}\dot{\boldsymbol{r}}_j\cdot\boldsymbol{\varepsilon}_{\boldsymbol{K},\lambda}\begin{pmatrix}\cos(\boldsymbol{K}\cdot\boldsymbol{r}_j)\\ \sin(\boldsymbol{K}\cdot\boldsymbol{r}_j)\end{pmatrix} \tag{12.28a}$$

$$\frac{\mathrm{d}p_{j,\alpha}}{\mathrm{d}t} = -\frac{\partial H}{\partial r_{j,\alpha}} = -\frac{e_j}{V^{\frac{1}{2}}}\sum_{\boldsymbol{K},\lambda} K_\alpha(\dot{\boldsymbol{r}}_j\cdot\boldsymbol{\varepsilon}_{\boldsymbol{K},\lambda})\,[Q_{\boldsymbol{K},\lambda,1}\sin(\boldsymbol{K}\cdot\boldsymbol{r}_j)+$$

$$-\,Q_{\boldsymbol{K},\lambda,2}\cos(\boldsymbol{K}\cdot\boldsymbol{r}_j)] - \frac{1}{4\pi\varepsilon_0}\frac{\partial}{\partial r_{j,\alpha}}\left[\sum_{j<j'}\frac{e_j e_{j'}}{r_{j,j'}}\right]. \tag{12.28b}$$

Equation (12.28b) together with (12.27b) gives a differential equation of motion of the particles which shows that

$$m_j(\mathrm{d}^2\boldsymbol{r}_j/\mathrm{d}t^2) = e_j(E+\dot{\boldsymbol{r}}\times\boldsymbol{B}). \tag{12.29}$$

Similarly it can be shown that within the limits already indicated, the changes in the field due to the particles are properly described.

3. Quantum mechanical treatment of the combined system

Emission, absorption, and scattering of radiation are brought about because of the interaction terms in the Hamiltonian function for the complete system. If there were no such interaction, there could be no transfer of energy between the matter and the radiation field, and the two parts of the system would exist entirely independent of each other. Since the interaction is in fact small, it can be treated as a perturbation for many purposes, and the energy eigenstates of the system without the interaction serve well as a set of basis functions. The interaction is then regarded as causing transitions, or quantum jumps, between these basis states.

Although the description of radiation, absorption, and scattering in terms of transitions between stationary states is useful and satisfactory for many purposes, it must also be remembered that the wave function or ket vector describing the system changes continuously with the time. The function represents a single pure state, but it is expressed as a sum of eigenstates of the Hamiltonian with the interaction omitted and it can be interpreted as giving the probability of finding the system in these various states used to describe it when the interaction is ignored. When there exists such a probability of the system being in a state different from the one in which it was at $t = 0$, that probability can be considered as the probability that the system has "jumped", during the time t, from the initial to the particular basis state. Thus, for example, the emission of radiation is treated by computing the probability that the system has jumped, during the time under consideration, from a state in which the material part has more than its minimum energy to a state in which the material part has less and the radiation field more energy than originally. It will appear as a result of the computation that such a probability has an appreciable value only in case the gain in energy by the field is just equal to the loss in energy by the material system, and furthermore that the gain in momentum by the field is equal to the loss in momentum by the material system. Thus the computation refers to interchanges of partial energies between different parts of the whole system, while the whole system itself is in a pure state which has developed continuously from the initial state. The state is not an energy eigenstate and the total energy, including the interaction energy, is not sharply defined. The momentum, however, is sharply defined and the interaction energy operator, H_1, commutes with the operator for the total momentum of the system, particles plus field.

3.1. *The interaction operator*

For treating the emission and absorption of radiation to the first approximation, only the cross product terms in the Hamiltonian operator are effective. For each particle (j) there is a term

$$\mathsf{H}_1^{(j)} = -\frac{e_j}{m_j V^{\frac{1}{2}}} \sum_{\boldsymbol{K},\lambda} \mathbf{p}_j \cdot \boldsymbol{\varepsilon}_{\boldsymbol{K},\lambda} [\mathsf{Q}_{\boldsymbol{K},\lambda,1} \cos(\boldsymbol{K}\cdot\mathbf{r}_j) + \mathsf{Q}_{\boldsymbol{K},\lambda,2} \sin(\boldsymbol{K}\cdot\mathbf{r}_j)]. \tag{12.30}$$

The quantities $\mathbf{p}_j$, $\mathbf{r}_j$, $\mathsf{Q}_{\boldsymbol{K},\lambda,1}$ and $\mathsf{Q}_{\boldsymbol{K},\lambda,2}$ are operators, but the only question about commutation arises in connection with $\mathbf{p}_j$ and $\mathbf{r}_j$. To be sure of having an operator that is Hermitian, both orders of all of the products can be taken. Also the $\mathsf{Q}_{\boldsymbol{K},\lambda,\mu}$ are replaced by

$$\mathsf{Q}_{\boldsymbol{K},\lambda,1} = -\mathrm{i}\left(\frac{\hbar}{2\varepsilon_0\omega_{\boldsymbol{K}}}\right)^{\frac{1}{2}} [\mathsf{R}^\dagger_{\boldsymbol{K},\lambda} - \mathsf{R}_{\boldsymbol{K},\lambda} + \mathsf{R}^\dagger_{-\boldsymbol{K},\lambda} - \mathsf{R}_{-\boldsymbol{K},\lambda}] \tag{12.31a}$$

$$\mathsf{Q}_{\boldsymbol{K},\lambda,2} = -\left(\frac{\hbar}{2\varepsilon_0\omega_{\boldsymbol{K}}}\right)^{\frac{1}{2}} [\mathsf{R}^\dagger_{\boldsymbol{K},\lambda} + \mathsf{R}_{\boldsymbol{K},\lambda} - \mathsf{R}^\dagger_{-\boldsymbol{K},\lambda} - \mathsf{R}_{-\boldsymbol{K},\lambda}]. \tag{12.31b}$$

Then

$$\mathsf{H}_1^{(j)} = \frac{\mathrm{i}}{2V^{\frac{1}{2}}} \frac{e_j}{m_j} \sum_{\boldsymbol{K},\lambda} \left(\frac{\hbar}{2\varepsilon_0\omega_K}\right)^{\frac{1}{2}} \times$$

$$\times [\mathsf{R}^\dagger_{\boldsymbol{K},\lambda}(\mathbf{p}_j \cdot \boldsymbol{\varepsilon}_{\boldsymbol{K},\lambda} \mathrm{e}^{-\mathrm{i}\boldsymbol{K}\cdot\mathbf{r}_j} + \mathrm{e}^{-\mathrm{i}\boldsymbol{K}\cdot\mathbf{r}_j}\mathbf{p}_j \cdot \boldsymbol{\varepsilon}_{\boldsymbol{K},\lambda}) +$$

$$- \mathsf{R}_{\boldsymbol{K},\lambda}(\mathbf{p}_j \cdot \boldsymbol{\varepsilon}_{\boldsymbol{K},\lambda} \mathrm{e}^{\mathrm{i}\boldsymbol{K}\cdot\mathbf{r}_j} + \mathrm{e}^{\mathrm{i}\boldsymbol{K}\cdot\mathbf{r}_j}\mathbf{p}_j \cdot \boldsymbol{\varepsilon}_{\boldsymbol{K},\lambda})]. \tag{12.32}$$

The sum over $\boldsymbol{K}$ in equation (12.32) is over the *whole* solid angle, and the terms with $-\boldsymbol{K}$ are not otherwise specifically indicated.

From the commutation relationships between $\mathbf{p}$ and $\mathbf{r}$ it can be shown that

$$\mathbf{p}_j \mathrm{e}^{\mathrm{i}\boldsymbol{K}\cdot\mathbf{r}_j} = \mathrm{e}^{\mathrm{i}\boldsymbol{K}\cdot\mathbf{r}_j}(\mathbf{p}_j + \hbar\boldsymbol{K}). \tag{12.33}$$

But since $\boldsymbol{K}\cdot\boldsymbol{\varepsilon}_{\boldsymbol{K},\lambda} = 0$ for both values of λ, the lack of commutation is unimportant here and we have

$$\mathsf{H}_1^{(j)} = \frac{\mathrm{i}}{V^{\frac{1}{2}}} \frac{e_j}{m_j} \sum_{\boldsymbol{K},\lambda} \left(\frac{\hbar}{2\varepsilon_0\omega_K}\right)^{\frac{1}{2}} (\mathsf{R}^\dagger_{\boldsymbol{K},\lambda} \mathrm{e}^{-\mathrm{i}\boldsymbol{K}\cdot\mathbf{r}_j} - \mathsf{R}_{\boldsymbol{K},\lambda} \mathrm{e}^{\mathrm{i}\boldsymbol{K}\cdot\mathbf{r}_j})\mathbf{p}_j \cdot \boldsymbol{\varepsilon}_{\boldsymbol{K},\lambda}. \tag{12.34}$$

3.2. *Radiation from an atom*

To illustrate the use of the operator of equation (12.34) consider a simple system of two particles, such as a hydrogen atom in the radiation field. Neglect the spin of the particles. Then let $\boldsymbol{R}$ be the vector coordinate of the center of mass and let $\boldsymbol{r}$ be the vector from particle 2 to particle 1. With these definitions

$$\boldsymbol{r}_1 = \boldsymbol{R} + \frac{m_2}{m_1+m_2}\boldsymbol{r}, \qquad \boldsymbol{p}_1 = \frac{m_0}{m_2}\boldsymbol{P}+\boldsymbol{p}$$

$$\boldsymbol{r}_2 = \boldsymbol{R} - \frac{m_1}{m_1+m_2}\boldsymbol{r}, \qquad \boldsymbol{p}_2 = \frac{m_0}{m_1}\boldsymbol{P}-\boldsymbol{p}. \tag{12.35}$$

$\boldsymbol{P}$ and $\boldsymbol{p}$ are the momenta conjugate to $\boldsymbol{R}$ and $\boldsymbol{r}$ respectively; and it will be convenient to define a reduced mass, $m_0 = m_1 m_2/(m_1+m_2)$. Then if $e_1 = -e$ and $e_2 = +e$, the operator for the interaction of the two particles with the field is the sum of two terms such as equation (12.34),

$$\begin{aligned} \mathsf{H}_1 = -\frac{ie}{V^{\frac{1}{2}}} \sum_{\boldsymbol{K},\lambda} \left(\frac{\hbar}{2\varepsilon_0\omega_{\boldsymbol{K}}}\right)^{\frac{1}{2}} \Bigg\{ \Bigg[&\mathsf{R}^\dagger_{\boldsymbol{K},\lambda} \exp\left(-i\boldsymbol{K}\cdot\boldsymbol{R} - i\frac{m_2}{m_1+m_2}\boldsymbol{K}\cdot\boldsymbol{r}\right) \\ &- \mathsf{R}_{\boldsymbol{K},\lambda} \exp\left(i\boldsymbol{K}\cdot\boldsymbol{R} + i\frac{m_2}{m_1+m_2}\boldsymbol{K}\cdot\boldsymbol{r}\right)\Bigg] \frac{1}{m_1}\left[\frac{m_0}{m_2}\boldsymbol{P}+\boldsymbol{p}\right]\cdot\varepsilon_{\boldsymbol{K},\lambda} \\ &- \Bigg[\mathsf{R}^\dagger_{\boldsymbol{K},\lambda} \exp\left(-i\boldsymbol{K}\cdot\boldsymbol{R} + i\frac{m_1}{m_1+m_2}\boldsymbol{K}\cdot\boldsymbol{r}\right) \\ &- \mathsf{R}_{\boldsymbol{K},\lambda} \exp\left(i\boldsymbol{K}\cdot\boldsymbol{R} - i\frac{m_1}{m_1+m_2}\boldsymbol{K}\cdot\boldsymbol{r}\right)\Bigg] \frac{1}{m_2}\left[\frac{m_0}{m_1}\boldsymbol{P}-\boldsymbol{p}\right]\cdot\varepsilon_{\boldsymbol{K},\lambda}\Bigg\}. \end{aligned} \tag{12.36}$$

Since the two particles are held together, in this case by their opposite charges and in the case of nuclear particles by other forces, the vector $\boldsymbol{r}$ will rarely be vary large. The quantity $\boldsymbol{K}\cdot\boldsymbol{r}$ will usually be less than one so the exponential in $\boldsymbol{K}\cdot\boldsymbol{r}$ can be expanded and only the first few terms retained. In this way the operator can be rearranged to be

$$\begin{aligned} \mathsf{H}_1 = -\frac{ie}{V^{\frac{1}{2}}} \sum_{\boldsymbol{K},\lambda} \left(\frac{\hbar}{2\varepsilon_0\omega_{\boldsymbol{K}}}\right)^{\frac{1}{2}} \Big\{ &[\mathsf{R}^\dagger_{\boldsymbol{K},\lambda} e^{-i\boldsymbol{K}\cdot\boldsymbol{R}} - \mathsf{R}_{\boldsymbol{K},\lambda} e^{i\boldsymbol{K}\cdot\boldsymbol{R}}] \frac{1}{m_0}\boldsymbol{p}\cdot\varepsilon_{\boldsymbol{K},\lambda} + \\ &- i[\mathsf{R}^\dagger_{\boldsymbol{K},\lambda} e^{-i\boldsymbol{K}\cdot\boldsymbol{R}} + \mathsf{R}_{\boldsymbol{K},\lambda} e^{i\boldsymbol{K}\cdot\boldsymbol{R}}]\boldsymbol{K}\cdot\boldsymbol{r}\left[\frac{\boldsymbol{P}}{m_1+m_2} + \frac{\boldsymbol{p}(m_2-m_1)}{m_1 m_2}\right]\cdot\varepsilon_{\boldsymbol{K},\lambda}\Big\}. \end{aligned} \tag{12.37}$$

The term containing $\boldsymbol{p} \cdot \varepsilon_{\boldsymbol{K},\lambda}$ gives rise to "dipole" radiation or absorption. The term in the next line containing $(\boldsymbol{K} \cdot \boldsymbol{r})(\boldsymbol{P} \cdot \varepsilon_{\boldsymbol{K},\lambda})$ might also be regarded as giving rise to dipole radiation since only the first power of $\boldsymbol{r}$ is present. However the presence of $\boldsymbol{P}$ indicates that the importance of this term depends on the motion of the center of mass of the system (atom). In most cases this is small compared with the internal motion described by $\boldsymbol{p}$ and the term can be ignored. The remaining term containing $(\boldsymbol{K} \cdot \boldsymbol{r})(\boldsymbol{p} \cdot \varepsilon_{\boldsymbol{K},\lambda})$ gives rise to quadrupole radiation or to magnetic dipole radiation. It is clearly zero when $m_1 = m_2$ but for atoms it may be significant, especially at the higher frequencies.

PROBLEM 3. Show that equation (12.33) follows from the commutation rules as well as from the Schroedinger representation of the momentum operator. Show also that $\exp(\mathrm{i}\boldsymbol{K}\cdot\boldsymbol{R})\,|\,\boldsymbol{P}'\rangle = |\,\boldsymbol{P}'+\hbar\boldsymbol{K}\rangle$.

3.3. *The dipole transitions*

To compute the probability of a dipole transition consider the initial state to be that in which the relative coordinate $\boldsymbol{r}$ is described by an energy eigenstate with eigenvalue E', the center of mass motion by an eigenfunction of the momentum $\boldsymbol{P}$ with eigenvalue $\boldsymbol{P}'$, and each mode of the radiation field by an energy eigenstate designated by $n'_{\boldsymbol{K},\lambda}$. The first step is then to evaluate the matrix elements between this initial state and the various "final" states described by other eigenvalues of the same operators. Then

$$\langle E'', \boldsymbol{P}'', n''_{\boldsymbol{K},\lambda}\,|\,\mathsf{H}_1\,|\,E', \boldsymbol{P}', n'_{\boldsymbol{K},\lambda}\rangle =$$

$$= -\frac{\mathrm{i}e}{m_0 V^{\frac{1}{2}}}\left(\frac{\hbar}{2\varepsilon_0\omega_K}\right)^{\frac{1}{2}} \times$$

$$\times\ \langle E'', \boldsymbol{P}'', n''_{\boldsymbol{K},\lambda}\,|\,(\mathsf{R}^{\dagger}_{\boldsymbol{K},\lambda}\,\mathrm{e}^{-\mathrm{i}\boldsymbol{K}\cdot\boldsymbol{R}} - \mathsf{R}_{\boldsymbol{K},\lambda}\,\mathrm{e}^{\mathrm{i}\boldsymbol{K}\cdot\boldsymbol{R}})\mathbf{p}\cdot\varepsilon_{\boldsymbol{K},\lambda}\,|\,E', \mathbf{P}', n'_{\boldsymbol{K},\lambda}\rangle$$

$$= -\frac{\mathrm{i}e}{m_0 V^{\frac{1}{2}}}\left(\frac{\hbar}{2\varepsilon_0\omega_K}\right)^{\frac{1}{2}} \times$$

$$\times\ [(n'_{\boldsymbol{K},\lambda}+1)^{\frac{1}{2}}\langle E'', \boldsymbol{P}'', n''_{\boldsymbol{K},\lambda}\,|\,\mathbf{p}\cdot\varepsilon\,|\,E', \boldsymbol{P}'-\hbar\boldsymbol{K}, n'_{\boldsymbol{K},\lambda}+1\rangle\ +$$

$$-\ (n'_{\boldsymbol{K},\lambda})^{\frac{1}{2}}\langle E'', \boldsymbol{P}'', n''_{\boldsymbol{K},\lambda}\,|\,\mathbf{p}\cdot\varepsilon\,|\,E', \boldsymbol{P}'+\hbar\boldsymbol{K}, n'_{\boldsymbol{K},\lambda}-1\rangle]. \qquad (12.38)$$

Only one value of $(\boldsymbol{K}, \lambda)$ is indicated in this equation, since if $n''_{\boldsymbol{K},\lambda}$ differs from $n'_{\boldsymbol{K},\lambda}$ for more than one $(\boldsymbol{K}, \lambda)$ the matrix element will vanish. Thus it is implied that $n''_{\boldsymbol{K},\lambda} = n'_{\boldsymbol{K},\lambda}$ in every case but one.

To evaluate these matrix elements one may note first that the eigenkets and eigenvalues of the states of relative motion of the two particles satisfy the equation

$$H(p, r) \mid E'\rangle = [(p^2/2m_0)+V(r)] \mid E'\rangle = E' \mid E'\rangle . \qquad (12.39)$$

Also

$$\mathsf{Hx}-\mathsf{xH} = (\mathsf{p}^2\mathsf{x}-\mathsf{x}\mathsf{p}^2)/2m_0 = (\hbar/\mathrm{i}m_0)\mathsf{p}_x \qquad (12.40\text{a})$$

so that

$$\langle E'' \mid \mathsf{Hx}-\mathsf{xH} \mid E'\rangle = (E''-E')\langle E'' \mid \mathsf{x} \mid E'\rangle = -(\mathrm{i}\hbar/m_0)\langle E'' \mid \mathsf{p}_x \mid E'\rangle$$

and

$$\langle E'' \mid \mathsf{p}_x \mid E'\rangle = -\mathrm{i}m_0\omega\langle E'' \mid \mathsf{x} \mid E'\rangle . \qquad (12.40\text{b})$$

Similar relations for y and z permit one to evalute the matrix elements for p in terms of those for r. Equation (12.38) then becomes

$$\langle E'', \boldsymbol{P}'', n''_{\mathbf{K},\lambda} \mid \mathsf{H}_1 \mid E', \boldsymbol{P}', n'_{\mathbf{K},\lambda}\rangle = -\mathrm{i}e\omega \left(\frac{\hbar}{2\varepsilon_0 V\omega_K}\right)^{\frac{1}{2}} \langle E'' \mid \mathbf{r}\cdot\boldsymbol{\varepsilon}_{\mathbf{K},\lambda} \mid E'\rangle$$

$$= [(n'_{\mathbf{K},\lambda}+1)^{\frac{1}{2}}\delta(\boldsymbol{P}''-\boldsymbol{P}'+\hbar\boldsymbol{K})\,\delta(n''_{\mathbf{K},\lambda}-n'_{\mathbf{K},\lambda}-1) +$$

$$+ n'^{\frac{1}{2}}_{\mathbf{K},\lambda}\delta(\boldsymbol{P}''-\boldsymbol{P}'-\hbar\boldsymbol{K})\,\delta(n''_{\mathbf{K},\lambda}-n'_{\mathbf{K},\lambda}+1)] \qquad (12.41)$$

where $\omega = (E'-E'')/\hbar$.

With equation (12.41) expressing the necessary matrix elements of H_1, one can evaluate the matrix elements of T after the manner used in equation (12.16). The ket vector at the time t is then the ket vector taken as the initial state plus all the other possible ket vectors, each multiplied by the appropriate element of the matrix for T.

At the time t the coefficient of a particular ket vector $\mid E'', \boldsymbol{P}'', n''_{\mathbf{K},\lambda}\rangle$, different from the initial one, is given by the corresponding matrix element of T

$$\langle E'', \boldsymbol{P}'', n''_{\mathbf{K},\lambda} \mid \mathsf{T} \mid E', \boldsymbol{P}', n'_{\mathbf{K},\lambda}\rangle =$$

$$= -\mathrm{i}\exp\left\{-\frac{\mathrm{i}}{\hbar}\langle E'', \boldsymbol{P}'', n''_{\mathbf{K},\lambda} \mid \mathsf{H}_0 \mid E'', \boldsymbol{P}'', n''_{\mathbf{K},\lambda}\rangle t\right\} \times$$

$$\times \frac{-\exp\{\mathrm{i}[(n''_{\mathbf{K},\lambda}-n'_{\mathbf{K},\lambda})\omega_K-\omega+\frac{1}{2}(\boldsymbol{P}''^2-\boldsymbol{P}'^2)/(m_1+m_2)\hbar]t\}}{\hbar[(n''_{\mathbf{K},\lambda}-n'_{\mathbf{K},\lambda})\omega_K-\omega+\frac{1}{2}(\boldsymbol{P}''^2-\boldsymbol{P}'^2)/(m_1+m_2)\hbar]} \times$$

$$\times\, e\omega\left(\frac{\hbar}{2\varepsilon_0 V\omega_K}\right)^{\frac{1}{2}} \langle E'' \mid \mathbf{r}\cdot\boldsymbol{\varepsilon}_{\mathbf{K},\lambda} \mid E'\rangle \times$$

$$\times\ [(n'_{\mathbf{K},\lambda}+1)^{\frac{1}{2}}\delta(\boldsymbol{P}''-\boldsymbol{P}'+\hbar\boldsymbol{K})\delta(n''_{\mathbf{K},\lambda}-n_{\mathbf{K},\lambda}-1)\ +$$

$$+\ n'^{\frac{1}{2}}_{\mathbf{K},\lambda}\delta(\boldsymbol{P}''_{''}-\boldsymbol{P}'-\hbar\boldsymbol{K})\delta(n''_{\mathbf{K},\lambda}-n'_{\mathbf{K},\lambda}+1)]\,. \tag{12.42}$$

The coefficient of the initial ket has the absolute value 1 and is merely $\exp\{-(\mathrm{i}/\hbar)\langle E', \boldsymbol{P}', n'_{\mathbf{K},\lambda} \mid \mathsf{H}_0 \mid E', \boldsymbol{P}', n'_{\mathbf{K},\lambda}\rangle t\}$. The initial state is an eigenstate of the Hamiltonian H_0 but it is not an eigenstate of the whole Hamiltonian when H_1 is included. The initial state is a state with a distribution of energy values around the mean value. However the distribution is narrow because the interaction term is small.

The distribution of energy over the eigenvalues of the total energy is constant in time. The mean remains constant and the mean square and higher means remain constant. However, as time goes on, the different partial energies may change, and the developing ket vector is interpreted in terms of "jumps" of the partial systems from one energy eigenvalue to another. The probability that the system has "jumped" from $\mid E', \boldsymbol{P}', n'_{\mathbf{K},\lambda}\rangle$ to $\mid E'', \boldsymbol{P}'' = \boldsymbol{P}'-\hbar\boldsymbol{K}, n''_{\mathbf{K},\lambda} = n'_{\mathbf{K},\lambda}+1\rangle$ is

$$\mid \langle E'', \boldsymbol{P}'-\hbar\boldsymbol{K}, n'_{\mathbf{K},\lambda}+1 \mid \mathsf{T} \mid E', \boldsymbol{P}', n'_{\mathbf{K},\lambda}\rangle \mid^2 =$$

$$= (n_{\mathbf{K},\lambda}+1)\,\frac{e^2\omega^2}{\varepsilon_0 V\hbar\omega_K}\,\frac{1-\cos\,[\omega_K-\omega-(\boldsymbol{P}'\cdot\boldsymbol{K}-\frac{1}{2}\hbar\boldsymbol{K}^2)/(m_1+m_2)]t}{[\omega_K-\omega-(\boldsymbol{P}'\cdot\boldsymbol{K}-\frac{1}{2}\hbar\boldsymbol{K}^2)/(m_1+m_2)]^2}\,\times$$
$$\times\ \mid \langle E'' \mid \mathbf{r}\cdot\boldsymbol{\varepsilon}_{\mathbf{K},\lambda} \mid E'\rangle \mid^2. \tag{12.43}$$

The argument of the cosine represents the energy difference (difference in H'_0) between the initial state and this one particular final state. $\hbar\omega_K$ represents the energy gained by the radiation field. $\hbar\omega = E'-E''$ is the energy lost by the relative motion of the particles and $\hbar(\boldsymbol{P}'\cdot\boldsymbol{K}-\frac{1}{2}\hbar\boldsymbol{K}^2)/(m_1+m_2)$ is the energy gained by the center of mass motion.

The square of the matrix element given in (12.43) is interpreted as the probability that the atom has lost $\hbar\omega$ of internal energy, that the center of mass momentum has changed from $\boldsymbol{P}'$ to $\boldsymbol{P}''$ with the corresponding change in energy, and that the $(\boldsymbol{K}, \lambda)$ mode of the electromagnetic field has increased in energy by $\hbar\omega_K$. Frequently, however, one is interested in the total probability that the atom has lost internal energy, regardless of just what modes of the electromagnetic field are excited. This total probability is given by a sum of terms such as equation (12.43) over all values of $(\boldsymbol{K}, \lambda)$. Since the volume V is taken very large, the sum over $\boldsymbol{K}$ can be replaced by an integral with the number of values of K per element of the $\boldsymbol{K}$ space as $(V/8\pi^3)\times \sin\vartheta\,\mathrm{d}\vartheta\mathrm{d}\varphi K^2\mathrm{d}K$. ϑ is the angle between $\boldsymbol{K}$ and an arbitrary direction chosen as the polar axis. As will be evident, it is convenient to take this axis along the direction of $\boldsymbol{P}'$, the initial direction of motion of the center of mass of the two particles.

The integrand as a function of K has a resonance denominator. Since $K = \omega_K c$ it becomes, using the above polar axis

$$\omega_K[1-(P'\cos\vartheta-\hbar\omega_K/2c)/2Mc]-\omega. \tag{12.44}$$

Setting the root of this equation equal to ω_K^0, the integral to give the total probability of emission of radiation in any direction is given by

$$\sum_\lambda \int |\langle E'', \boldsymbol{P}'-\hbar\boldsymbol{K}, n'_{\boldsymbol{K},\lambda}+1 \,|\, \mathsf{T} \,|\, E', \boldsymbol{P}', n'_{\boldsymbol{K},\lambda}\rangle|^2 \frac{V}{8\pi^3}\sin\vartheta\, \mathrm{d}\vartheta\, \mathrm{d}\varphi K^2\, \mathrm{d}K =$$

$$= \frac{e^2\omega^2}{8\pi^3\varepsilon_0 c^3}\int_0^\pi \sin\vartheta\,\mathrm{d}\vartheta \int_0^{2\pi}\mathrm{d}\varphi \sum_\lambda |\langle E''\,|\mathbf{r}\cdot\boldsymbol{\varepsilon}_{\boldsymbol{K},\lambda}\,|\,E'\rangle|^2(\bar{n}_{\boldsymbol{K},\lambda}+1)\times$$

$$\times\int_0^\infty \frac{1-\cos(\omega_K^0-\omega)t}{(\omega_K^0-\omega)^2}\,\omega_K\,\mathrm{d}\omega_K. \tag{12.45}$$

ω_K^0 is very close to ω_K, and over the small range in which $\omega_K^0 \approx \omega$, the difference may be ignored and the variable of integration taken to be ω_K^0. Also the quantity $\omega_K \approx \omega_K^0$ in the integrand may be taken outside the integral sign as equal to ω. The integral over $\mathrm{d}\omega_K$ then becomes

$$\int_0^\infty \frac{1-\cos(\omega_K^0-\omega)t}{(\omega_K^0-\omega)^2}\,\omega_K\,\mathrm{d}\omega_K = \omega t\int_{-\infty}^\infty \frac{1-\cos x}{x^2}\,\mathrm{d}x = \pi\omega t. \tag{12.46}$$

The total transition probability, up to the time t, is then

$$P(t) = \frac{e^2}{8\pi^2\varepsilon_0}\left(\frac{\omega}{c}\right)^3(\bar{n}_{\boldsymbol{K},\lambda}+1)\,t\int_0^\pi \sin\vartheta\,\mathrm{d}\vartheta\int_0^{2\pi}\mathrm{d}\varphi\sum_\lambda|\langle E''\,|\,\mathbf{r}\cdot\boldsymbol{\varepsilon}_{\boldsymbol{K},\lambda}\,|\,E'\rangle|^2. \tag{12.47}$$

The angles (ϑ, φ) define the direction of the vector $\boldsymbol{K}$ in a fixed coordinate system.

The scalar product $\mathbf{r}\cdot\boldsymbol{\varepsilon}_{\boldsymbol{K},\lambda}$ is clearly a function of the direction of $\boldsymbol{K}$ given by (ϑ, φ) as well as of the ket and bra vectors for the internal motion of the pair of particles. The ket and bra vectors permit evaluation of the matrix elements for the components of $\mathbf{r}$ which may be designated as $\langle E''\,|\,\mathsf{x}\,|\,E'\rangle$, etc. The sum of the squares of the scalar products of $\boldsymbol{r}$ with the two unit vectors $\boldsymbol{\varepsilon}_{\boldsymbol{K},\lambda,1}$ and $\boldsymbol{\varepsilon}_{\boldsymbol{K},\lambda,2}$ is the square of the projection of the vector $\langle E''\,|\,\mathbf{r}\,|\,E'\rangle$ on the plane perpendicular to $\boldsymbol{K}$ and therefore it is also the square of the vector product of $\langle E''\,|\,\mathbf{r}\,|\,E'\rangle$ with a unit vector in the direction of $\boldsymbol{K}$. The components of the unit vector in the direction of $\boldsymbol{K}$ are

$$K_{0x} = \sin\vartheta\cos\varphi, \quad K_{0y} = \sin\vartheta\sin\varphi, \quad K_{0z} = \cos\vartheta$$

and

$$\begin{aligned}
\sum_{\lambda} |\langle E''|\mathbf{r}\cdot\boldsymbol{\varepsilon}_{\mathbf{K},\lambda}|E'\rangle|^2 &= |\langle E''|\mathsf{x}|E'\rangle|^2[\sin^2\vartheta\sin^2\varphi + \cos^2\vartheta] \\
&+ |\langle E''|\mathsf{y}|E'\rangle|^2[\sin^2\vartheta\cos^2\Xi + \cos^2\vartheta] + |\langle E''|\mathsf{z}|E'\rangle|^2\sin^2\vartheta \\
&- \{[\langle E''|\mathsf{y}|E'\rangle^*\langle E''|\mathsf{z}|E'\rangle + \\
&\quad + \langle E''|\mathsf{y}|E'\rangle\langle E''|\mathsf{z}|E'\rangle^*]\sin\vartheta\cos\vartheta\sin\varphi \\
&+ [\langle E''|\mathsf{x}|E'\rangle^*\langle E''|\mathsf{z}|E''\rangle + \\
&\quad + \langle E''|\mathsf{x}|E'\rangle\langle E''|\mathsf{z}|E'\rangle^*]\sin\vartheta\cos\vartheta\cos\varphi \\
&+ [\langle E''|\mathsf{x}|E'\rangle^*\langle E''|\mathsf{y}|E'\rangle + \\
&\quad + \langle E''|\mathsf{x}|E'\rangle\langle E''|\mathsf{y}|E'\rangle^*]\sin\vartheta\cos\vartheta\sin\varphi\cos\varphi\}. \quad (12.48)
\end{aligned}$$

To get the total probability of transition with radiation in any direction, the expression in (12.48) must be integrated over the whole solid angle as indicated in (12.47). The final result is then

$$\begin{aligned}
P(t) = \frac{e^2 t}{3\varepsilon_0\pi\hbar}\left(\frac{\omega}{c}\right)^3(\bar{n}_{K,\lambda}+1)\,[|\langle E''|\mathsf{x}|E'\rangle|^2 + \\
+ |\langle E''|\mathsf{y}|E'\rangle|^2 + |\langle E''|\mathsf{z}|E'\rangle|^2]. \quad (12.49)
\end{aligned}$$

All of the terms in the second part of (12.48) integrate to zero. However, if one is interested in the transition probability only when radiation is emitted in some more limited solid angle the whole expression must be taken into account.

The situation is described by saying that equation (12.49) gives the probability that during the time t the two particle system (an atom) has lost the internal energy $(E'-E'')$, i.e. that it has "jumped" from the internal state $|E'\rangle$ to the state $|E''\rangle$. Most of this energy has appeared in the radiation field but there also may be a small increase or decrease of the translational energy of the center of mass.

From the above computations a number of conclusions can be drawn.

A. The emission of energy takes place in quanta. The change in the energy of a "mode of vibration" of the field is by one unit only. This is indicated by the $\delta(n''_{\mathbf{K},\lambda} - n'_{\mathbf{K},\lambda} - 1)$ in equation (12.42).

B. Energy is conserved in the radiation process. The amount of internal energy lost by the atom, $\hbar\omega$, is equal to the amount gained by the field

$\hbar\omega_K$, plus the amount gained by the atom in kinetic energy, $\hbar\omega_K \times (\boldsymbol{P}' \cos \vartheta - \hbar\omega_K/2c)/2Mc$.

C. Momentum is conserved in the process. The radiation field gains the momentum $\hbar\boldsymbol{K}$ and the center of mass motion of the atom loses the same amount. This follows from the $\delta(\boldsymbol{P}''-\boldsymbol{P}'+\hbar\boldsymbol{K})$ in equation (12.42).

D. The change in translational energy due to the recoil of the whole atom is

$$\boldsymbol{P}^2/2M = (\hbar^2\omega_K^2/2Mc^2)-(\hbar\omega_K \boldsymbol{P}' \cos \vartheta/Mc). \qquad (12.50)$$

This is usually quite small as evidenced by the terms in the denominator and can be closely approximated by the last term. It represents the Doppler effect which shows up in the difference between ω_K and ω_K^0.

E. The emission in a given direction is proportional to the square of the sine of the angle between $\langle E'' \mid \mathbf{r} \mid E' \rangle$ and the direction of emission. This is clear from the fact that the sum over λ of $| \langle E'' \mid \mathbf{r} \cdot \varepsilon_{\boldsymbol{K},\lambda} \mid E' \rangle |^2$ is the quantity appearing in (12.46).

F. The probability of emission depends on the radiation energy already in the field. This appears as the term $(\overline{n'_{\boldsymbol{K},\lambda}}+1)$ in the integrated form, and of $(n'_{\boldsymbol{K},\lambda}+1)$ before integration. The integration can be justified only if $n'_{\boldsymbol{K},\lambda}$ varies slowly enough with $\boldsymbol{K}$ to be replaced by a mean value, $\overline{n_{\boldsymbol{K}}^{\prime,\lambda}}$, taken outside the integral sign. If all the $n'_{\boldsymbol{K},\lambda}=0$ and there is no energy in the field, the procedure is obviously correct. If the radiation field represents a situation of thermal equilibrium, the mean value has a satisfactorily precise significance. But if some one, or a few, of the waves $\boldsymbol{K}$ are excited well above the others it is not true, and the probability of radiation into those modes is greatly enhanced. It is this property that accounts for the operation of lasers.

PROBLEM 4. Show that the change in energy associated with the recoil of the atom in emitting radiation represents the Doppler effect in the emitted radiation.

PROBLEMS 5. Compute the probability of the absorption of dipole radiation by a system of two particles.

3.4. *Interpretation of the wave function of the state vector*

The above treatment of the interaction of the radiation field with a system of particles, or an "atom", follows from a picture in which the state vector changes in a continuous fashion. At $t = 0$ it represents a state in which the internal motion of mass motion has a precisely defined momentum $\boldsymbol{P}'$ and the corresponding energy, and the radiation field has the energy and momentum associated with the values $n'_{\mathbf{K},\lambda}$. The state vector at the time t is, in addition to the initial vector itself, a sum of many state vectors each describable as was the initial state. In all of them the internal motion of the atom is different from its initial motion. In each of them some one wave in the radiation field has an additional quantum of energy, and the translational motion of the atom has changed in accordance with the conservation laws indicated in the analysis. But one may well ask, "What is *really* the state of the atom and the radiation field at the time t? Has the atom emitted a quantum of radiation, a photon, in the direction ($\boldsymbol{K}$) or has it not?". The answer is that the state of the total system at the time t cannot be described in such a precise form.

If an oscillating dipole is treated according to classical electrodynamics it appears to radiate in all directions at once, but not with uniform intensity. The intensity has the same angular distribution as the quantity

$$\Sigma_\lambda \,|\, \langle E'' \,|\, \mathbf{r} \cdot \boldsymbol{\varepsilon}_{\boldsymbol{K},\lambda} \,|\, E' \rangle \,|^2$$

which appears in equation (12.47). But methods for the detection of radiation in a particular direction ($\boldsymbol{K}$) will not detect less than one quantum of energy. Hence the function in equation (12.47) is interpreted as the probability that a quantum has been emitted in this particular direction and that the atom has lost the corresponding energy. An integral over all directions then gives the probability that the atom has emitted energy and the reciprocal is called the lifetime of the atomic state $| E' \rangle$.

3.5. *Emission of quadrupole radiation*

The probability of emission of dipole radiation was computed by picking out of the interaction part of the Hamiltonian operator those terms involving only the first power of $\boldsymbol{r}$, the separation between the two oppositely charged particles. The next terms in the series expansion of the exponentials give rise to the quadrupole radiation. Such radiation is important as resulting from transitions between atomic states when the dipole radiation is "forbidden", and also is dominant in cases where charges of only one sign are present such as in nuclei.

If one again neglects the term in $\boldsymbol{P} \cdot \boldsymbol{\varepsilon}_{\boldsymbol{K},\lambda}$ the part of the interaction Hamiltonian describing the quadrupole interaction can be recognized in equation (12.37):

$$[(m_2 - m_1)/m_1 m_2]\boldsymbol{K} \cdot \boldsymbol{r}\boldsymbol{p} \cdot \boldsymbol{\varepsilon}_{\mathbf{K},\lambda} = [(m_2 - m_1)/m_1 m_2]\boldsymbol{K} \cdot \boldsymbol{D} \cdot \boldsymbol{\varepsilon}_{\mathbf{K},\lambda}. \quad (12.51)$$

$\boldsymbol{D}$ is a tensor of the second rank, or a dyadic. The dyadic has nine components, $D_{ij} = r_i p_j$, which can be evaluated by means of the atomic wave functions.

The individual terms of such a tensor, or dyadic, depend on the coordinate system with respect to which they are expressed. However, the trace, $(D_{11}+D_{22}+D_{33})$ is an invariant. Also, any such tensor can be expressed as a sum of a symmetric and an antisymmetric tensor. Thus

$$D^{\mathrm{a}}_{ij} = \tfrac{1}{2}(D_{ij}-D_{ji}) \text{ and } D^{\mathrm{s}}_{ij} = \tfrac{1}{2}(D_{ij}+D_{ji}).$$

In this case the antisymmetric tensor is just half the angular momentum:

$$\tfrac{1}{2}(D_{ij}-D_{ji}) = \tfrac{1}{2}(r_i p_j - r_j p_i) = \tfrac{1}{2}L_k. \tag{12.52}$$

With this notation the part of the interaction Hamiltonian associated with the emission and absorption of quadrupole radiation is

$$\mathsf{H}_1'' = -\frac{e}{V^{\frac{1}{2}}}\frac{m_2-m_1}{m_1 m_2}\sum_{\boldsymbol{K},\lambda}\left(\frac{\hbar}{2\varepsilon_0\omega_K}\right)^{\frac{1}{2}}[\mathsf{R}^{\dagger}_{\boldsymbol{K},\lambda}\,\mathrm{e}^{-\mathrm{i}\boldsymbol{K}\cdot\boldsymbol{R}}+\mathsf{R}_{\boldsymbol{K},\lambda}\,\mathrm{e}^{\mathrm{i}\boldsymbol{K}\cdot\boldsymbol{R}}]\times$$

$$\times\,[\tfrac{1}{2}\boldsymbol{K}\times\boldsymbol{\varepsilon}_{\boldsymbol{K},\lambda}\cdot\boldsymbol{L}+\boldsymbol{K}\cdot\boldsymbol{D}^{\mathrm{s}}\cdot\boldsymbol{\varepsilon}_{\boldsymbol{K},\lambda}]. \tag{12.53}$$

The antisymmetric tensor part, proportional to $\boldsymbol{L}$, gives what is called the magnetic dipole because it contains the angular momentum multiplied by the charge e and divided by a mass. The symmetric part of the tensor gives the electric quadrupole radiation. The function of the masses shows that if $m_1 = m_2$ there is no contribution from H_1'' at all. However in atoms the nucleus is so much heavier than the electrons that the quadrupole and the magnetic dipole may both be significant.

By the procedure used in arriving at equation (12.40b) it can be shown that

$$\langle E''\mid \mathbf{r}_i\mathbf{p}_j+\mathbf{p}_i\mathbf{r}_j\mid E'\rangle = (\mathrm{i}m/\hbar)\,(E''-E')\,\langle E''\mid \mathbf{r}_i\mathbf{r}_j\mid E'\rangle. \tag{12.54}$$

In case the atomic states involved are isotropic, the diagonal terms of the tensor are all equal and give no contribution to the radiation probability since $\boldsymbol{K}\cdot\boldsymbol{\varepsilon}_{\boldsymbol{K},\lambda} = 0$. By evaluating the various terms of the tensors the selection rules and the transition probabilities for this radiation can be computed.

PROBLEM 6. Consider a hydrogen atom and neglect the electron spin. Compute the probability of spontaneous emission of dipole radiation when the atom is initially in an excited state. Derive the selection rules and indicate the necessary radial integrals.

PROBLEM 7. For an atom as in Problem 6 compute the probability of spontaneous emission of magnetic dipole and electric quadrupole interaction. Show that this is proportional to ω^5 compared with ω^3 for dipole radiation and hence is relatively more important for high frequencies.

PROBLEM 8. Consider a nucleus made up of a proton and a neutron held together by an arbitrary central force. Compute the probability of emission of dipole and quadrupole radiation when the center of mass is initially stationary.

3.6. *Scattering by a free electron*

It was recognized early in the development of quantum mechanics that a free electron cannot absorb energy from the radiation field. That it cannot radiate was obvious because radiation involves acceleration and this implies an external force. The impossibility of both radiation and absorption becomes obvious as soon as the quantum properties of the radiation field are recognized since it is impossible to conserve both energy and momentum in such an interaction. However, if the incident radiation is viewed as providing the acceleration which leads to radiation, such a process leads to scattering and a recoil of the electron. Classically treated, the process is one of radiation in all directions with the electron recoiling in the direction of the incident radiation. Treated quantum mechanically, the process is that of the Compton effect.

The part of the interaction Hamiltonian effective in single particle scattering is the square of the term in equation (12.26) representing the vector potential. Hence for a single electron let

$$\mathsf{H}_s = \frac{e^2}{2mV} \sum_{\boldsymbol{K},\lambda} \sum_{\boldsymbol{K}',\lambda'} [\mathsf{Q}_{\boldsymbol{K},\lambda,1} \cos(\boldsymbol{K}\cdot\boldsymbol{r}) + \mathsf{Q}_{\boldsymbol{K},\lambda,2} \sin(\boldsymbol{K}\cdot\boldsymbol{r})] \times$$
$$\times [\mathsf{Q}_{\boldsymbol{K}',\lambda',1} \cos(\boldsymbol{K}\cdot\boldsymbol{r}) + \mathsf{Q}_{\boldsymbol{K}',\lambda',2} \sin(\boldsymbol{K}\cdot\boldsymbol{r})] \boldsymbol{\varepsilon}_{\boldsymbol{K},\lambda} \cdot \boldsymbol{\varepsilon}_{\boldsymbol{K}',\lambda'}. \qquad (12.55)$$

In this expression the sums over $\boldsymbol{K}$ and $\boldsymbol{K}'$ cover only half of the solid angle. If the operators $\mathsf{Q}_{\boldsymbol{K},\lambda,\mu}$ are expressed in terms of $\mathsf{R}^{\dagger}_{\boldsymbol{K},\lambda}$ and $\mathsf{R}_{\boldsymbol{K},\lambda}$, and the sum is taken over the whole solid angle, the effective part of the Hamiltonian operator becomes

$$\mathsf{H}_s = -\left(\frac{e^2\hbar}{4\varepsilon_0 mV}\right)^{\frac{1}{2}} \sum_{\boldsymbol{K},\lambda} \sum_{\boldsymbol{K}',\lambda'} \left(\frac{1}{\omega_K \omega_{K'}}\right)^{\frac{1}{2}} [\mathsf{R}^{\dagger}_{\boldsymbol{K},\lambda} \mathrm{e}^{-\mathrm{i}\boldsymbol{K}\cdot\boldsymbol{r}} - \mathsf{R}_{\boldsymbol{K},\lambda} \mathrm{e}^{\mathrm{i}\boldsymbol{K}\cdot\boldsymbol{r}}] \times$$
$$\times [\mathsf{R}_{\boldsymbol{K}',\lambda'} \mathrm{e}^{-\mathrm{i}\boldsymbol{K}'\cdot\boldsymbol{r}} - \mathsf{R}_{\boldsymbol{K}',\lambda'} \mathrm{e}^{\mathrm{i}\boldsymbol{K}'\cdot\boldsymbol{r}}] \boldsymbol{\varepsilon}_{\boldsymbol{K},\lambda} \cdot \boldsymbol{\varepsilon}_{\boldsymbol{K}',\lambda'}$$

$$= -\left(\frac{e^2\hbar}{4\varepsilon_0 mV}\right)^{\frac{1}{2}} \sum_{K,\lambda}\sum_{K',\lambda'} \left(\frac{1}{\omega_K\omega_{K'}}\right)^{\frac{1}{2}} \times$$

$$\times [\mathsf{R}^\dagger_{K,\lambda}\mathsf{R}^\dagger_{K',\lambda'}\, e^{-i(K+K')\cdot r} - \mathsf{R}^\dagger_{K,\lambda}\mathsf{R}_{K',\lambda'}\, e^{-i(K-K')\cdot r}$$

$$- \mathsf{R}_{K,\lambda}\mathsf{R}^\dagger_{K',\lambda'}\, e^{i(K-K')\cdot r} + \mathsf{R}_{K,\lambda}\mathsf{R}_{K',\lambda'}\, e^{i(K+K')\cdot r}\quad \varepsilon_{K,\lambda}\cdot\varepsilon_{K',\lambda'}. \tag{12.56}$$

In this expression $\boldsymbol{r}$ represents the position of the electron. Since only one electron is present it needs no special designation but it must be noted that $\boldsymbol{r}$ represents an electron and not merely a point in the field. There need be no confusion with the $\boldsymbol{r}$ of the previous sections which represented the difference between the coordinates of the two particles.

For the initial state of the system let the only radiation present be $n'_{K,\lambda}$ quanta in the (K, λ) mode of the field, and let the electron be at rest, i.e. $p' = 0$. Such a state will be designated by the ket vector $|\,p', n'_{K,\lambda}\rangle$ with $p' = 0$. It is, of course, a highly idealized situation. The radiation is a single linearly polarized plane wave with an exactly defined frequency. It represents an energy density of $n'_{K,\lambda}\,\hbar\omega/V$ but is uniformly spread over the whole volume V. The specification that $p' = 0$ means that the electron is not localized at all. It is everywhere and can interact with the radiation all over the volume V.

The first and last terms in equation (12.56) are ineffective. They lead to a matrix element of H_s, but when used to give a matrix element of T the fact that they represent large changes of energy of the total system leads to negligible values.

With only the (K, λ) mode excited in the initial state, only the one $\mathsf{R}_{K,\lambda}$ operator can be effective, but it occurs twice because of the two sums in (12.56). Hence the important contributions are given by

$$\mathsf{H}_s\,|\,0, n'_{K,\lambda}\rangle = \frac{e^2\hbar}{2\varepsilon_0 mV}\, n'^{\frac{1}{2}}_{K,\lambda} \sum_{K',\lambda'} \left(\frac{1}{\omega_K\omega_{K'}}\right)^{\frac{1}{2}} \times$$

$$\times\ \varepsilon_{K,\lambda}\cdot\varepsilon_{K',\lambda'}\, e^{i(K-K')\cdot r}\,|\,0, n'_{K',\lambda'} = 1, n'_{K,\lambda}-1\rangle \tag{12.57}$$

and

$$\langle p', n''_{K',\lambda'} = 1, n'_{K,\lambda}-1\,|\,\mathsf{H}_s\,|\,0, n'_{K,\lambda'} = 0, n_{K,\lambda}\rangle =$$

$$= \frac{e^2\hbar}{2\varepsilon_0 mV}\, n'^{\frac{1}{2}}_{K,\lambda} \left(\frac{1}{\omega_K\omega_{K'}}\right)^{\frac{1}{2}} \varepsilon_{K,\lambda}\cdot\varepsilon_{K',\lambda'}\delta[(p' - \hbar(K-K')]\,. \tag{12.58}$$

The operators T and T_0 are, as usual, unit operators at $t = 0$. The off-diagonal terms of T_0 are then, to the first approximation,

$$\langle \hbar(\boldsymbol{K}-\boldsymbol{K}'), n'_{\boldsymbol{K},\lambda}-1, n''_{\boldsymbol{K}',\lambda'} = 1 \mid \mathsf{T}_0 \mid 0, n'_{\boldsymbol{K},\lambda}, n'_{\boldsymbol{K}',\lambda'}\rangle =$$

$$= \frac{e^2\hbar}{2\varepsilon_0 mV} n'^{\frac{1}{2}}_{\boldsymbol{K},\lambda} \left(\frac{1}{\omega_K\omega_{K'}}\right)^{\frac{1}{2}} \boldsymbol{\varepsilon}_{\boldsymbol{K}',\lambda'} \cdot \boldsymbol{\varepsilon}_{\boldsymbol{K},\lambda} \times$$

$$\times \frac{1-(\mathrm{i}/\hbar)\,\{[m^2c^4+\hbar^2(\boldsymbol{K}-\boldsymbol{K}')^2c^2]^{\frac{1}{2}}-mc^2+\hbar(\omega_{K'}\omega_K)\}t}{\{[m^2c^4+\hbar^2(\boldsymbol{K}-\boldsymbol{K}')^2c^2]^{\frac{1}{2}}-mc^2+\hbar(\omega_{K'}-\omega_K)\}} \tag{12.59}$$

for each value of $(\boldsymbol{K}', \lambda')$ except $(\boldsymbol{K}, \lambda)$. In the exponent and in the denominator, the relativistic expression for the energy of the electron is used even though the Hamiltonian function was formulated in the non-relativistic limit. The evaluation of the denominator is slightly simplified in this way.

The δ-function represents the conservation of momentum. In this special case, in which the initial state represents an electron at rest, the z axis may be taken along $\boldsymbol{K}$, and the x axis in the plane of $\boldsymbol{K}$ and $\boldsymbol{K}'$ perpendicular to $\boldsymbol{K}$. Then

$$p'_z = \hbar K - \hbar K' \cos\vartheta$$

$$p'_x = -\hbar K' \sin\vartheta \tag{12.60}$$

where ϑ is the angle between $\boldsymbol{K}$ and $\boldsymbol{K}'$. Thus the only components of the state vector which grow from their initial values of zero are those in which the motion of the electron and the direction of the wave in the electromagnetic field have the relationship given by (12.60). Squaring and adding the two equations (12.60) leads to

$$p'^2 = \hbar^2[K^2+K'^2-2KK'\cos\vartheta]. \tag{12.61}$$

With the restriction expressed by equation (12.61) the denominator in (12.59) is approximately

$$\hbar(\omega_K-\omega_{K'})-4(\hbar^2\omega_K\omega_{K'}/mc^2)\sin^2\tfrac{1}{2}\vartheta.$$

The square of the absolute value of the matrix element given in equation (12.59) is equal to the square of the corresponding element of T, which then becomes

$$|\langle \hbar(\boldsymbol{K}-\boldsymbol{K}'), n''_{\boldsymbol{K}',\lambda'} = 1, n'_{\boldsymbol{K},\lambda}-1 \mid \mathsf{T} \mid 0, n'_{\boldsymbol{K}',\lambda'} = 0, n'_{\boldsymbol{K},\lambda}\rangle|^2 =$$

$$= \frac{e^4}{\varepsilon_0^2 m^2 V^2 \omega_K\omega_{K'}} \frac{\sin^2[(\omega_K-\omega_{K'})-(4\hbar^2\omega_K\omega_{K'}/mc^2)\sin^2\frac{1}{2}\vartheta]\frac{1}{2}t}{[(\omega_K-\omega_{K'})-(4\hbar^2\omega_K\omega_{K'}/mc^2)\sin^2\frac{1}{2}\vartheta]^2} \times$$

$$\times (\boldsymbol{\varepsilon}_{\boldsymbol{K},\lambda}\cdot\boldsymbol{\varepsilon}_{\boldsymbol{K}',\lambda'})^2. \tag{12.62}$$

Since the maximum value of the sine is one, this matrix element is large only when the denominator is very small. To get the total observable quantity, the probability that radiation can be observed in an element of solid angle (ϑ, φ) of $\boldsymbol{K}'$ when the electron recoils with momentum $\boldsymbol{p}' = \hbar(\boldsymbol{K}-\boldsymbol{K}')$, it is necessary to sum (or integrate) over all values of ω_K, satisfying equation (12.61) and to sum over λ'. Hence

$$\pi(\vartheta, \varphi) = \frac{V}{8\pi^3}\int_0^\infty \sum_{\lambda'} |\langle\hbar(\boldsymbol{K}-\boldsymbol{K}'), n''_{\boldsymbol{K}',\lambda'} = 1, n'_{\boldsymbol{K},\lambda}-1 \,|\, \mathsf{T} \,|\, 0, n'_{\boldsymbol{K},\lambda'} = 0, n'_{\boldsymbol{K},\lambda}\rangle|^2 K'^2\,\mathrm{d}K'$$

$$= \frac{e^4}{8\pi^3\varepsilon_0^2 m^2 c^4}\,\frac{n'_{\boldsymbol{K},\lambda}c}{V}\sin^2\beta\,\frac{\omega_{K'}}{\omega\chi}\,t\int_{-\infty}^{\infty}\frac{\sin^2 x}{x^2}\,\mathrm{d}x$$

$$= \frac{(e^4/8\pi^2\varepsilon_0^2 m^2 c^4) I \sin^2\beta}{1+(2\hbar\omega_K/mc^2)\sin^2\frac{1}{2}\vartheta}\,t. \tag{12.63}$$

β is the angle between the electric vector of the incident radiation, which is the direction of $\boldsymbol{\varepsilon}_{\boldsymbol{K},\lambda}$, and the direction of the scattered radiation $\boldsymbol{K}'$. The sum over the two values of λ' for $\boldsymbol{\varepsilon}_{\boldsymbol{K}',\lambda'}$ of the scattered radiation gives the $\sin^2\beta$. The only significant contribution of the integral comes from those values of ω_K for which the denominator is near zero. Hence the $\omega_{K'}$ has been taken outside the integral and the ω_K in the denominator has been expressed in terms of the ω_K, for which the denominator vanishes. The term $(n'_{\boldsymbol{K},\lambda}c/V)$ represents the intensity of the incident radiation since $(n'_{\boldsymbol{K},\lambda}c/V)$ is the energy density and c is the velocity. It is replaced by I in the third line of (12.63).

The interpretation of the above analysis is that if a beam of polarized radiation is incident on a free electron most of the radiation will be unaffected. The unaffected radiation is represented by the constant diagonal terms of the T_0 matrix. However, some small amount of radiation will be observed at an angle (ϑ, φ) from the incident beam, and for each quantum detected there could be detected an electron recoiling with the momentum (p'_x, p'_z) given by equation (12.60) and an energy as indicated in equation (12.61). The probability of scattering is a maximum in the plane perpendicular to the electric vector of the incident radiation as indicated by the presence of $\sin^2\beta$ in equation (12.63).

Chapter 13

HYDROGEN-LIKE SPECTRA

The study of spectroscopy has proved to be one of the most fruitful methods for investigating the details of atomic structure, and it has been also of major importance in the development of quantum mechanics. Bohr's work on the hydrogen spectrum in 1913 gave much of the impetus which ultimately led to Schroedinger's work in 1925. From spectroscopy came also the ideas of electron spin and the Pauli exclusion principle. The exploratory work in atomic spectroscopy may now be said to be completed. There remains of course the effort to improve the methods of approximation to solutions of the equations. But the theory itself is of especial interest because of the way it illustrates many of the central features of quantum mechanics, and because of its application to atomic nuclei.

Spectroscopy involves the interaction between material systems and the radiation field. It is especially concerned with the emission and absorption of radiation. Since, however, the interaction is small, it is appropriately treated by the perturbation methods described in Chapter 12 and attention is directed in the first instance to the material system alone. In the "zero order approximation" then, the frequency of a spectral line is equal to the difference between two energy eigenvalues divided by h. An energy eigenvalue divided by hc is called a spectral term, and the difference between two terms gives the wave number, $\nu/c = 1/\lambda$, of a spectral line. Not all energy states combine in this way to give observed emission or absorption lines, and the calculation of the intensities (including zero intensity) is a part of the theoretical problem.

A completely satisfactory theory of atomic and molecular spectroscopy would make it possible to start with a given atomic or molecular system, specified by its Hamiltonian function, and to calculate all the energy eigenvalues and the probabilities of the emission or absorption of radiation connected with them. The complication inherent in the treatment of a many-particle system is such that complete attainment of these objectives is as yet impossible. Nevertheless, it has been possible to classify the energy levels in a useful manner, to make good estimates of their values, and to make fairly good estimates of the emission and absorption probabilities.

In the class of hydrogen-like spectra are included those due to atoms having only one electron outside of a closed shell, such as Li, Na, and K, as well as the truly hydrogen-like atoms such as He^+, Li^{++}, Be^{+++}, etc. Of course, with such atoms as Li and Na, the excitation may be so high that an electron is removed from the closed shell, and a different kind of spectrum is produced. There is usually little difficulty, however, in distinguishing the hydrogen-like spectrum from other spectra of the same atom.

The truly hydrogen-like spectra are shown by atomic hydrogen itself, by He^+, Be^{++}, etc.; but they represent a special case in that a relativistic treatment is necessary to understand the energy levels in detail. Such details are less important in spectra of atoms such as lithium or sodium because they are overshadowed by the departure of the effective central field from a Coulomb field. The relativistic treatment will not be given here, but a classification of the energy eigenvalues based on the analysis in Chapters 6 and 9 will suffice to illustrate the general features.

1. The spectrum of a single electron without spin

1.1. *General scheme of levels*

In Chapter 6 it was shown that the eigenvalues for the energy of an electron in the field of a heavy nucleus with charge Ze are

$$E_{n,l,m_l} = -\frac{Z^2\mu e^4}{8\varepsilon_0^2 h^2 n^2} = -\frac{\mathscr{R}Z^2hc}{n^2} \tag{13.1}$$

where $\mathscr{R}$ is the Rydberg constant and n is an integer. Each state is identified by two other quantum numbers in addition to the total quantum number n. They are the magnetic quantum number m_l and the azimuthal quantum number l, but the energy is dependent on n only. A state described by these three integers has the angular momentum $m_l\hbar$ about the polar axis, and the square of the total angular momentum is given by $l(l+1)\hbar^2$. For each value of n, all values of l from 0 up to and including $l = n-1$ are possible. For each value of l, m_l may take on all integral values such that $-l \leq m_l \leq l$.

For $n = 1$, l must be 0, and correspondingly $m_l = 0$. In this case there is just one state, which is the ground state according to equation (13.1). For $n = 2$, $l = 0$ or 1. When $l = 0$, $m_l = 0$, and when $l = 1$, $m_l = -1, 0,$

1. Thus a total of four states have the energy given by $n = 2$. Similarly for $n = 3$ there are nine states, one with $l = 0$, three with $l = 1$, and five with $l = 2$. For any value of n, the degeneracy, or the number of states having the same energy, is $\sum_{l=0}^{l=n} (2l+1) = n^2$. This degeneracy is changed when the electron spin is considered, when a magnetic field is applied, or when a central field different from a Coulomb field is used. The consideration of electron spin increases the number of states since an additional coordinate is involved. The other factors merely change the energies of states already considered.

With the energy values given in equation (13.1) the spectral wave numbers are given by the differences

$$(\nu/c)_{n',n} = [E(n', l', m_l') - E(n, l, m_l)]/hc = \mathscr{R}Z^2\left(\frac{1}{n^2} - \frac{1}{n'^2}\right). \quad (13.2)$$

With $Z = 1$ and $n = 2$ this is the famous Balmer formula for the series of lines in the visible spectrum emitted by atomic hydrogen. With $n = 1$ the frequencies of the Lyman series in the far ultraviolet are obtained, while $n = 3$ gives the Paschen series.

With $Z = 2$ the corresponding series of He^+ are obtained, and the higher integral values of Z correspond to Li^{++}, Be^{3+}, B^{4+}, C^{5+}, etc.

The spectrum of an electron in a Coulomb field has an especially simple form since the energy is dependent on n only, and not on l or m_l. If we are concerned with a system such as a lithium or a sodium atom, the single valence electron is in a Coulomb field only when far from the other electrons and the nucleus. Since, however, all but one of the electrons are held near the nucleus and only the one valence electron circulates around this inner electron core, a fair initial approximation is obtained by treating this valence electron as in a central but non-Coulomb field. Near the nucleus the potential energy will be $-Ze^2/r$. Far from the nucleus $V(r) \to -e^2/r$, and the transition from one form to the other can be approximated in various ways. The significant differences between such a spectrum and the hydrogen spectrum are due to the fact that in this kind of potential the energy depends on l.

All of the energies will be lower than in the case of hydrogen, because the potential energy is lower than in the Coulomb case with $Z = 1$. The terms with $l = 0$ will be lower than those with $l = 1$, and these latter will be lower than those with $l = 2$. However, because of the Pauli exclusion principle the total quantum number n cannot take on the value 1 in lithium or the values 1 or 2 in sodium. Such states are occupied by the core electrons.

Figure 13.1 gives a schematic illustration of the way in which the major

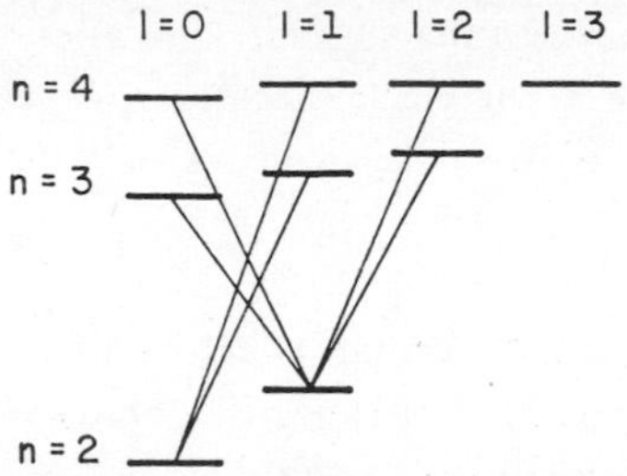

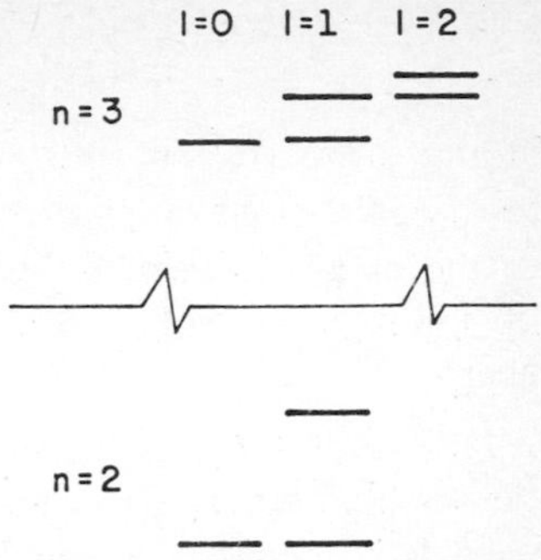

Fig. 13.1. Schematic arrangement of some energy levels of a single electron, without spin, moving in a non-Coulomb central field.

Fig. 13.2. Effect of electron spin in splitting the energy levels of an electron in a Coulomb field when the orbital motion is treated non-relativistically.

groups of energy levels are distributed in an atom such as lithium. With $n = 2$ there exist states with $l = 0$ and $l = 1$. With $n = 3$ there exist states with $l = 0, 1, 2$.

For historical reasons the states with $l = 0$ are called s states, with $l = 1$ are called p states, with $l = 2$, d states, $l = 3$, f states, etc. These terms are still useful, with lower-case letters designating states of single electrons, and capital letters designating states of more complex atoms.

Transitions from a series of states with a given value of l to a common lower state lead to a series of lines. Transitions from the series of s states to the state $n = 2$, $l = 1$ give rise to the "sharp" series. Those from the series of p states to the lowest state with $l = 0$ give rise to the principal series. These series names are much less common than formerly, and it is usually desirable to designate a series by giving the quantum numbers of the initial and final states.

PROBLEM 1. Assume that the potential of the central field in which an electron moves is given by

$$V(r) = -\frac{e^2}{4\pi\varepsilon_0}\left[\frac{1}{r} + (Z-1)\frac{e^{-\alpha r}}{r}\right]$$

and treat the second term as a perturbation. Find the first order correction to the eigenvalues of the energy for the levels with $n = 2$. Interpret the dependence of the result on the value of α.

1.2. *The Zeeman effect*

It was shown by P. Zeeman in 1896 that the spectral lines emitted by a luminous gas are split into several components when the gas is in a magnetic field. Some of the components are polarized with the electric vector parallel to the magnetic field. Others are circularly polarized if viewed along the direction of the field or transversely polarized if viewed at right angles to the field.

H. A. Lorentz gave a theoretical treatment of the Zeeman effect based on the simple oscillator picture of an atom current at that time. If a charged particle is bound to a center with a force proportional to the distance and the system is placed in a magnetic field, the motion can be described in terms of three frequencies. One is the frequency of the oscillator in the absence of the field and appears in a motion parallel to the field. The other two are, respectively, less than and greater than the undisturbed frequency and represent circular motions around the direction of the field.

The theory given by Lorentz predicts that a spectral line will split into three components. Except for negligible terms in the square of the magnetic field, these will be equally spaced, with the centre component at the position of the undisplaced line. The separation of the lines provides a means of measuring e/m since the spacing is proportional to this quantity.

A few spectral lines actually show such a normal triplet. Most, however, show a much more complicated structure referred to as the anomalous Zeeman effect, although it is much more common than the normal triplet. The anomalous Zeeman effect is due to the electron spin and its interaction with the orbital motion. When the spin is ignored, a quantum theoretical treatment also leads to a normal triplet.

Consider an electron in an arbitrary central field and a magnetic field along the polar axis. The Hamiltonian function is then

$$H = \frac{1}{2m}(p_x^2 + p_y^2 + p_z^2) - \frac{e}{2m}\boldsymbol{B}\cdot\boldsymbol{l} + V(\boldsymbol{r}) + \frac{e^2}{2m}A^2(\boldsymbol{r}). \qquad (13.3)$$

The Schroedinger equation in spherical coordinates is then

$$\nabla^2 u - \frac{ie}{\hbar}\boldsymbol{B}\frac{\partial u}{\partial \varphi} + \frac{2m}{\hbar^2}[E - V(\boldsymbol{r})]u = 0, \qquad (13.4)$$

when the term in A^2 is neglected. This neglect is usually justified in atomic spectroscopy since A is proportional both to the magnetic field $\boldsymbol{B}$ and to the distance from the nucleus, at which the vector potential can be set equal to zero. Equation (13.4) can be satisfied by a product of three functions

$$u(r, \vartheta, \varphi) = R_{n,l}(r)\, \Pi_l^{m_l}(\cos\vartheta) \frac{e^{im_l\varphi}}{(2\pi)^{\frac{1}{2}}}. \tag{13.5}$$

Substitution of the form (13.5) into equation (13.4), and the analysis used in Chapter 6, shows that the energy levels now depend on the magnetic quantum number m

$$E(n, l, m_l) = E(n, l) - m_l \mu_0 \hbar B, \tag{13.6}$$

where $\mu_0 = -|e|/2m$, and $E(n, l)$ includes the effect of the non-Coulomb field and depends on l. Each energy level is then split into $(2l+1)$ different levels corresponding to the $(2l+1)$ different values of m_l.

With this splitting of levels each spectral line will also be split. The observed frequencies will be

$$\begin{aligned}[E(n', l', m_l') - E(n, l, m_l)]/h &= [E(n', l') - E(n, l)]/h - (m_l' - m_l)\mu_0 B \\ &= \omega_0 - (m_l' - m_l)\mu_0 B.\end{aligned} \tag{13.7}$$

ω_0 represents the angular frequency emitted in the absence of a magnetic field. As will be shown in the next section, $(m_l' - m_l)$ can take on only the values 1, 0, -1; so equation (13.7) actually does represent a triplet of equally spaced lines.

PROBLEM 2. Write the Lagrangian function and the equations of motion for a charged particle in a uniform magnetic field and attracted toward the origin with a force proportional to the distance. Show that the equations are satisfied by a simple-harmonic motion along the direction of the field and two opposite circular motions in a plane perpendicular to the field.

PROBLEM 3. Show that equation (13.3) is the Hamiltonian function for the problem in question.

PROBLEM 4. Show that the energy eigenvalues are given by equation (13.6).

1.3. *Selection rules and intensities*

It was shown in Chapter 12 that the probability of the emission or absorption of dipole radiation is governed by the coordinate matrix element connecting two energy eigenstates. For radiation polarized with its electric vector along the z axis the matrix element

$$\langle n', l', m_l' \mid \mathbf{z} \mid n, l, m_l \rangle = \int u^*_{n', l', m'_l} r \cos\vartheta\, u_{n, l, m_l} r^2\, dr \sin\vartheta\, d\vartheta\, d\varphi \tag{13.8}$$

is required. The transition probability is proportional to the square of the absolute value of the matrix element, multiplied by the third power of the frequency. If this matrix element vanishes, the corresponding transition is said to be forbidden; it does not occur in the first approximation.

For a hydrogen-like atom, when the spin is neglected,

$$\langle n', l', m_l' \mid \mathbf{z} \mid n, l, m_l \rangle =$$
$$= \frac{1}{2\pi} \int_0^\infty R^*_{n',l'} R_{n,l} r^3 \, \mathrm{d}r \int_0^\pi \Pi_l^{m_l} \Pi_{l'}^{m'_l} \cos\vartheta \sin\vartheta \, \mathrm{d}\vartheta \int_0^{2\pi} \mathrm{e}^{\mathrm{i}(m_l - m'_l)\varphi} \, \mathrm{d}\varphi . \tag{13.9}$$

It can be seen immediately that this integral vanishes unless $m_l' = m_l$. Hence the magnetic quantum number does not change with the emission of radiation whose electric vector is along the z axis.

The integral over ϑ provides a selection rule on the quantum number l. One of the properties, listed in the Appendix, of the associated Legendre polynomials is

$$\cos\vartheta \, \Pi_l^m = \left[\frac{(l+m+1)(l-m+1)}{(2l+1)(2l+3)}\right]^{\frac{1}{2}} \Pi_{l+1}^m + \left[\frac{(l+m)(l-m)}{(2l+1)(2l-1)}\right]^{\frac{1}{2}} \Pi_{l-1}^m .$$

The integral over ϑ can thus be expressed as an integral of products of the angular functions Π_l^m. Because of the orthogonality, the integral is zero unless $l' = l \pm 1$, and in these two cases we have

$$\langle n', l+1, m_l \mid \mathbf{z} \mid n, l, m_l \rangle = \left[\frac{(l+m_l+1)(l-m_l+1)}{(2l+1)(2l+3)}\right]^{\frac{1}{2}} \int_0^\infty R^*_{n',l+1} R_{n,l} r^3 \, \mathrm{d}r \tag{13.10}$$

$$\langle n', l-1, m_l \mid \mathbf{z} \mid n, l, m_l \rangle = \left[\frac{(l+m_l)(l-m_l)}{(2l+1)(2l-1)}\right]^{\frac{1}{2}} \int_0^\infty R^*_{n',l-1} R_{n,l} r^3 \, \mathrm{d}r . \tag{13.10a}$$

The integral over r gives no selection rule on the quantum number n, but the transition probability can be evaluated for different cases when the wave functions are known.

The members of a Zeeman-effect triplet are close together in frequency, and so the relative intensities are given by the squares of the matrix elements. Transitions associated with radiation polarized parallel to the magnetic field then have the intensity

$$I_{\parallel, m_l} = I_0 \frac{(l+1)^2 - m_l^2}{(2l+1)(2l+3)} \tag{13.11}$$

where l is the smaller of the two l values. There will be $2l+1$ values of m_l

associated with this lower value of l, and hence there will be $2l+1$ transitions giving rise to the same undisplaced frequency. The total intensity of the central line will then be

$$I = I_0 \sum_{m=-l}^{m=l} \frac{(l+1)^2 - m_l^2}{(2l+1)(2l+3)} = \tfrac{1}{3}(l+1)I_0. \qquad (13.12)$$

A similar analysis shows that the matrix elements for the x coordinate are associated with transitions in which $m_l' = m_l \pm 1$ and that the sum of the transition probabilities corresponding to each of the displaced lines is equal to half the sum in equation (13.12). The triplet then consists of one undisplaced line polarized with its electric vector parallel to the magnetic field and two oppositely displaced lines of half the intensity with their electric vectors in the plane perpendicular to the magnetic field.

When the observed radiation is emitted in a direction perpendicular to the magnetic field, the lines are linearly polarized and the intensity is evenly divided between the two planes of polarization. When the observed radiation is that emitted in the direction of the field, the undisplaced line is missing altogether. The displaced lines are twice as strong as observed in the other direction, since both x and y polarization must be included, and they are oppositely circularly polarized.

PROBLEM 5. Work out the selection rules and the matrix elements for the coordinate x. Compute from them the intensity of the displaced lines in the normal triplet.

PROBLEM 6. The emission of circularly polarized light is governed by the matrix elements of $(x+iy)$ and $(x-iy)$. Show that the displaced components of the normal triplet are circularly polarized.

2. The spectrum of a single electron with spin

The preceding treatment of a single electron without spin is instructive but academic. An electron always has a spin, and the spin has a very significant influence on the spectrum.

2.1. *General scheme of levels*

As shown in Chapter 9 the inclusion of electron spin doubles all of the states, except those with $l = 0$. Each of the previous levels splits into two, one lQ above the original position and the other $(l+1)Q$ below. Figure 13.2 shows roughly the relative splitting of these levels.

The quantity Q can be evaluated, WALLER [1926], for the Coulomb-field wave functions and has the value*

$$Q_{n,l} = \frac{2\mathcal{R}\alpha^2 Z^4}{n^3 l(l+1)(2l+1)} \tag{13.13}$$

where $\alpha = e^2/\hbar c$ is the fine-structure constant.

The scheme of energy levels shown in fig. 13.2 is based on the electron motion in the Coulomb field and the electron spin. It is not exactly correct, however, to compute the energy of the electron motion nonrelativistically. When energy differences as small as $Q_{n,l}$ are important, account must be taken of the variation of electron mass with speed. This requires extensive modifications of the theory, outside of the range of this chapter. It is sufficient to note here that in atomic hydrogen the relativistic displacement of the levels associated with different values of l is just such as to bring certain pairs of the levels of fig. 13.2 into coincidence. The scheme of levels given by Dirac's relativistic theory, but classified in the terminology of this chapter, is then as shown in fig. 13.3. The spectrum associated with this system of levels is strongly dependent on the selection rules and the intensity relations of the permitted transitions. These will be taken up later.

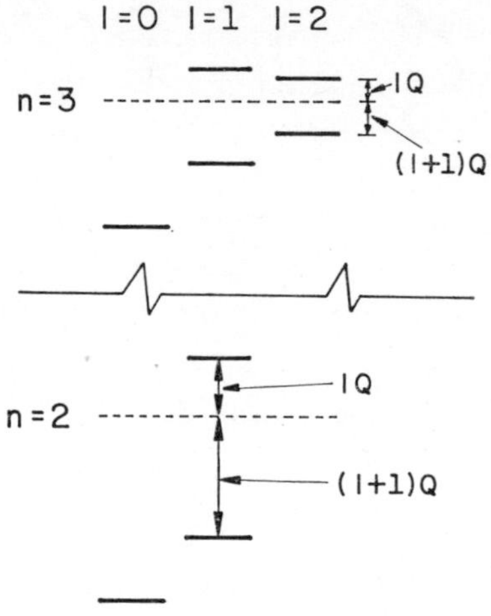

Fig. 13.3. The atomic energy levels in hydrogen when they are computed from the Dirac equation in which the spin appears as a consequence of relativity. The levels are arranged, however, according to the scheme of fig. 13.2.

* The expression (13.13) is not valid for l=0. For s states a relativistic treatment must be used to evaluate the spin effects.

This scheme of levels is in fair agreement with the observations, but careful spectroscopic measurements have shown it to be not quite exact. The observations show an additional displacement of the s levels relative to the others. With the advent of microwave techniques the displacement has been accurately determined, and modifications of the relativistic theory of an electron in a radiation field have given a good account of the best measurements; HOUSTON [1937], PASTERNAK [1938], LAMB and RETHERFORD [1947].

When the highest refinements of experimental and theoretical techniques are used, the hydrogen spectrum, which was the cornerstone of quantum theory in its early stages, remains one of the principal means for detailed quantitative comparison of theory and experiment.

In the case of atoms such as Li and Na a relativistic treatment of the valence electron is not necessary since the effect of the core electrons is only approximately known. The various series of lines are associated with transitions between double states except in the case of the sharp series and the principal series when one of the levels is single. Thus the principal series of Na is a series of doublets of which the well-known pair in the yellow is the first member. Other series are series of triplets since one of the possible transitions is forbidden by the selection rules.

2.2. *The anomalous Zeeman effect*

To calculate the splitting of the doublet levels in the presence of a magnetic field, the energy in the field and the spin-orbit interaction must be treated together. When the magnetic field is small and the Zeeman-effect separations are small compared with $Q_{n,l}$, the magnetic energy can be considered as an additional perturbation, after the spin-orbit interaction has been considered. This kind of procedure would give the anomalous Zeeman effect for small fields. If, however, both effects are treated together, the change in the Zeeman effect with increasing magnetic field can also be evaluated.

The part of the Hamiltonian operator to be treated as a perturbation is

$$\mathsf{H}_1 = (\mathsf{Q}/\hbar^2)\mathbf{l}\cdot\mathbf{s} - (\mu_0/\hbar)\boldsymbol{B}\cdot(\mathbf{l}+2\mathbf{s}), \tag{13.14}$$

where Q represents the operator from equation (9.43). Then placing the z axis along the magnetic field and following the notation of Chapter 9,

$$\mathsf{H}_1 \mid n, l, j, m\rangle = \\ = [(\mathsf{Q}/\hbar^2)\mathbf{l}\cdot\mathbf{s} - (\mu_0/\hbar)\boldsymbol{B}\cdot(\mathbf{l}+2\mathbf{s})] \sum_{m_l, m_s} a(n, l, m_l, m_s) \mid l, m_l, m_s\rangle \,. \tag{13.15}$$

The ket vector on the left is designated by the quantum numbers n and l of the wave functions treated in Chapter 6 and used in equation (13.5) above. The j and the m refer to total angular momentum, orbital and spin, and its component along the z-axis. On the right side the ket vectors are just those of the wave function in equation (13.5) multiplied by a spin eigenvector with eigenvalue along the polar axis of $m_s\hbar$. The n is omitted in writing the ket vector since we deal with only one value of n at a time. The inclusion of l is also really unnecessary but it helps to keep one reminded that the maximum value of m_l is l.

The operator can be expanded as in equation (9.48) and the result of its application is

$$\begin{aligned} \mathsf{H}_1 \mid n, l, j, m\rangle &= \sum_{m_l, m_s} a(l, m_l, m_s) \times \\ &\times \{Q[(l-m_l)^{\frac{1}{2}}(l+m_l+1)^{\frac{1}{2}}(\tfrac{1}{2}+m_s)^{\frac{1}{2}}(\tfrac{3}{2}-m_s)^{\frac{1}{2}} \mid l, m_l+1, m_s-1\rangle + \\ &+ (l+m_l)^{\frac{1}{2}}(l-m_l+1)^{\frac{1}{2}}(\tfrac{1}{2}-m_s)^{\frac{1}{2}}(\tfrac{3}{2}+m_s)^{\frac{1}{2}} \mid l, m_l-1, m_s+1\rangle + \\ &+ 2m_l s \mid l, m_l, m_s\rangle] - \mu_0 B(m_l+2m_s) \mid l, m_l, m_s\rangle\}\,. \end{aligned} \tag{13.16}$$

The scalar product of (13.16) with $\langle l, m_l', m_s' \mid$ leads to the secular equation

$$\begin{aligned} &\langle l, m_l', m_s' \mid E_1 - \mathsf{H}_1 \mid n, l, j, m\rangle = \\ &- Q(l-m_l'+1)^{\frac{1}{2}}(l+m_l')^{\frac{1}{2}}(\tfrac{3}{2}+m_s')^{\frac{1}{2}}(\tfrac{1}{2}-m_s')^{\frac{1}{2}} a(l, m_l'-1, m_s'+1) + \\ &+ [E_1 - 2m_l' m_s' Q + (m_l'+2m_s')\mu_0 B] a(l, m_l', m_s') + \\ &- Q(l+m_l'+1)^{\frac{1}{2}}(l-m_l')^{\frac{1}{2}}(\tfrac{3}{2}-m_s')^{\frac{1}{2}}(\tfrac{1}{2}+m_s')^{\frac{1}{2}} a(l, m_l'+1, m_s'-1) = 0\,. \end{aligned} \tag{13.17}$$

This equation differs from equation (9.60) only in containing the term in the magnetic field. But because of this term the eigenvalues depend on $m = m_l + m_s$. The equations (13.17) can be divided into pairs containing two coefficients each, as before, but the eigenvalues are different for the different pairs. One of these pairs is

$$\begin{aligned} &[E_1 - (m-\tfrac{1}{2})Q + (m+\tfrac{1}{2})\mu_0 B] a(l, m_l, \tfrac{1}{2}) + \\ &\qquad - Q(l+m+\tfrac{1}{2})^{\frac{1}{2}}(l-m+\tfrac{1}{2})^{\frac{1}{2}} a(l, m_l+1, -\tfrac{1}{2}) = 0 \\ &- Q(l+m+\tfrac{1}{2})^{\frac{1}{2}}(l-m+\tfrac{1}{2})^{\frac{1}{2}} a(l, m_l, \tfrac{1}{2}) + \\ &\qquad + [E_1 + (m+\tfrac{1}{2})Q + (m-\tfrac{1}{2})\mu_0 B] a(l, m_l+1, -\tfrac{1}{2}) = 0\,. \end{aligned} \tag{13.18}$$

In these equations the distinction between m and m_l must be carefully noted. $m\hbar = j_z' = (m_l+m_s)\hbar$, and each pair of equations contains only those coefficients $a(l, m_l, m_s)$ for which (m_l+m_s) has the same value. Also note that the second quantum number in the parenthesis of each $a(l, m_l, m_s)$ gives the value of m_l in the ket vector which it multiplies. Thus $a(l, m+1, -\frac{1}{2})$ is the coefficient of $| l, m_l+1, -\frac{1}{2}\rangle$.

For $m = \pm(l+\frac{1}{2})$ there is, again as before, just one equation leading to the eigenvalues

$$E_1 = lQ \pm (l+1)\mu_0 B. \tag{13.19}$$

For other values of m

$$E_1 = -\tfrac{1}{2}Q - m(\mu_0 B) \pm \tfrac{1}{2}[(2l+1)^2 Q^2 - 4m(\mu_0 B) + (\mu_0 B)^2]^{\frac{1}{2}}. \tag{13.20}$$

To follow the development of the energy level pattern from small values of B to very large ones it is convenient to define $\varepsilon = \mu_0 B\,(2l+1)Q$. Then equation (13.20) becomes

$$\frac{E_1}{Q} = -\tfrac{1}{2} - m(2l+1)\varepsilon \pm (l+\tfrac{1}{2}) \left[1 - \frac{4m}{2l+1}\varepsilon + \varepsilon^2\right]^{\frac{1}{2}}. \tag{13.21}$$

For small values of ε the two energy values are

$$E_1^{(1)} = lQ - m\left(\frac{2l+2}{2l+1}\right)\mu_0 B \tag{13.22a}$$

$$E_1^{(2)} = -(l+1)Q - m\left(\frac{2l}{2l+1}\right)\mu_0 B. \tag{13.22b}$$

By comparison with the results in Chapter 9 where the magnetic field was not included it is clear that $E_1^{(1)}$ refers to the states for which $j = l+\frac{1}{2}$ and $E_1^{(2)}$ to the states for which $j = l-\frac{1}{2}$. Thus $(2l+2)/(2l+1) = (2j+1)/(2l+1)$ and $2l/(2l+1) = (2j+1)/(2l+1)$ in each case. The quantity $g_{j,l} = (2j+1)/(2l+1)$ is called the Landé g-factor, with which it is possible to express the splitting of each level as proportional to the magnetic field for small values of the field.

On the other hand, when $\varepsilon \gg 1$, the whole set of six energy values becomes a set of five equally spaced levels if the two close together are considered to be coincident. Figure 13.4 indicates the way in which these energy values, for the case $l = 1$, change from two groups with 2 and 4 components respectively to a set of 5. In the limit of large ε the effect of the magnetic field is predominant. The orbital angular momentum and the spin must be considered separately and their effect added.

The equally spaced levels at high fields give rise to spectral lines which are essentially normal triplets, and the transformation of the spectrum from the relatively complicated "anomalous" Zeeman effect at low fields toward a

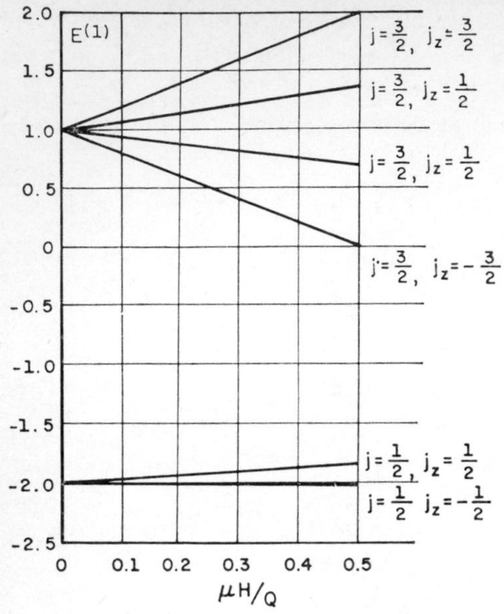

Fig. 13.4. Effect of a "small" magnetic field on the energy levels of a single electron with spin in a central field. This distribution of levels leads to the anomalous Zeeman effect. The magnetic field is "small" since the splitting it produces is small compared with that due to the spin-orbit interaction.

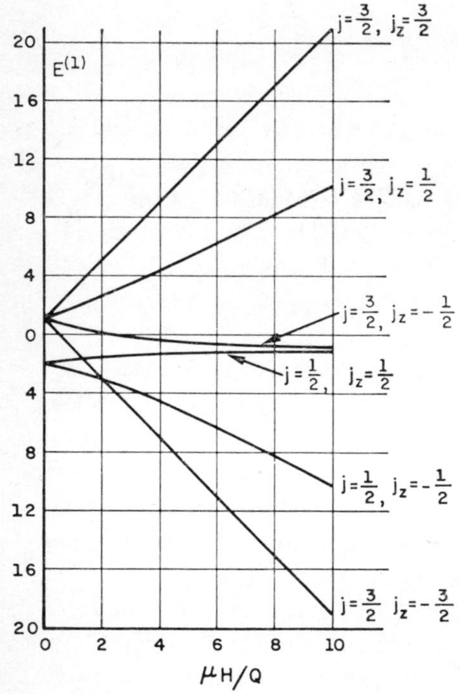

Fig. 13.5. Effect of a "large" magnetic field leading to the Paschen-Back effect, an approximation to a normal triplet.

normal triplet at high fields is known as the Paschen-Back effect. It is observed in spectra having small enough multiplet separations.

2.3. *Selection rules and intensities*

As will be shown later, most selection rules and intensity ratios can be evaluated on the basis of the commutation rules only. However, the absolute intensities, or probabilities of transition, must be evaluated by evaluating such matrix elements as are indicated in Chapter 12. It is also instructive to consider this evaluation in some of the simple cases as illustrations of the complete procedure.

For the case of low magnetic fields, $\varepsilon \to 0$, the ket vectors are as given in Chapter 9:

$$| n, l, l+\tfrac{1}{2}, m_l+\tfrac{1}{2}\rangle = \left(\frac{l+m_l+1}{2l+1}\right)^{\frac{1}{2}} | l, m_l, \tfrac{1}{2}\rangle + \left(\frac{l-m_l}{2l+1}\right)^{\frac{1}{2}} | l, m_l+1, -\tfrac{1}{2}\rangle \tag{13.23a}$$

$$| n, l, l-\tfrac{1}{2}, m_l+\tfrac{1}{2}\rangle = \left(\frac{l-m_l}{2l+1}\right)^{\frac{1}{2}} | l, m_l, \tfrac{1}{2}\rangle - \left(\frac{l+m_l+1}{2l+1}\right)^{\frac{1}{2}} | l, m_l+1, -\tfrac{1}{2}\rangle . \tag{13.23b}$$

For the given l only the two values of j are possible.

Consider the matrix element of the coordinate z between two states $| n', l', j', m'\rangle$ and $| n, l, j, m\rangle$. Since z is independent of the spin, only ket vectors with the same spin will lead to a bracket different from zero. Thus

$$\begin{aligned} \langle n', l', l'+\tfrac{1}{2}, m' \,|\, \mathbf{z} \,|\, n, l, l+\tfrac{1}{2}, m\rangle &= \int R^*_{n', l'}(r) R_{n, l}(r) r^3 \,\mathrm{d}r \times \\ \times \left(\frac{(l'+m_l'+1)\,(l+m_l+1)}{(2l'+1)\,(2l+1)}\right)^{\frac{1}{2}} &\int_0^\pi \cos\vartheta \sin\vartheta \,\mathrm{d}\vartheta \int_0^{2\pi} Y_{l'}^{m_l'*}(\vartheta, \varphi) Y_l^{m_l}(\vartheta, \varphi)\,\mathrm{d}\varphi \\ + \left(\frac{(l'-m_l')\,(l-m_l)}{(2l'+1)\,(2l+1)}\right)^{\frac{1}{2}} &\int_0^\pi \cos\vartheta \sin\vartheta \,\mathrm{d}\vartheta \int_0^{2\pi} Y_{l'}^{m_l'+1*}(\vartheta, \varphi) Y_l^{m_l+1}(\vartheta, \varphi)\,\mathrm{d}\varphi . \end{aligned} \tag{13.24}$$

The integrals over φ are clearly zero unless $m_l' = m_l$, and over ϑ are zero unless $l' = l \pm 1$. Using the values of the integrals that do not vanish

$$\begin{aligned} \langle n', l+1, l+\tfrac{3}{2}, m \,|\, \mathbf{z} \,|\, n, l, l+\tfrac{1}{2}, m\rangle &= \mathscr{R}(n', l+1; n, l) \times \\ \times \left[\frac{(l+m_l+2)^{\frac{1}{2}}(l+m_l+1)\,(l-m_l+1)^{\frac{1}{2}}}{(2l+3)\,(2l+1)} \right. &+ \left. \frac{(l-m_l+1)^{\frac{1}{2}}(l-m_l)\,(l+m_l+2)^{\frac{1}{2}}}{(2l+3)\,(2l+1)}\right] \end{aligned} \tag{13.25}$$

for the case $l' = l+1$. $\mathscr{R}(n', l+1; n, l)$ represents the radial integral. For the other case, in which $l' = l-1$

$$\langle n', l-1, l+\tfrac{1}{2}, m \mid \mathbf{z} \mid n, l, l+\tfrac{1}{2}, m\rangle = \mathscr{R}(n', l-1; n, l) \times$$
$$\times \left[\frac{(l+m_l+1)^{\frac{1}{2}}(l+m_l)(l-m_l)^{\frac{1}{2}}}{(2l+1)(2l-1)} + \frac{(l+m_l+1)^{\frac{1}{2}}(l-m_l-1)(l-m_l)^{\frac{1}{2}}}{(2l+1)(2l-1)}\right]. \tag{13.26}$$

These can be more succinctly written in terms of j and m. In equation (13.25) $j = l+\frac{1}{2}$ and $m = m_l+\frac{1}{2}$,

$$|\langle n', l+1, j+1, m \mid \mathbf{z} \mid n, l, j, m\rangle|^2 = \mathscr{R}^2[(j+1)^2-m^2]/4(j+1)^2. \tag{13.25a}$$

In equation (13.26) where $j = l-\frac{1}{2}$ and $m = m_l+\frac{1}{2}$,

$$|\langle n', l-1, j-1, m \mid \mathbf{z} \mid n, l, j, m\rangle|^2 = \mathscr{R}^2(j^2-m^2)/4j^2. \tag{13.26a}$$

The same results in terms of j and m are obtained with equation (13.23b) so the relative transition probabilities are expressible in terms of j and m without reference to the l involved. Of course the l appears in the $\mathscr{R}$ term where its effect depends upon the departure of the central field from a Coulomb field.

Transitions between the two different kinds of states in (13.23a) and (13.23b) lead to

$$\langle n', l+1, l+\tfrac{1}{2}, m \mid \mathbf{z} \mid n, l, l+\tfrac{1}{2}, m\rangle =$$
$$= \mathscr{R}(n', l+1; n, l)\left[\frac{(l+m_l+1)(l-m_l+1)}{(2l+1)(2l+3)} - \frac{(l+m_l+2)(l-m_l)}{(2l+1)(2l+3)}\right]$$
$$= \mathscr{R}(n', l+1; n, l)\frac{m}{2j(j+1)} \tag{13.27}$$

for the case $l' = l+1$. The matrix element is zero for $l' = l-1$.

These results show that in addition to the selection rule on l there is a limitation on changes in j. j can change only by ± 1 or by 0.

In the absence of a magnetic field the energy of a state is independent of m. In a gas one may expect the states with different values of m to be present in equal numbers and so the intensity of a spectral line polarized with the electric vector along the z axis will be given by the sum over all values of m, of the above intensities. For the three cases

$$j' = j+1 : I \sim \mathscr{R}^2 \sum_{m=-j}^{m=j} \frac{(j+1)^2-m^2}{4(j+1)^2} = \mathscr{R}^2\frac{(2j+1)(2j+3)}{12(j+1)} \tag{13.28a}$$

$$j' = j : I \sim \mathscr{R}^2 \sum_{m=-j}^{m=+j} \frac{m^2}{4j^2(j+1)^2} = \mathscr{R}^2 \frac{(2j+1)}{12j(j+1)} \tag{13.28b}$$

$$j' = j-1 : I \sim \mathscr{R}^2 \sum_{m=-j}^{m=j} \frac{j^2-m^2}{4j^2} = \mathscr{R}^2 \frac{(2j+1)(2j-1)}{12j}. \tag{13.28c}$$

It can be shown by a similar process that identical results are obtained by using matrix elements of x or y. This must be so since the z axis is in no way different from the others in the absence of a magnetic or other field to fix a direction in space.

From equations (13.28) can be obtained the relative intensities of the lines in the multiplets of one-electron spectra; and to the extent that the radial integrals can be evaluated, the absolute intensities can be obtained also. Since, in many cases, the different members of a multiplet have very nearly the same frequency, the factor ν^4 can be ignored and the intensities will be proportional to the expressions in (13.28).

PROBLEM 7. Work out the Zeeman effect pattern for transitions between a p state and an s state when the magnetic field is small.

PROBLEM 8. Evaluate the coefficients from equations (13.18) when $\varepsilon \gg 1$ and then evaluate the transition probabilities and the Zeeman effect pattern for transitions between a p state and an s state.

PROBLEM 9. The "diffuse" series of a hydrogen-like spectrum is associated with transitions between a series of d states and a low-lying p state. Show that this is a series of triplets and find the intensity ratios of the members.

PROBLEM 10. Work out selection rules on j and m for radiation polarized along the x and y axes.

CHAPTER 14

TWO-ELECTRON SPECTRA

The study of a system consisting of two electrons moving in the central field of a fixed nucleus provides illustrations of many of the important results of spectroscopy without introducing undue complications. For this reason such a system will be treated at length. The calculations for two-electron systems are, of course, strictly applicable only to such atoms as He, Li^+, Be^{++}, etc., but practically they are useful for all cases in which there are two electrons outside of a closed shell, such as Be, B^+, C^{++}, Mg, Ca, etc. The approximation in these cases is similar to that in which the treatment of a one-electron system is applied to the alkali metals.

1. The Hamiltonian function

The Hamiltonian operator to be used for the two-electron system is

$$\mathsf{H} = \sum_{i=1}^{2} \left[\frac{\mathsf{P}_i^2}{2m} + V(r_i)\right] + \frac{e^2}{4\pi\varepsilon_0 r_{1,2}} + \frac{\mu}{mc^2} \sum_{i=1}^{2} \left[\frac{1}{r_i}\frac{\partial V(r_i)}{\partial(r_i)}\right] \mathbf{l}_i \cdot \mathbf{s}_i . \tag{14.1}$$

The first sum represents the kinetic energy and the potential energy of the electrons in the central field. The form of this central field is not specified, since in the case of two electrons outside of a closed shell, it will depend on the nature of the closed shell. At large distances, however, $V(r) \to -2e^2/4\pi\varepsilon_0 r$. On account of the identity of the electrons they are both taken as moving in the same central field. The term $e^2/4\pi\varepsilon_0 r_{1,2}$ is the electrostatic repulsion between the electrons. When the mutual repulsion is small compared with the energy in the central field, such as in the case of C^{++}, the approximations to be described are fairly good. When it is of the same order of magnitude, as in He, the approximate results are much poorer.

The last term in the Hamiltonian function represents the interaction of the spin of each electron with its own orbital motion. $\mu = -|e|/m$, and $\mathbf{l}_1$ and

$\mathbf{l}_2$ represent the angular momentum operators of the two electrons. Strictly speaking this is not the whole Hamiltonian function. There should be included terms representing the interaction of each spin with the orbital motion of the other electron, and also the mutual interaction of the spins. For most spectroscopic problems these latter terms can be omitted without seriously affecting the results, although, in a few cases such as He and Li^+, they are of importance in treating the fine structure. Furthermore, when dealing with the finest details of a spectrum, the possibility that relativistic effects are important must be taken into account.

The usual procedure for treating this problem is to regard the last two terms of the Hamiltonian function as perturbations. The solutions for the first part are assumed to be known. When the term $e^2/4\pi\varepsilon_0 r_{1,2}$ is more important than the spin-orbit interaction, it is applied first, and the spin-orbit interaction is applied as a second perturbation to the result thus obtained. This case is called Russell-Saunders coupling because of the pioneer work of Russell and Saunders in analyzing spectra of this kind. The opposite case, in which the spin-orbit interaction is the more important and is applied first, is called the case of *jj* coupling. In this case the spin of each electron is first combined with its own orbital motion to give a total angular momentum j for each electron. The electrons are then coupled together by means of the electrostatic interaction. It is also possible to apply both perturbations simultaneously. This will be done to obtain information about the intermediate cases, but the customary term designations are justified only in one or the other limit.

The perturbation treatment is considerably simplified by an important theorem, which is only a slight extension of theorems already mentioned in other connections. Consider two Hermitian operators, A and B, which commute with each other, and a set of eigenkets $| b'\rangle$ of the operator B with the real eigenvalues b'. Then

$$b' \langle b' | \mathsf{A} | b'' \rangle = \langle b' | \mathsf{BA} | b'' \rangle = \langle b' | \mathsf{AB} | b'' \rangle = b'' \langle b' | \mathsf{A} | b'' \rangle$$

and it follows that

$$\langle b' | \mathsf{A} | b'' \rangle = 0 \quad \text{unless} \quad b' = b''. \tag{14.2}$$

Because of this theorem it is convenient to find operators which commute with the perturbation to be applied, in this case the electrostatic interaction, and then to select, as a starting point, groups of eigenfunctions of these operators. Such functions will be linear combinations of the energy eigenfunctions of the Hamiltonian without the last two terms. Equation (14.2) then shows that in finding the correct zero-approximation functions for the application of the perturbation, only those in a single group need be considered. In some cases the groups can be made so small as to contain one function only.

2. The electrostatic interaction

The solutions of the unperturbed problem will be products of the one-electron functions, characteristic of the energy when only the central field is taken into account. These products require eight quantum numbers for their specification,

$$\left| \begin{matrix} n_1, l_1, m_{l_1}, m_{s_1} \\ n_2, l_2, m_{l_2}, m_{s_2} \end{matrix} \right\rangle = \frac{1}{2\pi} R_{n_1,l_1}(r_1) R_{n_2,l_2}(r_2) \Pi_{l_1}^{m_{l1}}(\cos \vartheta_1) \Pi_{l_2}^{m_{l2}}(\cos \vartheta_2)$$

$$\times \exp\{\mathrm{i}(m_{l_1}\varphi_1 + m_{l_2}\varphi_2)\} \mid m_{s_1}, m_{s_2}\rangle . \tag{14.3}$$

On the right-hand side, this ket vector is expressed as a wave function for the orbital part but with the ket vector notation for the spin part. The energy of the unperturbed system depends only on n_1, l_1, n_2 and l_2, so that the zero-approximation functions must be formed out of the $4(2l_1+1)(2l_2+1)$ functions which have the same energy, by summing over m_{l_1} and m_{l_2}, and using the four combinations of m_{s_1} and m_{s_2}. If the perturbation were applied directly in the usual way it would lead to $4(2l_1+1)(2l_2+1)$ simultaneous equations. It is possible to avoid this complication by selecting, in advance, suitable linear combinations of the unperturbed functions and applying the perturbation to them. To select such states we make use of commutation relations between the operators of which the states are eigenstates.

2.1. *Operators that commute with the electrostatic interaction*

The operators which commute with the electrostatic interaction include

(i) *Any function of the spin only.* This is obvious since the electrostatic interaction is entirely independent of the spin variables.

(ii) $L_z = (\hbar/\mathrm{i})(\partial/\partial\varphi_1 + \partial/\partial\varphi_2)$. To show this, the electrostatic interaction can be expressed in terms of the arguments of the wave function. One of the properties of Legendre polynomials is that

$$\frac{1}{r_{1,2}} = \sum_l \frac{r_\mathrm{a}^l}{r_\mathrm{b}^{l+1}} P_l(\cos \vartheta)$$

where r_a is the smaller and r_b is the larger of the distances r_1 and r_2. ϑ is the angle between these two radii. The Legendre polynomials of this angle can be expressed in terms of the associated polynomials of ϑ_1 and ϑ_2 to give

$$\frac{1}{r_{1,2}} = \sum_{l,m} \frac{r_{\mathrm{a}}^{l}}{r_{\mathrm{b}}^{l+1}} \frac{\Pi_l^m(\cos\vartheta_1)\Pi_l^m(\cos\vartheta_2)}{[\Pi_l(1)]^2} \mathrm{e}^{im(\varphi_1-\varphi_2)}. \qquad (4.14)$$

From the dependence of this expression on φ_1 and φ_2, it follows that the operator L_z commutes with $1/r_{1,2}$.

(iii) $\mathsf{L}^2 = \mathsf{L}_x^2+\mathsf{L}_y^2+\mathsf{L}_z^2$. Since L_z commutes with $1/r_{1,2}$, L_x and L_y do likewise and as a consequence so does L^2. The reason L^2 is of importance rather than L_x and L_y themselves is that there are no eigenstates of both L_x and L_z. On the other hand, L^2 commutes with L_z, and there can be states characteristic of both at the same time. The letter L is used here instead of J as in Chapter 9 since it refers to orbital angular momentum only.

(iv) *The operator* P *which represents the interchange of the four coordinates of the two electrons.* Since $1/r_{1,2}$ is symmetrical in these coordinates, it clearly commutes with P.

PROBLEM 1. Illustrate the theorem of equation (14.2) by showing that $\langle n, l', m_l \mid \mathsf{l}_z \mid n, l, m_l\rangle = 0$ when $l' \neq l$.

2.2. *Eigenfunctions of the operators that commute with* $1/r_{1,2}$

The problem is now to find combinations of the wave functions, or ket vectors, (14.3) which are eigenfunctions of as many of the above operators as possible. Since all functions of the spin operators commute with $1/r_{1,2}$ any function of the spin variables serves as well as any other until the spin-orbit interaction is taken into account. The four possible combinations of m_{s_1} and m_{s_2} serve the purpose satisfactorily.

The function of equation (14.3) can well be written as a ket vector, since it is specified by eigenvalues of the various one-electron operators. The operator L_z operating on it gives

$$\mathsf{L}_z \left| \begin{matrix} n_1, l_1, m_{l_1}, m_{s_1} \\ n_2, l_2, m_{l_2}, m_{s_2} \end{matrix} \right\rangle = (m_{l_1}+m_{l_2})\hbar \left| \begin{matrix} n_1, l_1, m_{l_1}, m_{s_1} \\ n_2, l_2, m_{l_2}, m_{s_2} \end{matrix} \right\rangle \qquad (14.5)$$

so it is already an eigenvector of L_z with the eigenvalue $(m_1+m_2)\hbar$. Those ket vectors with the same value of m_1+m_2 may be designated by $M_{\mathrm{L}} = m_1+m_2$, a "magnetic" quantum number for the combined system. Since the maximum values of m_{l_1} and m_{l_2} are l_1 and l_2, the maximum value of the sum will be l_1+l_2, and there will be but one function (with given values of $n_1, n_2, l_1, l_2, m_{s_1}, m_{s_2}$) with this maximum value. For $M_{\mathrm{L}} = l_1+l_2-1$ there will be two functions, one with $m_{l_1} = l_1$, $m_{l_2} = l_2-1$, and one with $m_{l_1} = l_1-1$, $m_{l_2} = l_2$. For $M = l_1+l_2-s$ there will be $(s+1)$ functions when

$s \leq 2l_2$, $(l_2 \leq l_1)$. When $s > 2l_2$ there will be just $(2l_2+1)$ functions for all the remaining positive values of M_L, including zero. For negative values of M_L there will be just as many functions as for the corresponding positive values. For a pair of electrons one of which is in a d state and the other in a p state, $l_1 = 2$ and $l_2 = 1$. For $M_L = \pm 3$ there is one function, for $M_L = \pm 2$ there are two, for $M_L = \pm 1$ there are three, and for $M_L = 0$ there are again three.

The next problem is to select from each of these groups of eigenfunctions of L_z, those linear combinations which are eigenfunctions of L^2. Because L^2 commutes with L_z, equation (14.2) shows this is possible. The direct method of making the selection is to insert an undetermined linear combination into the eigenvalue equations and determine the coefficients. This procedure can be illustrated in two simple cases. Consider first the state with $M_L = l_1+l_2$. There is only one ket vector of this kind; there are no coefficients to be determined. It remains only to show that this state vector is really an eigenstate of the square of the total angular momentum and to find the eigenvalue. The operator may be expressed in terms of the operators on the individual particles:

$$\mathsf{L}^2 = (\mathsf{l}_{x_1}+\mathsf{l}_{x_2})^2+(\mathsf{l}_{y_1}+\mathsf{l}_{y_2})^2+(\mathsf{l}_{z_1}+\mathsf{l}_{z_2})^2 = \mathsf{l}_1^2+\mathsf{l}_2^2+2\mathsf{l}_1\cdot\mathsf{l}_2. \tag{14.6}$$

Then

$$\mathsf{L}^2 \left| \begin{matrix} n_1, l_1, l_1, m_{s_1} \\ n_2, l_2, l_2, m_{s_2} \end{matrix} \right\rangle = [l_1(l_1+1)\hbar^2+l_2(l_2+1)\hbar^2+2\mathsf{l}_1\cdot\mathsf{l}_2] \left| \begin{matrix} n_1, l_1, l_1, m_{s_1} \\ n_1, l_1, l_1, m_{s_2} \end{matrix} \right\rangle$$

$$= (l_1+l_2)(l_1+l_2+1)\hbar^2 \left| \begin{matrix} n_1, l_1, l_1, m_{s_1} \\ n_2, l_2, l_2, m_{s_2} \end{matrix} \right\rangle$$

$$= L(L+1)\hbar^2 \left| \begin{matrix} n_1, l_1, m_{s_1}, n_2, l_2, m_{s_2} \\ L = M_L = l_1+l_2 \end{matrix} \right\rangle. \tag{14.6a}$$

The operator l_i^2 operates on only the angular function $Y(\vartheta_i\varphi_i)$ and results in multiplying it by $l_i(l_i+1)$. The product $2\mathsf{l}_1\cdot\mathsf{l}_2$ can be expressed as in previous chapters but the only effective term is $2\mathsf{l}_{z_1}\mathsf{l}_{z_2}$ which results in multiplication by $2l_1l_2$. This ket vector is thus properly designated by the eigenvalues L and $M_L = L = l_1+l_2$.

In the next case with $M_L = l_1+l_2-1$ there are two ket vectors and two equations. Let

$$\left| \begin{matrix} n_1, l_1, m_{s_1}, n_2, l_2, m_{s_2} \\ L, M = l_1+l_2-1 \end{matrix} \right\rangle = a \left| \begin{matrix} n_1, l_1, l_1, m_{s_1} \\ n_2, l_2, l_2-1, m_{s_2} \end{matrix} \right\rangle + b \left| \begin{matrix} n_1, l_1, l_1-1, m_{s_1} \\ n_2, l_2, l_2, m_{s_2} \end{matrix} \right\rangle. \tag{14.7}$$

Operating on this linear combination by L^2 and making use of the orthogonality of the ket vectors leads to two simultaneous equations which can be satisfied with two possible eigenvalues of L^2. They are given by $L^2 = L(L+1)$ with $L = l_1+l_2$ and $L = l_1+l_2-1$. The corresponding values of the coefficients a and b can be determined to give the functions corresponding to these eigenvalues. The coefficients are, in fact, the Clebsch-Gordon coefficients discussed in Chapter 9. The process can then be continued for all the allowed values of L and M_L.

PROBLEM 2. Evaluate the coefficients a and b of equation (14.7).

PROBLEM 3. Find the wave functions, or ket vectors, that are eigenfunctions of L^2 with $M_L = l_1+l_2-2$.

2.3. *Eigenvalues from the operator relationships*

As shown in Chapter 8 for the harmonic oscillator, and in Chapter 9 for the angular momentum, it is often possible to learn something about the eigenvalues of an operator without knowing all about the wave functions or the state vectors. It was shown in problem 3 of Chapter 9 that if any operator K commutes with all three components of the angular momentum $\mathbf{L}$, it will have matrix elements only between states of the same L and M_L. It can be shown further that these matrix elements will be independent of M_L.

Let $|\gamma, L, M\rangle$ be an eigenket of L^2, of L_z, and of other operators whose eigenvalues are lumped together as γ. Then

$$\begin{aligned}\mathsf{L}_+\mathsf{K}\,|\,\gamma, L, M_L\rangle &= \mathsf{L}_+ \sum_{\gamma'} \langle\gamma', L, M_L\,|\,\mathsf{K}\,|\,\gamma, L, M_L\rangle\,|\,\gamma', L, M_L\rangle \\ &= \sum_{\gamma'} \hbar[L(L+1)-M_L^2-M_L]^{\frac{1}{2}}\times \\ &\quad\times\langle\gamma', L, M_L\,|\,\mathsf{K}\,|\,\gamma, L, M_L\rangle\,|\,\gamma', L, M_L+1\rangle.\end{aligned}$$

Also

$$\begin{aligned}\mathsf{K}\mathsf{L}_+\,|\,\gamma, L, M_L\rangle &= \mathsf{K}[L(L+1)-M_L^2-M_L]^{\frac{1}{2}}\,\hbar\,|\,\gamma, L, M_L+1\rangle \\ &= \sum_{\gamma'} \hbar[L(L+1)-M_L^2-M_L]^{\frac{1}{2}}\times \\ &\quad\times\langle\gamma', L, M_L+1\,|\,\mathsf{K}\,|\,\gamma, L, M_L+1\rangle\,|\,\gamma', L, M_L+1\rangle.\end{aligned}$$

Comparison of these two series indicates that

$$\langle\gamma', L, M_L+1\,|\,\mathsf{K}\,|\,\gamma, L, M_L+1\rangle = \langle\gamma', L, M_L\,|\,\mathsf{K}\,|\,\gamma, L, M_L\rangle. \quad (14.8)$$

This theorem can be applied immediately to the operator L^2 itself, and makes it possible to determine the eigenvalues of L^2 without solving equations such as would be obtained from functions of the form (14.7). As shown in equation (14.6)

$$\left\langle \begin{matrix} n_1, l_1, m_{s_1}, n_2, l_2, m_{s_2} \\ M_L = L = l_1+l_2 \end{matrix} \middle| \mathsf{L}^2 \middle| \begin{matrix} n_1, l_1, m_{s_1}, n_2, l_2, m_{s_2} \\ M_L = L = l_1+l_2 \end{matrix} \right\rangle = (l_1+l_2)(l_1+l_2+1). \tag{14.9}$$

The above theorem then indicates that

$$\left\langle \begin{matrix} n_1, l_1, m_{s_1}, n_2, l_2, m_{s_2} \\ M_L = L-1 = l_1+l_2-1 \end{matrix} \middle| \mathsf{L}^2 \middle| \begin{matrix} n_1, l_1, m_{s_1}, n_2, l_2, m_{s_2} \\ M_L = L-1 = l_1+l_2-1 \end{matrix} \right\rangle$$

also is equal to $(l_1+l_2)(l_1+l_2+1)$. However we do not immediately have the ket vector with $M_L = l_1+l_2-1$. We know only that it is a linear combination of two ket vectors as indicated in equation (14.7). The pair of simultaneous equations for determining the coefficients in (14.7) is consistent only when the determinant of its coefficients is equal to zero. But the sum of the roots of such a determinantal equation is equal to the sum of the diagonal terms. The diagonal terms are relatively easy to evaluate since

$$\left\langle \begin{matrix} n_1, l_1, m_{l_1}, m_{s_1} \\ n_2, l_2, m_{l_2}, m_{s_2} \end{matrix} \middle| (\mathbf{l}_1^2+\mathbf{l}_2^2+2\mathbf{l}_1\cdot\mathbf{l}_2) \middle| \begin{matrix} n_1, l_1, m_{l_1}, m_{s_1} \\ n_2, l_2, m_{l_2}, m_{s_2} \end{matrix} \right\rangle =$$

$$= l_1(l_1+1)+l_2(l_2+1)+2m_{l_1}m_{l_2}. \tag{14.10}$$

Hence for the case of $M_L = m_{l_1}+m_{l_2} = l_1+l_2-1$ the two diagonal terms are

$$l_1(l_1+1)+l_2(l_2+1)+2l_1(l_2-1) \quad \text{and} \quad l_1(l_1+1)+l_2(l_2+1)+2(l_1-1)l_2$$

whose sum is

$$2l_1(l_1+1)+2l_2(l_2+1)+4l_1l_2-2l_1-2l_2. \tag{14.11}$$

This is equal to the value in equation (14.8) plus the second value of $L(L+1)$. It follows, then, that this second value is $(l_1+l_2-1)(l_1+l_2)$. Such a process can be carried on to show that the number of values of L is equal to the smaller of the two values of l_i, and are the successive integers (l_1+l_2), (l_1+l_2-1), (l_1+l_2-2), etc.

PROBLEM 4. Show that the combination of two angular momenta $\boldsymbol{l}_i$ will lead to eigenfunctions of the total angular momentum $\boldsymbol{L}^2$ with values of L given by the integral solutions of the vector equation

$$\boldsymbol{L} = \boldsymbol{l}_1+\boldsymbol{l}_2.$$

This can be done by evaluating the diagonal sums for different values at M_L. This result is the basis of the "vector model" often used in spectroscopy.

2.4. *Spin variables and the Pauli exclusion principle*

The functions of equation (14.3) have thus far been divided into $(2l_1+1)\times(2l_2+1)$ groups of four functions each. These four functions differ only in the spin variables, and it was shown in Chapter 10 that these spin functions can be arranged in such a way as to be eigenfunctions of S^2 and S_z. The eigenfunctions with $S = 1$ are symmetric and do not change sign if S_1 and S_2 are interchanged. On the other hand the function for $S = 0$ is antisymmetric. One must then form suitable symmetric and antisymmetric combinations of the orbital functions which can be combined with the spin functions to form states satisfying the Pauli exclusion principle. Thus the states formed from combinations of the ket vectors of equation (14.3) can be designated by the eigenvalues L, M_L, S, M_S, representing operators which commute with the electrostatic interaction.

Because there are three functions with $S = 1$, and because they all have the same energy when the electrostatic interaction, but not the spin-orbit interaction, is considered they are called a triplet. The other state, with $S = 0$, is separated from the triplet by the electrostatic interaction and is called a singlet. Thus the energy eigenstates of a two electron atom are singlets or triplets.

In the conventional notation a state is described by the capital letters S, P, D, F, etc. corresponding to $L = 0, 1, 2, 3$, etc. The multiplicity is indicated by a preceding superscript. Thus ^{1}S signifies a singlet state with $L = 0$, ^{3}P a triplet state with $L = 1$. The electron configuration is expressed by means of lower case letters indicating the single particle functions or ket vectors that appear in the products. If $l_1 = 0$ and $l_2 = 1$, the configuration is called an sp configuration. If $l_1 = 1$ and $l_2 = 1$ with $n_1 \neq n_2$ it is called a pp configuration. If $n_1 = n_2$ also, it is called p^2.

2.5. *Case of equivalent electrons*

If the two electrons are equivalent, if $n_1 = n_2$ and $l_1 = l_2$, there will be a reduction in the number of states, since some of the ket vectors described above will turn out to be zero. The orbital functions alone will be either symmetric or antisymmetric and so can be combined with only one type of spin function. For two equivalent p electrons, the configuration p^2, the permitted states are ^{1}D, ^{3}P, ^{1}S. For d^2 they are ^{1}G, ^{3}F, ^{1}D, ^{3}P, ^{1}S.

PROBLEM 5. Show that for a p^2 configuration there is no ^{1}P state.

PROBLEM 6. Write out the wave functions for the states in an sp configuration that are eigenstates of L^2, L_z, S^2 and S_z.

Problem 7. Write out the wave functions for the ^{1}F, ^{3}F, ^{1}D, and ^{3}D states with $M_L = 3$ and $M_L = 2$ for a pd configuration.

2.6. *Application of the electrostatic interaction*

The previous discussion has shown that in the case of two electron systems it is possible to organize the states of a single configuration in such a way that each one is completely described by eigenvalues of four operators which commute with the electrostatic interaction. These are the zero order states, and no solution of a secular equation is necessary. The first order correction to the energy is then just the mean value of the electrostatic interaction for each of these states.

As already stated in equation (14.4) the electrostatic interaction can be expressed in terms of the spherical coordinates of the two electrons:

$$\frac{e^2}{r_{12}} = e^2 \sum_{l,m} \frac{r_a^l}{r_b^{l+1}} \frac{\Pi_l^m(\cos\vartheta_1)\Pi_l^m(\cos\vartheta_2)}{[\Pi_l^0(1)]^2} e^{im(\varphi_1-\varphi_2)}, \tag{14.12}$$

where r_a represents the smaller of r_1 and r_2 while r_b represents the larger. This permits the integration to be carried out.

The integrals involved in applying this interaction are of two kinds. The first is of the form

$$\begin{aligned}\frac{e^2}{\pi 4^2}\sum_{l,m}\int_0^\infty r_1^2\,dr_1\int_0^\infty r_2^2\,dr_2\,|R_{n_1l_1}(r_1)|^2\,|R_{n_2l_2}(r_2)|^2\,\frac{r_a^l}{r_b^{l+1}}\times\\ \times\frac{1}{[\Pi_l^0(1)]^2}\int_0^\pi[\Pi_{l_1}^{m_1}(\cos\vartheta_1)]^2\,\Pi_l^m(\cos\vartheta_1)\,e^{im\varphi_1}\sin\vartheta_1\,d\vartheta_1\,d\varphi_1\times\\ \times\int_0^\pi[\Pi_{l_2}^{m_2}(\cos\vartheta_2)]^2\Pi_l^m(\cos\vartheta_2)\,e^{-im\varphi_2}\sin\vartheta_2\,d\vartheta_2\,d\varphi_2.\end{aligned} \tag{14.13}$$

It is clear that the integrals over φ_1 and over φ_2 vanish unless $m = 0$, so the whole expression reduces to

$$\begin{aligned}e^2\sum_l F\begin{pmatrix}n_1,l_1\\ n_2,l_2\end{pmatrix}\int_0^\pi[\Pi_{l_1}^0(\cos\vartheta_1)]^2\Pi_l^0(\cos\vartheta_1)\sin\vartheta_1\,d\vartheta_1\times\\ \times\int_0^\pi[\Pi_{l_2}^0(\cos\vartheta_2)]^2\Pi_l^0(\cos\vartheta_2)\sin\vartheta_2\,d\vartheta_2\end{aligned} \tag{14.13a}$$

where the radial integrals are combined in $F\binom{n_1,l_1}{n_2,l_2}$. These can be evaluated only after the radial functions $R_{n,l}$ are known. These functions depend on the form assumed for $V(r)$ in order to take proper account of the inner shells of electrons.

The other integrals are those which appear when the roles of the electrons are interchanged in one of the functions. They have the form

$$\frac{e^2}{4\pi^2}\sum_{l,m}\int_0^\infty r_1^2\,\mathrm{d}r_1\int_0^\infty r_2^2\,\mathrm{d}r_2 R^*_{n_1l_1}(r_1)R_{n_2l_2}(r_1)R^*_{n_2l_2}(r_2)R_{n_1l_1}(r_2)\,\frac{r_\mathrm{a}^l}{r_\mathrm{b}^{l+1}}\times$$

$$\frac{1}{[\Pi_l^0(1)]^2}\int_0^\pi \Pi_{l_1}^{m_1}(\cos\vartheta_1)\Pi_{l_2}^{m_2}(\cos\vartheta_1)\Pi_l^m(\cos\vartheta_1)\,\mathrm{e}^{\mathrm{i}(m_2-m_1+m)\varphi_1}\sin\vartheta_1\,\mathrm{d}\vartheta_1\,\mathrm{d}\varphi_1\times$$

$$\int_0^\pi \Pi_{l_2}^{m_2}(\cos\vartheta_2)\Pi_{l_1}^{m_1}(\cos\vartheta_2)\,\Pi_l^m(\cos\vartheta_2)\,\mathrm{e}^{-\mathrm{i}(m_2-m_1+m)\varphi_2}\sin\vartheta_2\,\mathrm{d}\vartheta_2\,\mathrm{d}\varphi_2\,. \quad (14.14)$$

In this case the integrals clearly vanish unless $(m+m_2-m_1) = 0$. The requirement that $m = 0$ in the first type of integral is merely a special case of this condition.

The value of the integral of the product of three spherical harmonics can be written out as a finite sum but it is rather complicated. For the simpler cases, with values of l_1 and l_2 that are not too great, it is sufficient to work out each individual case, subject to the general rule that the integral is zero unless (l_1+l_2+l) is an even number and no one of the three l's is greater than the sum of the other two.

The radial integrals in this second case may be designated by $G_l\binom{n_1,l_1}{n_2,l_2}$, and the correction to the energy can be expressed as a sum of a relatively few terms,

$$\frac{e^2}{4\pi^2}\sum_l G_l\begin{pmatrix}n_1, l_1\\ n_2, l_2\end{pmatrix}\left[\int_0^\pi \Pi_{l_1}^{m_1}(\cos\vartheta)\Pi_{l_2}^{m_2}(\cos\vartheta)\Pi_l^m(\cos\vartheta)\sin\vartheta\,\mathrm{d}\vartheta\right]^2 \quad (14.14\mathrm{a})$$

subject to the condition that

$$m_1 = m_2+m\,.$$

2.7. *An illustration of the electrostatic interaction*

Let $W\left(\begin{smallmatrix}n_1,l_1\\ n_2,l_2\end{smallmatrix}\middle|\begin{smallmatrix}M_L\\ L\end{smallmatrix}\right)$ represent that linear combination of orbital functions with the eigenvalues of $L(L+1)\hbar^2$ for the square of the total angular momentum and $M_L\hbar$ for the total angular momentum about the z axis. The single particle function designated by n_1 and l_1 is a function of $(r_1, \vartheta_1, \varphi_1)$ and the other of $(r_2, \vartheta_2, \varphi_2)$. The positions of the various symbols in the argument of W indicates their significance. The upper right-hand symbol will always represent the z-component of the *orbital* angular momentum. The upper left-hand symbol will give the one-particle function of the coordinates $(r_1, \vartheta_1, \varphi_1)$ and the lower left-hand symbol will represent a function of $(r_2, \vartheta_2, \varphi_2)$.

Consider an sp configuration. The only value of L is 1 and the values of M_L are $(1, 0, -1)$:

$$W\begin{pmatrix} n_1, 0 & 1 \\ n_2, 1 & 1 \end{pmatrix} = R_{n_1,0}(r_1)R_{n_2,1}(r_2)Y_0^0(\vartheta_1\varphi_1)Y_1^1(\vartheta_2\varphi_2). \qquad (14.15)$$

The spin functions ar not included in W. This function contains only one term and there remains only the interchange of the two electrons to form symmetric and antisymmetric functions to be combined with the corresponding spin functions. The complete functions are then

$$U(^3\mathrm{P}) = (\tfrac{1}{2})^{\frac{1}{2}}\left[W\begin{pmatrix} n_1, 0 & 1 \\ n_2, 1 & 1 \end{pmatrix} - W\begin{pmatrix} n_2, 1 & 1 \\ n_1, 0 & 1 \end{pmatrix}\right] | S = 1\rangle$$

$$U(^1\mathrm{P}) = (\tfrac{1}{2})^{\frac{1}{2}}\left[W\begin{pmatrix} n_1, 0 & 1 \\ n_2, 1 & 1 \end{pmatrix} + W\begin{pmatrix} n_2, 1 & 1 \\ n_1, 0 & 1 \end{pmatrix}\right] | S = 0\rangle. \qquad (14.16)$$

The electrostatic interaction for these two states is then given by

$$\sum_l F_l\begin{pmatrix} n_1, 0 \\ n_2, 1 \end{pmatrix}\int_0^\pi [\Pi_0^0(\cos\vartheta_1)]^2\Pi_l^0(\cos\vartheta_1)\sin\vartheta_1\,\mathrm{d}\vartheta_1 \times$$

$$\times\int_0^\pi [\Pi_1^0(\cos\vartheta_2)]^2\Pi_l^0(\cos\vartheta_2)\sin\vartheta_2\,\mathrm{d}\vartheta_2$$

$$\pm\sum_l G_l\begin{pmatrix} n_1, 0 \\ n_2, 1 \end{pmatrix}\left[\int_0^\pi \Pi_1^1(\cos\vartheta)\Pi_1^0(\cos\vartheta)\Pi_l^{-1}(\cos\vartheta)\sin\vartheta\,\mathrm{d}\vartheta\right]^2 \qquad (14.17)$$

where the $(+)$ refers to the singlet state and the $(-)$ to the triplet. Hence

$$\left\langle {}^3\mathrm{P}\left|\frac{e^2}{r_{12}}\right|{}^3\mathrm{P}\right\rangle = F_0\begin{pmatrix} n_1, 0 \\ n_2, 1 \end{pmatrix} - \tfrac{1}{3}G_1\begin{pmatrix} n_1, 0 \\ n_2, 1 \end{pmatrix}$$

$$\left\langle {}^1\mathrm{P}\left|\frac{e^2}{r_{12}}\right|{}^1\mathrm{P}\right\rangle = F_0\begin{pmatrix} n_1, 0 \\ n_2, 1 \end{pmatrix} + \tfrac{1}{3}G_1\begin{pmatrix} n_1, 0 \\ n_2, 1 \end{pmatrix}. \qquad (14.18)$$

The separation of the singlet and triplet states is given, to this approximation, by $\frac{2}{3}\ G_1\binom{n_1,0}{n_2,1}$. The comparison of such a result with observed spectral levels is not very significant since the radial functions used in evaluating the F_0 and G_1 integrals depend on the approximate central field adopted to account for the inner shells of electrons. Only in the case of helium is the central field a simple Coulomb field.

In the case of a pp configuration one may again look first at the state with maximum M_L, $M_L = 2$,

$$W\begin{pmatrix} n_1, 1 & 2 \\ n_2, 1 & 2 \end{pmatrix} = R_{n_1,1}(r_1)R_{n_2,1}(r_2)Y_1^1(\vartheta_1, \varphi_1)Y_1^1(\vartheta_2, \varphi_2). \qquad (14.19)$$

In this case an interchange of electrons appears only in the radial functions so

$$\begin{aligned} U(^3\mathrm{D}) &= (\tfrac{1}{2})^{\frac{1}{2}}\,[R_{n_1,1}(r_1)R_{n_2,1}(r_2)-R_{n_1,1}(r_2)R_{n_2,1}(r_1)]\times \\ &\quad\times\; Y_1^1(\vartheta_1\varphi_1)Y_1^1(\vartheta_2\varphi_2)\,|\,S=1\rangle \\ U(^1\mathrm{D}) &= (\tfrac{1}{2})^{\frac{1}{2}}\,[R_{n_1,1}(r_1)R_{n_2,1}(r_2)+R_{n_1,1}(r_2)R_{n_2,1}(r_1)]\times \\ &\quad\times\; Y_1^1(\vartheta_1\varphi_1)Y_1^1(\vartheta_2\varphi_2)\,|\,S=0\rangle . \end{aligned} \tag{14.20}$$

The electrostatic interaction is then

$$\begin{aligned} E^{(1)}(L=2) &= e^2\sum_l\left[F_l\begin{pmatrix}n_1, 0\\ n_2, 1\end{pmatrix}\mp G_l\begin{pmatrix}n_1, 0\\ n_2, 1\end{pmatrix}\right]\times \\ &\quad\times\int_0^\pi [\Pi_1^1(\cos\vartheta)]^2\Pi_l^0(\cos\vartheta)\sin\vartheta\,\mathrm{d}\vartheta]^2 \\ &= F_0+\tfrac{1}{25}F_2\mp G_0\mp\tfrac{1}{25}G_2 . \end{aligned} \tag{14.21}$$

For the cases with $M=1$, appropriate linear combinations could be formed for $L=1$ as well as $L=2$. However it is also possible to use the theorem of equation (14.8) since the electrostatic interaction commutes with all three components of L.

The states with $M=1$ are $|\,{}^{n_1,1,0}_{n_2,1,1}\rangle$ and $|\,{}^{n_1,1,1}_{n_2,1,0}\rangle$, where the ket vectors without the spin are designated as $|\,{}^{n_1,l_1,m_{l1}}_{n_2,l_2,m_{l2}}\rangle$. Each of these can be used to form both a symmetric and an antisymmetric function and they both give the same value for the interaction. The sum is then

$$E^{(1)}(L=2)+E^{(1)}(L=1) = 2F_0-\tfrac{4}{25}F_2\mp 2G_0\mp\tfrac{4}{25}G_2 .$$

Since $E^{(1)}(L=2)$ is known, it follows that

$$E^{(1)}(L=1) = F_0-\tfrac{5}{25}F_2\mp G_0\mp\tfrac{5}{25}G_2 . \tag{14.22}$$

Similarly, for $M=0$ there are three combinations of m_1 and m_2. Two of these give $F_0+\frac{1}{25}F_2\mp(G_0+\frac{1}{25}G_2)$. The other gives $(F_0+\frac{4}{25}F_2)\mp(G_0\mp\frac{4}{25}G_2)$. The result for $L=0$ is then

$$E^{(1)}(L=0) = (F_0+\tfrac{10}{25}F_2)\mp(G_0+\tfrac{10}{25}G_2) . \tag{14.23}$$

PROBLEM 7. Work out the values of the integrals for a pp configuration as described above.

PROBLEM 8. Work out the linear combinations of wave functions corresponding to $L=2$, $L=1$ and $L=0$ for the case $M_\mathrm{L}=0$, in an sp configuration.

The energy values given above all contain the integral $F_0\binom{n_1,1}{n_2,1}$ and so the separations of the six energy levels can be expressed in terms of three integrals. A number of such spectra are known but they do not conform at all well to the indicated relationships. It must be concluded that this zero approximation set of state functions is inadequate to give much information about the energy levels.

The case of two equivalent p electrons may be treated by noting that the F and G integrals then are identical. The energy corrections are then

$$^1\mathrm{D}:\ E^{(1)} = 2F_0 + \tfrac{2}{25}F_2$$

$$^3\mathrm{P}:\ E^{(1)} = 2F_0 - \tfrac{10}{25}F_2 \qquad \frac{^1\mathrm{S} - {}^1\mathrm{D}}{^1\mathrm{D} - {}^3\mathrm{P}} = 1.50.$$

$$^1\mathrm{S}:\ E^{(1)} = 2F_0 + \tfrac{20}{25}F_2$$

Data with which to test this relationship are available and some may be tabulated as follows:

Atom	Configuration	Ratio
Theory	$n\mathrm{p}^2$	1.50
C	$2\mathrm{p}^2$	1.13
N^+	$2\mathrm{p}^2$	1.14
O^{++}	$2\mathrm{p}^2$	1.14
Si	$3\mathrm{p}^2$	1.48
Ge	$4\mathrm{p}^2$	1.50
Sn	$5\mathrm{p}^2$	1.39

Here again the agreement with the approximate calculation is satisfactory in only a few cases, and they seem to be accidental. It is evident that the interaction with other configurations must be taken into account. In the first three cases in the table the 2p 3p configuration may be of importance.

3. Spin-orbit interaction in *LS* coupling

In the extreme case of *LS*, or Russell-Saunders, coupling, the spin-orbit interaction is applied to those states which have been separated in energy by the electrostatic interaction. The fact that the spin-orbit interaction does not commute with all the quantities L^2, L_z, S_2, S_z, which have been used to characterize these states, and furthermore does not commute with the operators for the radial part of the kinetic energy, indicates that such a procedure gives only an average value for the correction to the energy. Nevertheless, when the energy of the spin-orbit interaction is small compared with the energy-level separation brought about by the electrostatic interaction, it can be applied as though the states were exact solutions of the problem with the electrostatic interaction included. For isolated states the first-order correction to the energy is just the average value of the perturbation. In case there is still degeneracy and several states have the same energy, these must be considered as a group.

Since the electrostatic interaction has separated those states with different values of L and S, the only degeneracy that remains is that of M_L and M_S. For the singlet states each value of L is represented by $(2L+1)$ values of M_L, and for the triplets each value of L and M_L corresponds to three values of M_S, making $3(2L+1)$ states altogether. The natural procedure is then to try to reduce these groups of states to smaller groups by considering operators which commute with the spin-orbit interaction.

Let the spin-orbit interaction be represented by H_2.
Then

$$H_2 = f(r_1)\mathbf{l}_1 \cdot \mathbf{s}_1 + f(r_2)\mathbf{l}_2 \cdot \mathbf{s}_2 . \tag{14.24}$$

3.1. *Operators commuting with* H_2

1. An operator which commutes with H_2 is

$$\mathsf{J}_z = \mathsf{L}_z + \mathsf{S}_z . \tag{14.25}$$

This commutes with L^2, with S^2, and with the electrostatic interaction, since each part commutes with them. To show that J_z commutes with H_2, it is necessary to consider only one electron at a time, since all operators for one electron commute with those for the other. Thus

$$\begin{aligned}(\mathsf{l}_z+\mathsf{s}_z)(\mathsf{l}_x\mathsf{s}_x+\mathsf{l}_y\mathsf{s}_y+\mathsf{l}_z\mathsf{s}_z)-(\mathsf{l}_x\mathsf{s}_x+\mathsf{l}_y\mathsf{s}_y+\mathsf{l}_z\mathsf{s}_z)(\mathsf{l}_z+\mathsf{s}_z)\\ = (\mathsf{l}_z\mathsf{l}_x-\mathsf{l}_x\mathsf{l}_z)\mathsf{s}_x+(\mathsf{l}_z\mathsf{l}_y-\mathsf{l}_y\mathsf{l}_z)\mathsf{s}_y+\mathsf{l}_x(\mathsf{s}_z\mathsf{s}_x-\mathsf{s}_x\mathsf{s}_z)+\mathsf{l}_y(\mathsf{s}_z\mathsf{s}_y-\mathsf{s}_y\mathsf{s}_z)\\ = i\hbar(\mathsf{l}_y\mathsf{s}_x-\mathsf{l}_x\mathsf{s}_y+\mathsf{l}_x\mathsf{s}_y-\mathsf{l}_y\mathsf{s}_x) = 0 .\end{aligned}$$

2. Another such operator is

$$\begin{aligned}\mathsf{J}^2 &= (\mathsf{L}_x+\mathsf{S}_x)^2+(\mathsf{L}_y+\mathsf{S}_y)^2+(\mathsf{L}_z+\mathsf{S}_z)^2\\ &= \mathsf{L}^2+\mathsf{S}^2+2\mathbf{L}\cdot\mathbf{S} .\end{aligned} \tag{14.26}$$

The fact that J^2 commutes with L^2, S^2 and H_2 can easily be established using the commutation rules for the component parts.

The functions, or ket vectors, to be treated now must always include the spin. They may be designated by L, M_L, S and M, where $M\hbar = J'_z$. Each state will be characterized by a value of M, and because J_z commutes with H_2, it will be independent of the other states in its response to H_2. For singlet states the value of S is always zero, so the value of J'_z is that of $M_L\hbar$.

For singlet states no additional classification is necessary for the application of H_2 since there is but one state in each group, but it is still of interest to investigate the value of J^2 also. One has, from equation (14.26)

$$\mathsf{J}^2 \left| \begin{matrix} M_\mathrm{L} & M \\ L & S=0 \end{matrix} \right\rangle = \hbar^2 L(L+1) \left| \begin{matrix} M_\mathrm{L} & M \\ L & S=0 \end{matrix} \right\rangle .$$

The ket symbol does not include the value of J. It is only for this particular one that the ket is an eigenket of J^2 as well as of L^2. This is true because the operator L^2 gives $\hbar^2 L(L+1)$ times the eigenket of L^2 and the operator S^2 gives zero for all singlet states. Furthermore

$$\mathsf{L}_z\mathsf{S}_z \left| \begin{matrix} M_\mathrm{L} & M \\ L & S=0 \end{matrix} \right\rangle = 0$$

since the operator S_z is equivalent in this case to multiplication by zero. The remaining operators $(\mathsf{L}_x+\mathrm{i}\mathsf{L}_y)(\mathsf{S}_x-\mathrm{i}\mathsf{S}_y)$ and $(\mathsf{L}_x-\mathrm{i}\mathsf{L}_y)(\mathsf{S}_x+\mathrm{i}\mathsf{S}_y)$ also give zero. Consequently the value of J for a singlet state is that of L.

3.2. *Application of* H_2 *as a perturbation*

Consider first the singlet function of the spin variables. This has the form

$$(\tfrac{1}{2})^{\frac{1}{2}} [|\tfrac{1}{2}(1), -\tfrac{1}{2}(2)\rangle - |-\tfrac{1}{2}(1), \tfrac{1}{2}(2)\rangle] \tag{14.27}$$

where $\frac{1}{2}(i)$ indicates that the ket is an eigenket of s_{zi} with the eigenvalue $\frac{1}{2}\hbar$. The spin-orbit interaction for each electron can be expressed in the form

$$\mathsf{l}_i \cdot \mathsf{s}_i = \tfrac{1}{2}\mathsf{l}_{+i}\mathsf{s}_{-i} + \tfrac{1}{2}\mathsf{l}_{-i}\mathsf{s}_{+i} + \mathsf{l}_{zi}\mathsf{s}_{zi} . \tag{14.28}$$

The application of s_{-i} to the ket (14.27) leads to a ket vector orthogonal to (14.27) for either value of i. The same thing is true of s_{+i} and s_{zi}. It is thus possible to conclude from the spin functions alone that the singlet terms are not affected by the spin-orbit interaction in this approximation.

In the case of the triplets there may be as many as three functions for one value of J_z'. The maximum value of J_z' is $(M_\mathrm{L}+1)\hbar$ for the maximum value of M_L, which is L. For this $J_z' = (L+1)\hbar$ there is but one state

$$\left| \begin{matrix} M_\mathrm{L} = L, S_z = 1, M = L+1 \\ L, S = 1 \end{matrix} \right\rangle =$$

$$= (\tfrac{1}{2})^{\frac{1}{2}} \left[W \left(\begin{matrix} n_1, l_1 \\ n_2, l_2 \end{matrix} \middle| \begin{matrix} L \\ L \end{matrix} \right) - W \left(\begin{matrix} n_2, l_2 \\ n_1, l_1 \end{matrix} \middle| \begin{matrix} L \\ L \end{matrix} \right) \right] |\tfrac{1}{2}(1), \tfrac{1}{2}(2)\rangle . \tag{14.29a}$$

For $J_z' = \hbar L$ there are two states

$$\left| \begin{matrix} M_\mathrm{L} = L, S_z = 0 \\ L, S = 1, M = L \end{matrix} \right\rangle = \tfrac{1}{2} \left[W \left(\begin{matrix} n_1, l_1 \\ n_2, l_2 \end{matrix} \middle| \begin{matrix} L \\ L \end{matrix} \right) - W \left(\begin{matrix} n_2, l_2 \\ n_1, l_1 \end{matrix} \middle| \begin{matrix} L \\ L \end{matrix} \right) \right] \times$$

$$\times [|\tfrac{1}{2}(1), -\tfrac{1}{2}(2)\rangle + |-\tfrac{1}{2}(1), \tfrac{1}{2}(2)\rangle], \tag{14.29b}$$

$$\left| \begin{matrix} M_L = L-1, S_z = +1 \\ L, S = 1, M = L \end{matrix} \right\rangle =$$

$$= (\tfrac{1}{2})^{\frac{1}{2}} \left[W \begin{pmatrix} n_1, l_1 & L-1 \\ n_2, l_2 & L \end{pmatrix} - W \begin{pmatrix} n_2, l_2 & L-1 \\ n_1, l_1 & L \end{pmatrix} \right] | \tfrac{1}{2}(1), \tfrac{1}{2}(2) \rangle . \tag{14.29c}$$

For $J_z' = \hbar(L-1)$ there are three states from which three orthogonal linear combinations can be formed. They are

$$\left| \begin{matrix} M_L = L-2, S_z = 1 \\ L, S = 1, M = L-1 \end{matrix} \right\rangle, \left| \begin{matrix} M = L-1, S_z = 0 \\ L, S = 1, M = L-1 \end{matrix} \right\rangle, \left| \begin{matrix} M = L, S_z = -1 \\ L, S = 1, M = L-1 \end{matrix} \right\rangle . \tag{14.29d}$$

Three is the maximum number of linearly independent functions, which can be eigenfunctions of J_z for a fixed L in a two electron system. This is because there are only three possible values of S_z. Corresponding to these functions are three values of J_z'. All of the states characterized by a single value of J will give the same first order value for the perturbation H_2.

It is possible, of course, to find the linear combinations corresponding to the different values of J by the usual method of applying the operator J^2. It is also possible to evaluate the energy displacements by the diagonal-sum rule as was done in the treatment of the electrostatic interaction.

PROBLEM 9. Evaluate the coefficients in the linear combinations of the functions in (14.29b) and (14.29c) and those in (14.29d) that constitute eigenfunctions of the square of the total angular momentum J^2.

PROBLEM 10. Write the set of ket vectors that correspond to a pd configuration, and show that for each value of J_z' there are enough independent ket vectors to provide the $(5-J_z'/\hbar)$ values of J.

3.3. *Configuration* sl

This combination of an s state with any other one-electron state leads to singlet and triplet states with $L = l$. Since there is only one value of L, there is no problem of finding the correct linear combination to form an eigenfunction of L^2. The maximum value of $J_z' = (L+1)\hbar$ and the corresponding state vector is

$$\left| \begin{matrix} M_L = l, M_S = 1, M = l+1 \\ L = l, S = 1 \end{matrix} \right\rangle =$$

$$= (\tfrac{1}{2})^{\frac{1}{2}} \left[W \begin{pmatrix} n_1, l & l \\ n_2, 0 & l \end{pmatrix} - W \begin{pmatrix} n_2, 0 & l \\ n_1, l & l \end{pmatrix} \right] | \tfrac{1}{2}(1), \tfrac{1}{2}(2) \rangle . \tag{14.30}$$

The operator H_2 may be written in terms of the raising and lowering operators already defined:

$$\mathsf{H}_2 = \tfrac{1}{2} \sum_{i=1}^{2} f(r_i)\,[\mathsf{l}_{+i}\mathsf{s}_{-i} + \mathsf{l}_{-i}\mathsf{s}_{+i} + 2\mathsf{l}_{zi}\mathsf{s}_{zi}]\,. \tag{14.31}$$

The operators s_{-i} and s_{+i} applied to the spin ket in (14.30) give ket vectors orthogonal to it, so only the term $\sum_{i=1}^{2} f(r_i)\mathsf{l}_{zi}\mathsf{s}_{zi}$ will affect the first order energy correction. Hence

$$\left\langle \begin{matrix} M_L = l, M_S = 1, M = l+1 \\ L = l, S = 1 \end{matrix} \middle| \mathsf{H}_2 \middle| \begin{matrix} M_L = l, M_S = 1, M = l+1 \\ L = l, S = 1 \end{matrix} \right\rangle = l\hbar Q(n_1, l) \tag{14.32}$$

where

$$Q(n_1, l) = \tfrac{1}{2} \int \left| W \begin{pmatrix} n_1, l & l \\ n_2, 0 & l \end{pmatrix} \right|^2 f(r_1) r_1^2\, \mathrm{d}r_1 r_2^2\, \mathrm{d}r_2\,. \tag{14.33}$$

By the arguments already used it follows that the energy correction will be independent of J_z' or of M, and will be the same for all states for which $J = L+1$.

For $M = L = l$ there are two functions,

$$\begin{aligned} &\left| \begin{matrix} M_L = L, M_S = 0, M = l \\ L = l, S = 1 \end{matrix} \right\rangle = \\ &= \tfrac{1}{2} \left[W \begin{pmatrix} n_1, l & l \\ n_2, 0 & l \end{pmatrix} - W \begin{pmatrix} n_2, 0 & l \\ n_1, l & l \end{pmatrix} \right] [|\tfrac{1}{2}(1), -\tfrac{1}{2}(2)\rangle + |-\tfrac{1}{2}(1), \tfrac{1}{2}(2)\rangle], \\ &\left| \begin{matrix} M_L = L-1, M_S = 1, M = l \\ L = l, S = 1 \end{matrix} \right\rangle = \\ &= (\tfrac{1}{2})^{\frac{1}{2}} \left[W \begin{pmatrix} n_1, l & l-1 \\ n_2, 0 & l \end{pmatrix} - W \begin{pmatrix} n_2, 0 & l-1 \\ n_1, l & l \end{pmatrix} \right] |\tfrac{1}{2}(1), \tfrac{1}{2}(2)\rangle\,. \end{aligned} \tag{14.34}$$

When applied to the first of these ket vectors the spin operators all give zero expectation value. The whole contribution to the diagonal sum comes from the second state vector, where the $(\mathsf{l}_z\mathsf{s}_z)$ terms give $(l-1)Q = (L-1)Q$.

Of the three functions with $M = L-1$ only those with $M_S = \pm 1$ contribute to the expectation value. They contribute $-2Q$. Thus there are the three diagonal sums

$$\begin{aligned} &E_2(J = L+1) = LQ \\ &E^{(1)}(J = L+1) + E^{(1)}(J = L) = (L-1)Q \\ &E^{(1)}(J = L+1) + E^{(1)}(J = L) + E^{(1)}(J = L-1) = -2Q\,. \end{aligned}$$

The first order corrections to the energy due to H_2 are then

$$\begin{aligned} &E^{(1)}(J = L+1) = LQ\,, \qquad E^{(1)}(J = L) = -Q\,, \\ &E^{(1)}(J = L-1) = -(L+1)Q\,. \end{aligned} \tag{14.35}$$

From these follows the Landé interval rule

$$\frac{E(L+1)-E(L)}{E(L)-E(L-1)} = \frac{L+1}{L}. \tag{14.36}$$

This rule is fairly well followed since in a number of cases the spin-orbit interaction energy is quite small compared with the other separations of energy values. Two examples are as follows:

	Configuration 2s2p		
Element	J	Term value	Difference
Mg I	0	39821.3	
			19.9
	1	39801.4	
			40.9
	2	39760.5	
Al II	0	114468.4	
			61.8
	1	114406.6	
			125.5
	2	114281.1	

3.4. *General two-electron configuration*

Here again we may start with the maximum value of J_z', which is $(l_1+l_2+1)\hbar$:

$$\left| \begin{matrix} M_L = l_1+l_2, M_S = 1, M = L+1 \\ L = l_1+l_2, S = 1 \end{matrix} \right\rangle =$$

$$= (\tfrac{1}{2})^{\frac{1}{2}} \left[W\begin{pmatrix} n_1, l_1 & l_1+l_2 \\ n_2, l_2 & l_1+l_2 \end{pmatrix} - W\begin{pmatrix} n_2, l_2 & l_1+l_2 \\ n_1, l_1 & l_1+l_2 \end{pmatrix} \right] | \tfrac{1}{2}(1), \tfrac{1}{2}(2)\rangle . \tag{14.37}$$

As before, the only terms in H_2 which contribute to the mean value energy are $\mathsf{l}_{z_1}\mathsf{s}_{z_1}$ and $\mathsf{l}_{z_2}\mathsf{s}_{z_2}$

$$\mathsf{l}_{z_1}\mathsf{s}_{z_1} \left| \begin{matrix} M_L = l_1+l_2, M_S = 1, M = L+1 \\ L = l_1+l_2, S = 1 \end{matrix} \right\rangle =$$

$$= \frac{\hbar^2}{8^{\frac{1}{2}}} \left[l_1 W\begin{pmatrix} n_1, l_1 & l_1+l_2 \\ n_2, l_2 & l_1+l_2 \end{pmatrix} - l_2 W\begin{pmatrix} n_2, l_2 & l_1+l_2 \\ n_1, l_1 & l_1+l_2 \end{pmatrix} \right] | \tfrac{1}{2}(1), \tfrac{1}{2}(2)\rangle, \tag{14.38a}$$

$$\mathsf{l}_{z_2}\mathsf{s}_{z_2} \left| \begin{matrix} M_L = l_1+l_2, M_S = 1, M = L+1 \\ L = l_1+l_2, S = 1 \end{matrix} \right\rangle =$$

$$= \frac{\hbar^2}{8^{\frac{1}{2}}} \left[l_2 W\begin{pmatrix} n_1, l_1 & l_1+l_2 \\ n_2, l_2 & l_1+l_2 \end{pmatrix} - l_1 W\begin{pmatrix} n_2, l_2 & l_1+l_2 \\ n_1, l_1 & l_1+l_2 \end{pmatrix} \right] | \tfrac{1}{2}(1), \tfrac{1}{2}(2)\rangle . \tag{14.38b}$$

This mean value is the energy correction for this state and can be written

$$E^{(1)}(J = L+1) = l_1 Q(n_1, l_1) + l_2 Q(n_2, l_2)$$
$$= (l_1+l_2)\frac{l_1 Q(n_1, l_1) + l_2 Q(n_2, l_2)}{(l_1+l_2)} = LQ_0 . \tag{14.39}$$

For the next value of J_z' which is $(l_1+l_2)\hbar$ the two functions are

$$\left|\begin{matrix} M_L = l_1+l_2, M_S = 0, M = L \\ L = l_1+l_2, S = 1 \end{matrix}\right\rangle = \tfrac{1}{2}\left[W\begin{pmatrix} n_1, l_1 & l_1+l_2 \\ n_2, l_2 & l_1+l_2 \end{pmatrix}\right.$$

$$\left. - W\begin{pmatrix} n_2, l_2 & l_1+l_2 \\ n_1, l_1 & l_1+l_2 \end{pmatrix}\right] [|\tfrac{1}{2}(1), -\tfrac{1}{2}(2)\rangle + |-\tfrac{1}{2}(1), \tfrac{1}{2}(2)\rangle], \quad (14.40a)$$

$$\left|\begin{matrix} M_L = l_1+l_2-1, M_S = 1 \\ L = l_1+l_2, S = 1, M = L \end{matrix}\right\rangle = (\tfrac{1}{2})^{\frac{1}{2}}\left[W\begin{pmatrix} n_1, l_1 & l_1+l_2-1 \\ n_2, l_2 & l_1+l_2 \end{pmatrix}\right.$$

$$\left. - W\begin{pmatrix} n_2, l_2 & l_1+l_2-1 \\ n_1, l_1 & l_1+l_2 \end{pmatrix}\right] |\tfrac{1}{2}(1), \tfrac{1}{2}(2)\rangle. \quad (14.40b)$$

The form of $W\binom{n_1,l_1|l_1+l_2}{n_2,l_2|l_1+l_2}$ has already been used, but $W\binom{n_1,l_1|l_1+l_2-1}{n_2,l_2|l_1+l_2}$ must be determined. This can be done by applying the operator $\mathsf{L}_x - i\mathsf{L}_y = (\mathsf{l}_{x_1}+\mathsf{l}_{x_2}) - i\,(\mathsf{l}_{y_1}+\mathsf{l}_{y_2})$ to the known function according to equation (9.22b)

$$\mathsf{L}_- W\begin{pmatrix} n_1, l_1 & l_1+l_2 \\ n_2, l_2 & l_1+l_2 \end{pmatrix} = (2L)^{\frac{1}{2}}\hbar W\begin{pmatrix} n_1, l_1 & l_1+l_2-1 \\ n_2, l_2 & l_1+l_2 \end{pmatrix}. \quad (14.41)$$

The operator L_- is composed of l_{1-} and l_{2-}. These applied individually to the component parts of W lead to

$$(\mathsf{l}_{x_1} - i\mathsf{l}_{y_1})R_{n_1,l_1}(r_1)R_{n_2,l_2}(r_2)Y_{l_1}^{l_1}(\vartheta_1, \varphi_1)Y_{l_2}^{l_2}(\vartheta_2, \varphi_2) =$$
$$= \hbar(2l_1)^{\frac{1}{2}}R_{n_1,l_1}(r_1)R_{n_2,l_2}(r_2)Y_{l_1}^{l_1-1}(\vartheta_1, \varphi_1)Y_{l_2}^{l_2}(\vartheta_2, \varphi_2). \quad (14.42)$$

The application of $(\mathsf{l}_{x_2} - i\mathsf{l}_{y_2})$ is similar so that

$$L^{\frac{1}{2}}W\begin{pmatrix} n_1, l_1 & l_1+l_2-1 \\ n_2, l_2 & l_1+l_2 \end{pmatrix} = R_{n_1,l_1}(r_1)R_{n_2,l_2}(r_2)\times$$
$$\times[l_1^{\frac{1}{2}}Y_{l_1}^{l_1-1}(\vartheta_1, \varphi_1)Y_{l_2}^{l_2}(\vartheta_2, \varphi_2) + l_2^{\frac{1}{2}}Y_{l_1}^{l_1}(\vartheta_1, \varphi_1)Y_{l_2}^{l_2-1}(\vartheta_2, \varphi_2)]. \quad (14.43)$$

The spin operators give an average value of zero for the function (14.40a) as before but each one multiplies (14.40b) by $\frac{1}{2}\hbar$. Also, as before, the terms $l_{z_1}s_{z_1}$ and $l_{z_2}s_{z_2}$ are the effective ones so that

$$\left\langle\begin{matrix} M_L = l_1+l_2-1, M_S = 1 \\ L = l_1+l_2, S = 1, M = L \end{matrix}\right| \mathsf{H}_2 \left|\begin{matrix} M_L = l_1+l_2-1, M_S = 1 \\ L = l_1+l_2, S = 1, M = L \end{matrix}\right\rangle =$$
$$= \frac{l_1(l_1+l_2-1)Q(n_1, l_1) + l_2(l_1+l_2-1)Q(n_2, l_2)}{l_1+l_2}. \quad (14.44)$$

It follows that

$$E^{(1)}(J = L) = -\frac{l_1Q_1 + l_2Q_2}{l_1+l_2} = -Q_0. \quad (14.45)$$

By continuing the process it follows that the energy correction for the other state is

$$E^{(1)}(J = L-1) = -(l_1+l_2+1)Q_0 = -(L+1)Q_0 \qquad (14.46)$$

which shows that the Landé interval rule holds for triplets in all kinds of two electron configurations.

The Landé interval rule was first derived from the observed spectral terms and holds in a wide variety of cases. The above treatment shows that it may be expected to hold when the triplet splitting is small compared with the difference between the singlet and the triplet and if there are no other configurations with the same L and J values too near in unperturbed energy.

4. Spin-orbit interaction in arbitrary coupling

The case of Russell-Saunders, or LS, coupling is the limiting case in which the spin-orbit interaction is negligible compared with the separation of the multiplets due to the electrostatic interaction. In general, of course, this limiting case is not realized, and the coupling is more properly spoken of as intermediate. Occasionally the electrostatic interaction is so small compared with the multiplet splitting that the coupling approaches the other limiting case, known as jj coupling. This implies that the spin-orbit interaction is to be applied to each individual electron and that the resulting one-electron states, which are classified according to their j values, are the states to which the electrostatic interaction is applied as a perturbation.

Instead of treating the jj coupling as a special limiting case we shall treat the general case. This means that the electrostatic and the spin-orbit interactions will be applied together. Again the problem is so to classify the states into groups that the application of the perturbation can be made without solving too large a secular equation. As has been shown in the previous discussion, this involves classification of the states according to the quantum numbers M and J. Instead of trying to work out a general result we shall treat a special case, the case of an sl configuration. As usual this is the simplest configuration to treat. It gives rise to one singlet and one triplet. For the singlet $J = L$, and for the triplet $J = L+1, L, L-1$. The singlet and triplet with $J = L$ must be considered together. Out of them was shall form two linear combinations which are strictly neither singlet nor triplet but are the functions suitable for the application of the whole perturbation $\mathsf{H}_1+\mathsf{H}_2$.

The maximum value of M is $L+1$. Such a state has $J = L+1$ and $S = 1$,

$$\left| \begin{matrix} M_L = l, M_S = 1, M = l+1 \\ L = l, S = 1, J = l+1 \end{matrix} \right\rangle =$$

$$= (\tfrac{1}{2})^{\frac{1}{2}} \left[W\begin{pmatrix} n_1, l & l \\ n_2, 0 & l \end{pmatrix} - W\begin{pmatrix} n_2, 0 & l \\ n_1, l & l \end{pmatrix} \right] | \tfrac{1}{2}(1), \tfrac{1}{2}(1)\rangle . \tag{14.47}$$

For $M = J-1 = L$ there are three functions

$$\left| \begin{matrix} M_L = l, M_S = 0, \\ L = l, S = 1, \end{matrix} M = L \right\rangle =$$

$$= \tfrac{1}{2} \left[W\begin{pmatrix} n_1, l & l \\ n_2, 0 & l \end{pmatrix} - W\begin{pmatrix} n_2, 0 & l \\ n_1, l & l \end{pmatrix} \right] [| \tfrac{1}{2}(1) - \tfrac{1}{2}(2)\rangle + | -\tfrac{1}{2}(1), \tfrac{1}{2}(2)\rangle], \tag{14.48a}$$

$$\left| \begin{matrix} M_L = l-1, M_S = 1, \\ L = l, S = 1, \end{matrix} M = L \right\rangle =$$

$$= (\tfrac{1}{2})^{\frac{1}{2}} \left[W\begin{pmatrix} n_1, l & l-1 \\ n_2, 0 & l \end{pmatrix} - W\begin{pmatrix} n_2, 0 & l-1 \\ n_1, l & l \end{pmatrix} \right] | \tfrac{1}{2}(1), \tfrac{1}{2}(2)\rangle, \tag{14.48b}$$

$$\left| \begin{matrix} M_L = l, M_S = 0, \\ L = l, S = 0, \end{matrix} M = L \right\rangle =$$

$$= \tfrac{1}{2} \left[W\begin{pmatrix} n_1, l & l \\ n_2, 0 & l \end{pmatrix} + W\begin{pmatrix} n_2, 0 & l \\ n_1, l & l \end{pmatrix} \right] [| \tfrac{1}{2}(1), -\tfrac{1}{2}(2)\rangle - | -\tfrac{1}{2}(1), \tfrac{1}{2}(2)\rangle]. \tag{14.48c}$$

A straightforward procedure would be to apply the perturbation H_2 and then to solve the secular equation to find the proper linear combinations of these three functions. Another procedure would be to find first the two linear combinations of (14.48a) and (14.48b) which are eigenfunctions of J^2. Since H_2 commutes with J^2 it would then be necessary to treat together only the two functions with the same value of J instead of all three. This reduction by means of J^2 could be carried out by applying the operator J^2 and solving the secular equation in the usual way, but it is instructive to consider another method for doing the same thing.

It was shown in Chapter 9 that

$$(\mathsf{J}_x - i\mathsf{J}_y) | \gamma, J, M\rangle = [J(J+1) - M^2 + M]^{\frac{1}{2}}\hbar | \gamma, J, M-1\rangle . \tag{14.49}$$

Then since

$$(\mathsf{J}_x - i\mathsf{J}_y) = (\mathsf{L}_x - i\mathsf{L}_y) + (\mathsf{S}_x - i\mathsf{S}_y)$$

and a relation similar to (14.49) holds for both the orbital and the spin parts,

$$(\mathsf{J}_x - i\mathsf{J}_y) \left| \begin{matrix} M_L = l, M_S = 1, M = l+1 \\ L = l, S = 1, J = l+1 \end{matrix} \right\rangle =$$

$$= \hbar[(L+1)(L+2) - L(L+1)]^{\frac{1}{2}} \left| \begin{matrix} M_L = l, M_S = 1, M = l \\ L = l, S = 1, J = l+1 \end{matrix} \right\rangle$$

$$
\begin{aligned}
&= \hbar L^{\frac{1}{2}} \left[W \begin{pmatrix} n_1, l & l-1 \\ n_2, 0 & l \end{pmatrix} - W \begin{pmatrix} n_2, 0 & l-1 \\ n_1, l & l \end{pmatrix} \right] |\, \tfrac{1}{2}(1), \tfrac{1}{2}(1)\rangle + \\
&+ \frac{\hbar}{2^{\frac{1}{2}}} \left[W \begin{pmatrix} n_1, l & l \\ n_2, 0 & l \end{pmatrix} - W \begin{pmatrix} n_2, 0 & l \\ n_1, l & l \end{pmatrix} \right] [|\, \tfrac{1}{2}(1), -\tfrac{1}{2}(2)\rangle + |-\tfrac{1}{2}(1), \tfrac{1}{2}(2)\rangle].
\end{aligned}
\tag{14.50}
$$

It then follows that

$$
\begin{aligned}
\left| \begin{matrix} S = 1, M = l \\ L = l, J = l+1 \end{matrix} \right\rangle &= \left(\frac{2}{2L+2} \right)^{\frac{1}{2}} \left| \begin{matrix} M_L = l, M_S = 0, \\ L = l, S = 1, \end{matrix} \; M = L \right\rangle + \\
&+ \left(\frac{2L}{2L+2} \right)^{\frac{1}{2}} \left| \begin{matrix} M_L = l-1, M_S = 1, \\ L = l, S = 1, \end{matrix} \; M = L \right\rangle .
\end{aligned}
\tag{14.51}
$$

The other function for which $J = L$, is the combination orthogonal to (14.51), or

$$
\begin{aligned}
\left| \begin{matrix} S = 1, M = l \\ L = l, J = l \end{matrix} \right\rangle &= \left(\frac{2L}{2L+2} \right)^{\frac{1}{2}} \left| \begin{matrix} M_L = l, M_S = 0, \\ L = l, S = 1, \end{matrix} \; M = L \right\rangle + \\
&- \left(\frac{2}{2L+2} \right)^{\frac{1}{2}} \left| \begin{matrix} M_L = l-1, M_S = 1, \\ L = l, S = 1, \end{matrix} \; M = L \right\rangle .
\end{aligned}
\tag{14.52}
$$

The four states now include two with $J = L+1$, equations (14.47) and (14.51), and two with $J = L$, equations (14.52) and (14.47c). The two states with $J = L+1$ are distinguished by the two values of M, but the states with $J = L$ both have $M = L$.

PROBLEM 11. Show that the state of equation (14.48c) is an eigenket of J^2 and find the eigenvalue.

The application of H_2 as a perturbation now requires the determination of the coefficients in the linear combination

$$
\left| \begin{matrix} M = L \\ J = L \end{matrix} \right\rangle = \alpha \left| \begin{matrix} S = 1, M = l \\ L = l, J = l \end{matrix} \right\rangle + \beta \left| \begin{matrix} S = 0, M = l \\ L = l, J = l \end{matrix} \right\rangle .
\tag{14.53}
$$

These coefficients must satisfy the equations

$$
\begin{aligned}
&[E^{(1)} - \langle S = 1 \,|\, \mathsf{H}_2 \,|\, S = 1 \rangle]\alpha - \langle S = 1 \,|\, \mathsf{H}_2 \,|\, S = 0 \rangle \beta = 0, \\
&-\langle S = 0 \,|\, \mathsf{H}_2 \,|\, S = 1 \rangle \alpha + [E^{(1)} - X - \langle S = 0 \,|\, \mathsf{H}_2 \,|\, S = 0 \rangle]\beta = 0.
\end{aligned}
\tag{14.54}
$$

In these equations the state vectors of (14.53) have been designated only by their values of S, and X represents the difference between the singlet and triplet terms due to the electrostatic interaction,

$$
X = \int W \begin{pmatrix} n_1, l & l \\ n_2, 0 & l \end{pmatrix} \frac{e^2}{r_{12}} W \begin{pmatrix} n_2, 0 & l \\ n_1, l & l \end{pmatrix} \mathrm{d}v_1 \, \mathrm{d}v_2 .
\tag{14.55}
$$

The two values of $E^{(1)}$ for which the equations (14.54) have non-zero solutions are the first order corrections to the energies when both the electrostatic interaction and the spin-orbit interaction are taken into account. Both solutions are eigenstates of J^2 and are distinguished only by their different energies.

The evaluation of the matrix elements follows as in examples already worked out and the secular equation is

$$\begin{vmatrix} [E^{(1)}+Q] & -[L(L+1)]^{\frac{1}{2}}Q \\ -[L(L+1)]^{\frac{1}{2}}Q & [E^{(1)}-X] \end{vmatrix} = E^{(1)2}+(Q-X)E^{(1)}-QX^{-L(L+1)Q^2} = 0 \tag{14.56}$$

and

$$E^{(1)} = \tfrac{1}{2}(X-Q) \pm \tfrac{1}{2}[(X+Q)^2+4L(L+1)]^{\frac{1}{2}}. \tag{14.57}$$

Equation (14.56) was obtained by treating the case of $M = L$ but it is valid for all values of M from $+L$ to $-L$. It could have been obtained by treating a determinant set up for an arbitrary value of M. The value of $E^{(1)}$ for $J = L+1$ is LQ and it has already been shown that for $J = L-1$, $E^{(1)} = -(L+1)Q$. The first order approximation corrections to the energies, in the absence of a magnetic field, may then be tabulated

L	S	J	$E^{(1)}$
L	0	L	$\frac{1}{2}(X-Q)+\frac{1}{2}[(X+Q)^2+4L(L+1)Q^2]^{\frac{1}{2}}$
L	1	$L+1$	LQ
L	1	L	$\frac{1}{2}(X-Q)-\frac{1}{2}[(X+Q)^2+4L(L+1)Q^2]^{\frac{1}{2}}$
L	1	$L-1$	$-(L+1)Q$

When $X \gg Q$, the division into singlet and triplet is quite clear and the triplet obeys the interval rule. This is true whether X is positive or negative. In this limiting case the assignment of the quantum number S has real significance, since the functions are essentially of one type or the other. When X and Q are of the same order of magnitude, the distinction between singlets and triplets is obliterated and there are merely four states, of which two have the same value of J. When $X \ll Q$, there are two doublets for the limiting case of jj coupling.

Problem 13. Show how the triplet interval ratio obtained from equation (14.57) departs from the Landé rule as Q/X becomes larger.

Problem 14. In the mercury spectrum one group of term values is as follows

$$^1P_1 = -30112.8 \qquad ^3P_1 = -44768.9$$
$$^3P_2 = -40138.3 \qquad ^3P_0 = -46536.2.$$

From these determine the quantities Q and X and the extent to which equation (14.57) represents the level separations.

PROBLEM 15. Work out the Zeeman effect for an sp configuration.

5. The parity operator

Let P be the operator which reverses the sign of all of the position coordinates of all of the electrons. It does not affect the spin coordinates. Then

$$\begin{aligned} \mathsf{P}U(x, y, z, s) &= U(-x, -y, -z, s) \\ \mathsf{P}U(r, \vartheta, \varphi, s) &= U(r, \pi-\vartheta, \pi+\varphi, s). \end{aligned} \tag{14.58}$$

The effect of this operator on the functions of one electron in a central field has already been noted,

$$\begin{aligned} \mathsf{P}R_{n,l}(r)\Pi_l^m(\cos\vartheta)\,\mathrm{e}^{\mathrm{i}m\varphi} &= R_{n,l}(r)\Pi_l^m(-\cos\vartheta)\,\mathrm{e}^{\mathrm{i}m\pi}\,\mathrm{e}^{\mathrm{i}m\varphi} \\ &= (-1)^l R_{n,l}(r)\Pi_l^m(\cos\vartheta)\,\mathrm{e}^{\mathrm{i}m\varphi}. \end{aligned} \tag{14.59}$$

The functions of the one-electron problem are eigenfunctions of this operator with the eigenvalues ± 1, and it is clear that there are no other eigenvalues since $P^2 = 1$. Those functions with the eigenvalue $+1$ are called even states, and those with -1 are called odd states. A state of one electron is even or odd according as the azimuthal quantum number l is even or odd. That this property is a constant of the motion follows from the fact that P commutes with the Hamiltonian function.

The operator P commutes with the Hamiltonian function for any number of electrons, and hence the states of an atom can be strictly classified as even or odd. Since the unperturbed states are products of one-electron states, their parity can be obtained as the parity of the sum of the azimuthal quantum numbers. Thus the parity is a property of an electron configuration, rather than of one of the states in that configuration. In case two or more electron configurations are considered together in applying the perturbation, only those of the same parity will need to be considered. No combination will ever occur between states of different parity. This operator then divides the states of an atom strictly into two groups between which no combinations will occur when applying the Hamiltonian function.

6. Summary of operators

In the course of this discussion we have classified the states of the atomic system according to a variety of operators of which they are characteristic.

Some of these commute with the complete Hamiltonian function and so strictly characterize the energy states. Others commute with only part of the Hamiltonian and are suitable for characterization only when the remainder is neglected. These may be summarized as follows:

Operator	Quantum number
L^2	L
L_z	M_L
S^2	S
S_z	M_S
J^2	J
J_z	M
P	± 1
P_{ji}	-1

The first four operators commute with the unperturbed Hamiltonian function and with the electrostatic interaction but not with the spin-orbit interaction. They have a precise significance only in extreme Russell-Saunders coupling. In other cases they have only an approximate meaning.

The last four operators commute with the complete Hamiltonian function and so have a precise significance in all couplings.

7. Selection rules

As was shown in Chapter 13 the probability of the transition of an atom from a state $|\,a\rangle$ to a state $|\,b\rangle$ is determined by the matrix elements of various functions of the electron coordinates $\boldsymbol{r}_i$ between these two states. Thus for dipole radiation the vector matrix elements $\langle b\,|\sum_i \boldsymbol{r}_i\,|\,a\rangle$ are important. In particular, if all components of this matrix element are zero, the transition is said to be forbidden. In case dipole radiation is forbidden there is still the possibility of quadrupole radiation which is determined by matrix elements of quadratic functions of the electron coordinates.

Some examples of selection rules have already been treated by evaluating the matrix elements between specific state vectors. A more general formulation can be given by considering the commutation rules.

There are numerous commutation rules which follow from the basic rules for the commutation of coordinates and momenta. Among them are the following:

$$L_i R_j - R_j L_i = J_i R_j - R_j L_i = \mathrm{i}\hbar R_k e_{ijk} \tag{14.60a}$$

$$L_i R_i - R_i L_i = J_i R_i - R_i J_i = 0 \tag{14.60b}$$

$$J_i^2 R_j - R_j J_i^2 = i\hbar(J_i R_k - R_k J_i) e_{ijk} \tag{14.60c}$$

$$J_i^2 R_i - R_i J_i^2 = 0 . \tag{14.60d}$$

The capital letters are used to indicate that each operator represents the sum of the operators for the individual electrons. e_{ijk} is $+1$ when (i, j, k) are in cyclical order such as (x, y, z) or (y, z, x), and is -1 when one pair is interchanged, such as (x, z, y).

Also the parity operator "anticommutes" with the coordinates

$$\mathsf{P} R_i + R_i \mathsf{P} = 0 . \tag{14.61}$$

From these commutation rules there follow a variety of selection rules including the following.

1. *Transitions accompanied by dipole radiation can take place only between states of different parity.* This is because the parity of a coordinate is negative. Then, since the application of the parity operator to all the elements in a bracket expression is only an inversion of the coordinate system it will not change the value of the integral and

$$\langle b \mid \sum_i \boldsymbol{r}_i \mid a \rangle = \langle b \mid \mathsf{P}^{-1} [\mathsf{P} \sum_i \boldsymbol{r}_i] \mathsf{P} \mid a \rangle = -\langle b \mid \mathsf{P}^{-1} [\sum_i \boldsymbol{r}_i] \mathsf{P} \mid a \rangle . \tag{14.62}$$

Hence unless the operator P operating on $\mid a \rangle$ together with $\mathsf{P}^{-1} = \mathsf{P}^\dagger$ operating on the bra vector $\langle b \mid$ provide a minus sign to compensate the effect of P operating on $\sum_i \boldsymbol{r}_i$, the bracket is zero.

2. *Transitions accompanied by dipole radiation polarized parallel to the z axis can take place only between states of the same M, or the same eigenvalue of* J_z. Since $\mathsf{J}_z \mathsf{z} - \mathsf{z} \mathsf{J}_z = 0$

$$\langle J', M' \mid \mathsf{J}_z \mathsf{z} - \mathsf{z} \mathsf{J}_z \mid J, M \rangle = \\ = M'\hbar \langle J', M' \mid \mathsf{z} \mid J, M \rangle - \langle J', M' \mid \mathsf{z} \mid J, M \rangle M\hbar = 0$$

and

$$(M' - M) \langle J', M' \mid \mathsf{z} \mid J, M \rangle = 0 . \tag{14.63}$$

3. *For dipole radiation with the electric vector in the x-y plane the eigenvalue of* J_z *must change by one unit.* Since

$$\mathsf{J}_z(\mathsf{x} \pm i\mathsf{y}) - (\mathsf{x} \pm i\mathsf{y})\mathsf{J}_z = \pm\hbar(\mathsf{x} \pm i\mathsf{y})$$

$$M' \langle J', M' \mid (\mathsf{x} \pm i\mathsf{y}) \mid J, M \rangle - M \langle J', M' \mid (\mathsf{x} \pm i\mathsf{y}) \mid J, M \rangle = \\ = \pm \langle J, M' \mid (\mathsf{x} \pm i\mathsf{y}) \mid J, M \rangle$$

and

$$(M' - M \pm 1) \langle J', M' \mid \mathsf{x} \pm i\mathsf{y} \mid J, M \rangle = 0 . \tag{14.64}$$

4. *For dipole radiation with the electric vector along the z axis J must change by* ± 1 *or* 0. The proof of this statement requires the additional commutation rules:

$$\begin{aligned} \mathrm{J}^2\mathrm{x}-\mathrm{x}\mathrm{J}^2 &= -2\,i\hbar(\mathrm{J}_y\mathrm{z}-\mathrm{y}\mathrm{J}_z) \\ \mathrm{J}^2\mathrm{y}-\mathrm{y}\mathrm{J}^2 &= 2\,i\hbar(\mathrm{J}_x\mathrm{z}-\mathrm{x}\mathrm{J}_z) \\ \mathrm{J}^2\mathrm{z}-\mathrm{z}\mathrm{J}^2 &= 2\,i\hbar(\mathrm{J}_y\mathrm{x}-\mathrm{y}\mathrm{J}_x). \end{aligned} \tag{14.65}$$

With these and repeated uses of the other commutation rules can be derived the operator equation

$$\mathrm{J}^4\mathrm{z}-2\mathrm{J}^2\mathrm{z}\mathrm{J}^2+\mathrm{z}\mathrm{J}^4-2\hbar^2(\mathrm{J}^2\mathrm{z}+\mathrm{z}\mathrm{J}^2)+4\hbar^2(\mathbf{J}\cdot\mathbf{R})\mathrm{J}_z = 0. \tag{14.66}$$

If now one forms the matrix elements of this equation between eigenstates of J^2 and J_z which differ only in the eigenvalues of J^2 there results the equation

$$[(J'+J+1)^2-1]\,[(J'-J)^2-1]\,\langle J', M \mid \mathrm{z} \mid J, M\rangle = 0. \tag{14.67}$$

This involves the fact that the term $(\mathbf{J}\cdot\mathbf{R})\mathrm{J}_z$ makes no contribution to the equation when $J' \neq J$ since $(\mathbf{J}\cdot\mathbf{R})$ commutes with J^2. The first bracket in (14.67) cannot be zero since neither J nor J' can be negative. Hence, when $J' \neq J$, $J' = J\pm 1$ is a condition for the non-vanishing of $\langle J', M \mid \mathrm{z} \mid J, M\rangle$. The case $J' = J$ is not forbidden by this analysis so there are three possible changes in J.

The above selection rules expressed in terms of the eigenvalues of J, J_z and P are exact. They hold for all free atoms. Similar but only approximate conclusions can be drawn from the commutation rules of other operators. The selection rules on L and L_z are the same as on J and J_z insofar as the states involved are eigenstates of L and L_z. On the other hand all of the coordinates commute with S^2 and S_z so that insofar as the states are eigenstates of S^2 and S_z the quantum numbers S and M_S do not change with the emission of radiation. This is a reason for the convenience of dividing the spectral terms into sets of different multiplicity. In the extreme Russell-Saunders limit there are no transitions between states of different S.

Quantum mechanics has been applied very extensively to atomic and molecular spectroscopy, to the computation of energy levels and the estimation of transition probabilities. The qualitative understanding of the problems is entirely satisfactory and the quantitative results give no indication of deficiencies in the theory. As is clear, the treatment in these two chapters is intended merely to indicate the nature of the approach and not to give anything like a complete treatment.

CHAPTER 15

QUANTUM MECHANICAL SCATTERING THEORY

The work of Rutherford in analyzing the scattering of alpha particles initiated the present day view of atomic structure. The experimental results, interpreted by means of classical mechanics and the Coulomb law of force between particles, showed the existence of a heavy atomic nucleus. On this basis there developed the whole theory of atomic and molecular structure. It is important, therefore, to apply the methods of quantum mechanics to this scattering process if only to see that the original conclusions are still valid. In addition, however, most of our knowledge of nuclear properties is now based on the scattering of one nuclear particle in collision with another, and such scattering experiments must be interpreted by means of quantum mechanics.

The treatment of quantum mechanical problems where two or more particles collide and interact requires some considerations not necessary in handling the preceding topics. The principal reason for this difference is that some of the simultaneous observables (for example, energy and momentum) have continuous ranges of eigenvalues. This fact introduces some conceptual as well as some computational differences. An allied difference is that the eigenfunctions may not be quadratically integrable in the same sense as are the wavefunctions that represent bound states.

As in the discussions above, use will be made of the various equivalent formulations of the theory. Time dependent arguments will be employed to develop the structure of the theory and to establish some general theorem, as these methods more closely follow classical ideas; to discuss practical calculations use will be made of the Schroedinger representation, the wave mechanics, as this is usually more productive of numerical results. In all of the presentation, an appeal will be made to physical intuition and the relationship of physical observation and theory emphasized.

The subject of scattering includes a wide variety of phenomena. Among these are the collisions of complex systems such as atoms or molecules, and collisions in which additional particles such as mesons are produced. Here, however, we shall treat only the simplest cases of collisions between two simple stable particles such as electrons, protons, neutrons, etc. and shall ignore any possibilities of internal structure and intrinsic spin.

1. Separation of the center of mass

As in other two-body problems, it is convenient to separate the center of mass and to distinguish its motion from the relative motion of the two particles. Let X, Y, Z, be the coordinates of the center of mass, and let $x = x_1 - x_2$, $y = y_1 - y_2$, $z = z_1 - z_2$ be the relative coordinates. Then, ψ, the wave-function of the system is given by

$$\psi = \Omega(X, Y, Z, t)\,\gamma(x, y, z, t),$$

where Ω and γ must satisfy the equations

$$-\frac{\hbar}{\mathrm{i}}\frac{\partial\Omega}{\partial t} = -\frac{\hbar^2}{2(m_1+m_2)}\nabla^2\Omega \tag{15.1}$$

and

$$-\frac{\hbar}{\mathrm{i}}\frac{\partial\gamma}{\partial t} = -\frac{\hbar^2}{2\mu}\nabla^2\gamma + V(x, y, z)\gamma\,. \tag{15.2}$$

The reduced mass μ is given by $m_1 m_2/(m_1+m_2)$.

Equation (15.1) describes the free motion of the center of mass. The results obtained from equation (15.2) describe the collision with reference to a system of coordinates moving with the center of mass and must be transformed appropriately to give the motion of the particles in the system of coordinates fixed in the laboratory.

The function $\Omega(X, Y, Z, t)$ is, of course, not influenced by the scattering, since the interaction is a function of the relative coordinates only. The motion of the center of mass is given by the initial momentum of the moving particle if one particle is initially at rest. Let v be the initial velocity of the particle whose mass is m_1, and let it be directed along the z axis. The other particle is assumed stationary. $m_1 v$ is the total momentum of the system, and if it is taken exactly along the z axis, the function Ω will have the form

$$\Omega(X, Y, Z, t) \sim \mathrm{e}^{(\mathrm{i}/\hbar)m_1 vZ}\mathrm{e}^{-(\mathrm{i}/\hbar)(m_1\mu/2m_2)v^2 t}\,. \tag{15.3}$$

This function represents the uniform motion of the center of mass along the

Z axis. The energy assigned to this motion is that of a particle of mass (m_1+m_2) moving with the velocity $m_1 v/(m_1+m_2)$.

It is the function γ which will be the object of study in this chapter. Its asymptotic behavior at large distances can be written down directly if the forces of interaction described by $V(x, y, z)$ have a suitable short range. At large distances the motion of the effective particle, whose mass is $\mu = m_1 m_2/(m_1+m_2)$, will move as a free particle and can be treated as such.

2. Transformation to laboratory coordinates

Observations are usually made in the laboratory system of coordinates. Usually one of the particles is initially at rest in the laboratory system, so the center of mass is moving. The transformation between these two systems is important. The scattering in the two systems is quite different, except when the mass of the incident particle is very small compared with the mass of the particle initially at rest.

The transformation can be made by consideration of the conservation of momentum. The velocity in the x, y, z system, both before and after collision (for an elastic collision), is v. Then for the particle scattered in the direction (θ, φ) the components are

$$\dot{x} = v \sin\theta \cos\varphi, \quad \dot{y} = v \sin\theta \sin\varphi, \quad \dot{z} = v \cos\theta.$$

If we now insert the values of $\dot{x}, \dot{y}, \dot{z}$ in terms of $\dot{x}_1, \dot{y}_1, \dot{z}_1, \dot{x}_2, \dot{y}_2, \dot{z}_2$, it follows for the case in which m_2 is initially at rest that

$$m_1 \dot{z}_1 + m_2 \dot{z}_2 = m_1 v; \quad \dot{z}_1 - \dot{z}_2 = v \cos\theta,$$

and

$$(m_1+m_2)\dot{z} = m_1 v + m_2 v \cos\theta,$$

$$(m_1+m_2)\dot{z}_2 = m_1 v - m_1 v \cos\theta.$$

Expressions for $\dot{x}_1, \dot{x}_2, \dot{y}_1$ and y can be obtained in a similar way.

If θ_0 is the angle in the laboratory coordinates at which the incident particle is scattered

$$\tan \theta_0 = \frac{\sin\theta}{(m_1/m_2)+\cos\theta}. \tag{15.4}$$

If θ_r is the angle at which the stationary particle recoils

$$\tan \theta_r = \frac{\sin\theta}{1-\cos\theta}. \tag{15.5}$$

From equation (15.4) it is apparent that $\theta_0 \to \theta$ as $m_1/m_2 \to 0$. Furthermore if $m_1 = m_2$, the maximum value of θ_0 is 90°. If $m_1 > m_2$, there is a maximum value of θ_0 less than 90° beyond which no particles are scattered and each value of θ_0 below the maximum corresponds to two values of θ. For $m_1 = 2m_2$, $\theta = 0$ and $\theta = \pi$ both give $\theta_0 = 0$; $\theta = \frac{2}{3}\pi$ gives the maximum value of $\theta_0 = \frac{1}{6}\pi$. Values of θ between 0 and $\frac{2}{3}\pi$ give values of θ_0 between 0 and $\frac{1}{6}\pi$. Values of θ between $\frac{2}{3}\pi$ and π cover again the values of θ_0 between $\frac{1}{6}\pi$ and 0.

It is also interesting to note from equation (15.5) that the angle of recoil is independent of the masses of the particles.

The transformation for the scattering, as a function of the angle θ to that as a function of θ_0, must also take into account the differing solid angles in the two systems of coordinates. Since the solid angles are $\sin\theta_0 \, d\theta_0 \, d\varphi_0$ and $\sin\theta \, d\theta \, d\varphi$, respectively, we have the equation

$$I(\theta_0) \sin\theta_0 \, d\theta_0 = I(\theta) \sin\theta \, d\theta . \tag{15.6}$$

$\varphi_0 = \varphi$ and $d\varphi_0 = d\varphi$. Then from equation (15.4) it follows that

$$\sin\theta_0 \, d\theta_0 = \frac{\pm(1+(m_1/m_2)\cos\theta)}{[1+2(m_1/m_2)\cos\theta+(m_1/m_2)^2]^{\frac{3}{2}}} \sin\theta \, d\theta ,$$

and consequently

$$I(\theta_0) = \frac{[1+2(m_1/m_2)\cos\theta+(m_1/m_2)^2]^{\frac{3}{2}}}{[1+(m_1/m_2)\cos\theta]} I(\theta) , \tag{15.7}$$

when the angles θ_0 and θ are related by equation (15.4).

These and other conclusions, drawn from the conservation of momentum and the conservation of energy only, can be carried over into quantum mechanics without change.

PROBLEM 1. If the scattering cross section in the center-of-mass coordinates, $I(\theta)$, is a constant independent of θ, find the cross section in the laboratory coordinates, $I(\theta_0)$. Take the cases with $m_1/m_2 \to 0$, $m_1/m_2 = 1$, and $m_1/m_2 = 2$.

PROBLEM 2. Treat the cases of the above problem when $I(\theta) = \cos^2\theta$, and compute the fraction of the particles for which $\theta_0 < 45°$ and for which $\theta_0 < 90°$.

3. General time dependent operator treatment of scattering; time propagators and Lippmann-Schwinger equation

In a scattering problem two particles collide and are scattered from each other. A typical experimental arrangement defines the energy or momentum of incoming particles (the beam) which are allowed to strike an initially almost stationary target particle. If the energy of the projectile particle is well defined, there are some conceptual difficulties in the problem of normalization. These have been met before, and have been treated by a limiting process. A slightly different normalizing process will be treated later in this chapter. However, because of this difficulty, the problem will first be considered in terms of two colliding wave packets, representing the beam and the target particles respectively, and each of them may be visualized as initially contained in a finite spatial volume. Also, since the treatment of two interacting particles can always be discussed in terms of the motion of their center of mass, and the motion of one "particle" about the center of mass, the problem will be treated as the scattering of a single particle of mass μ by a stationary center of force. The two particles will be taken as non-identical and spinless, and the center of force will be assumed to be limited in spatial extent so that effects beyond a certain finite range may be neglected. This excludes, of course, a treatment of Coulomb scattering.

3.1. *The initial state*

The initial state of the system $|A, t\rangle$ for $t < 0$ will be supposed to have as a Schroedinger representation a wave packet, normalized ($\langle A | A\rangle = 1$) and located in some suitable volume of space to represent one particle, of mass μ, incident on the (single, fixed) scattering center. For these general considerations the exact shape of the wave packet is unimportant, although some specific shapes may be used as illustrations.

Let the Hamiltonian function in the center of mass coordinates be H_S. For $t < 0$ it may be assumed that no scattering has occurred and that the wave packet is entirely outside the range of the force. Then for $t < 0$ the effective part of the Hamiltonian is H_0 and the corresponding part of the time development operator, T_S, is T_0. H_0 and T_0 are the free particle Hamiltonian and time displacement operators respectively. One may note that the time displacement operator is often referred to as the propagator since it describes the propagation of the initial state. At any time t, H_S and T_S are related by the Schroedinger equation

$$-\frac{\hbar}{i}\dot{\mathsf{T}}_S = \mathsf{H}_S\mathsf{T}_S. \tag{15.8}$$

In a typical experiment the wave packet as a whole will move in a given (average) direction and with a given (average) momentum which may be specified by P, but will of course be composed of a range of momenta p in the neighborhood of P. A compact wave packet will require a large spread in p but if the location in space need not be sharply defined, the range of p need not be large. In any case (to within a normalization factor) the initial state for times $0 > t > t_0$ may be described as

$$\begin{aligned} | A, t\rangle &= \int \langle \bar{p} \mid A, t\rangle \mid \bar{p}\rangle\, \mathrm{d}^3\boldsymbol{p} \\ &= \int \langle \bar{p} \mid \mathsf{T}_0(t-t_0) \mid A, t_0\rangle \mid \bar{p}\rangle\, \mathrm{d}^3\boldsymbol{p} \\ &= \int \exp\{-\mathrm{i}\mathsf{H}_0\,(t-t_0)/\hbar\}\, \langle \bar{p} \mid A, t_0\rangle \mid \bar{p}\rangle\, \mathrm{d}^3\boldsymbol{p} \\ &= \int \exp\{-\mathrm{i}p^2(t-t_0)/2\mu\hbar\}\, \langle \bar{p} \mid A, t_0\rangle \mid \bar{p}\rangle\, \mathrm{d}^3\boldsymbol{p}. \end{aligned} \tag{15.9}$$

Here the kets $| \bar{p}\rangle$ are free particle eigenkets of momentum. The symbol $\mathrm{d}^3\boldsymbol{p}$ indicates integrations over the three dimensional vector $\bar{\boldsymbol{p}}$, and $p^2/2\mu$ is the energy of the effective free particle. In equation (15.9) the Hamiltonian H_0 in the exponent is, of course, an operator which operates on the bra vector $\langle p \,|$. Since $| A\rangle$ is normalized it follows that

$$\int | \langle \bar{p} \mid A, t_0\rangle |^2\, \mathrm{d}^3\boldsymbol{p} = 1. \tag{15.10}$$

As is evident from the form of (15.9) the normalization is maintained for $t < 0$.

3.2. *The scattering*

Some time after $t = 0$ it may be supposed that the wave packet comes under the influence of the scattering center and is affected by it. The Hamiltonian H_0 and the propagator T_0 are then no longer adequate and one must use H_S and T_S. It may be supposed that the Hamiltonian suddenly changes from H_0 to H_S at time $t = 0$, and that T_S "jumps" from T_0 to T_S. However since the particle, by hypothesis, has been outside the range of the scattering center for $t < 0$ this sudden change introduces no difficulties. In fact the whole Hamiltonian and propagator could have been used from the beginning since those parts other than H_0 and T_0 would have had no effect. Nevertheless we may define

$$\mathsf{H}_S = \mathsf{H}_0 + \mathsf{V}, \tag{15.11}$$

where $\mathsf{V}(t < 0)$ may be taken as zero and $\mathsf{V}(t > 0)$ as non-zero but independent of the time. Similarly T_S may be divided into two parts

$$\mathsf{T}_S = \mathsf{T}_0 + \mathsf{T}, \tag{15.12}$$

where T is taken to be zero for $t < 0$.

The advantage of using a T which is zero for $t > 0$ is that it permits the definition of a Laplace transform

$$\mathscr{T}(s) = (\mathrm{i}/\hbar) \int_0^\infty \mathsf{T}(t)\, \mathrm{e}^{-st}\, \mathrm{d}t. \tag{15.13}$$

The Laplace transform provides a means of studying solutions of the Schroedinger equation in a form different from that used previously but which has advantages in some cases. For an elementary discussion of the properties of the Laplace transform see CHURCHILL (1944). The Laplace transform will normally be designated by a script letter for the function of which it is the transform. The factor $(\mathrm{i}/\hbar)$ in (15.13) departs from the usual convention but is convenient in the applications of this chapter.

In many similar time-dependent developments of scattering theory the Fourier transform, rather than the Laplace transform, is used. These methods are equivalent. The latter method has been chosen here because in the definition of Laplace-transformable functions $\mathsf{T}(t)$, (15.13), it is supposed that $\mathsf{T}(t)$ is zero for $t < 0$. This allows a "roughness" for $\mathsf{T}(t)$, at $t \geq 0$, and easily allows the insertion of initial conditions into the calculations. In a Fourier treatment the symmetry for $\pm t$ is sometimes confusing; the "roughness" allowed by the Laplace transform at $t = 0$ accounts physically for the difference between "prior" scattering and "post" scattering.

In using equations such as (15.13) it must be remembered that the integrand is an operator. Since operators can be added, they can be integrated. One may also consider the integral of a matrix representation of the operators, and integrate each matrix element as a function of the time to get the Laplace transform of that element. Furthermore $\mathscr{T}(s)$ will be an analytic function of s except for certain singularities, because of its definition as an integral.

All of the properties of the scattering are contained in the operator $\mathsf{T}(t)$

and consequently in the transformed operator $\mathscr{T}(s)$. If something can be learned about $\mathscr{T}(s)$, the consequences for $\mathsf{T}(t)$ follow directly. The inverse Laplace transform is

$$\mathsf{T}(t) = -(\hbar/2\pi) \int_{\gamma} e^{st}\, \mathscr{T}(s)\, ds\,, \tag{15.14}$$

where γ is a contour along a vertical line in the complex s-plane taken to the right of all the singularities of $\mathscr{T}(s)$, and extending from $-i\infty$ to $+i\infty$. This integral may be closed to the left on a very large semi-circle since the integrand becomes negligible for large negative real values of s when $t > 0$ because of the exponential factor. From the Cauchy theorem of residues it follows that the operator $\mathsf{T}(t)$ will depend only on the properties of the singularities of $\mathscr{T}(s)$ inside the closed contour of integration. From the physical situation it is clear that all of these poles must be on or to the left of the imaginary axis, since otherwise $\mathsf{T}(t)$ for $t \to \infty$ would not converge. Since $\mathsf{T}(t)$ for $t < 0$ must be zero, the integral must vanish when closed with a large contour on the positive side of the imaginary axis. The location of the poles and the residues at the poles are determined by the scattering potential, via the Lippmann-Schwinger equation to be discussed below, and all the information about the scattering is contained in the locations of these poles and their residues. Alternatively the locations of poles and their residues can be inferred from scattering experiments.

Since the properties of $\mathsf{T}(t)$ will be deduced from $\mathscr{T}(s)$ by means of performing contour integrals in the s-plane, the Laplace transformation rules of integration for equation (15.14), make clear the required nature of the contour. This is a further reason for using the Laplace transform.

Although T_S and T_0 are not really zero for $t < 0$, their behavior during this time is such as to merely move the wave packet up to a place where the scattering center can become effective. The wave packet at the edge of the range of the scattering center and at $t = 0$ may be taken as the initial condition, and all operators may be transformed by equation (15.13), which involves treating everything as starting at $t = 0$. Thus

$$\begin{aligned} \mathscr{T}_0(s) &= (i/\hbar) \int_0^{\infty} \exp(-i\mathsf{H}_0 t/\hbar)\, e^{-st}\, dt \\ &= (i/\hbar)/[s + (i/\hbar)\mathsf{H}_0] = 1/[\mathscr{T}_0(s)]^{-1}. \end{aligned} \tag{15.15}$$

The last equality assumes, of course, the existence of an inverse of the operator $\mathscr{T}_0(s)$.

The significance of an operator in the denominator of an equation such as (15.15) must be carefully considered. Equation (15.15) appears as a result of purely formal manipulation, but its further use must take account of the operator nature of H_0. If $s = -(\mathrm{i}/\hbar)E$ equation (15.15) shows that

$$[\mathscr{T}_0(-\mathrm{i}E/\hbar)]^{-1} = [\mathsf{H}_0 - E], \tag{15.16}$$

This is just the operator for the Schroedinger equation for the energy eigenvalues E. It can be represented by a matrix based on any complete set of ket vectors and is a diagonal matrix if the basic vectors are eigenkets of H_0.

The matrix representations of $\mathscr{T}_0(s)$ can be obtained as the Laplace transforms of the corresponding matrix representations of $\mathsf{T}_0(t)$. If these are based on eigenkets of H_0, the corresponding matrices are diagonal. For use in scattering, however, it is more useful to use the values of the vector linear momentum as a basis. Then

$$\langle \bar{p}_2 \mid \mathsf{T}_0 \mid \bar{p}_1 \rangle = \langle \bar{p}_2 \mid \mathrm{e}^{-\mathrm{i}E_1 t/\hbar} \mid \bar{p}_1 \rangle = \delta(\boldsymbol{p}_2 - \boldsymbol{p}_1)\, \mathrm{e}^{-\mathrm{i}E_1 t/\hbar}, \tag{15.17}$$

and correspondingly

$$\langle \bar{p}_2 \mid \mathscr{T}_0(-\mathrm{i}E_0/\hbar) \mid \bar{p}_1 \rangle = \delta(\bar{\boldsymbol{p}}_2 - \boldsymbol{p}_1)/(E_1 - E_0). \tag{15.17a}$$

In these expressions the δ-function requires that the vectors $\bar{\boldsymbol{p}}_2$ and $\bar{\boldsymbol{p}}_1$ are equal, even though the energy depends on p^2 only. In (15.17a) there are really three energies, $p_2/2\mu$, $p_1^2/2\mu$ and E_0.

In a similar way it follows from (15.12) and the Laplace transformation that

$$\mathscr{T}_s(s = -\mathrm{i}E/\hbar) = \frac{1}{\mathsf{H}_0 + \mathsf{V} - E} = \frac{1}{[\mathscr{T}_0(s)]^{-1} + \mathsf{V}}. \tag{15.18}$$

The operator V in the denominator must be properly interpreted, and the basic significance of a quantity in the denominator is that its multiplication by itself must give one, or the unit operator. Hence the significance of equation (15.18) is that

$$\{[\mathscr{T}_0(s)]^{-1} + \mathsf{V}\}\, \mathscr{T}_s(s) = \mathscr{T}_s(s)\, \{[\mathscr{T}_0(s)]^{-1} + \mathsf{V}\} = \mathsf{I}. \tag{15.19}$$

From equations (15.12) and (15.14) the propagator can be written

$$\mathsf{T}_s(t) = \mathsf{T}_0(t) - (\hbar/2\pi) \int_\gamma \mathscr{T}(s)\, \mathrm{e}^{st}\, \mathrm{d}s \tag{15.20}$$

if $\mathscr{T}(s)$ is known. Since

$$\mathsf{T}_S(t) = \exp\{-\mathrm{i}(\mathsf{H}_0+\mathsf{V})t/\hbar\}$$
$$= \exp\{-\mathrm{i}\mathsf{H}_0 t/\hbar\} + [\exp\{-\mathrm{i}(\mathsf{H}_0+\mathsf{V})t/\hbar\} - \exp\{-\mathrm{i}\mathsf{H}_0 t/\hbar\}],$$

one may write

$$\mathscr{T}(s) = \frac{\mathrm{i}}{\hbar}\int_0^\infty [\exp\{-\mathrm{i}(\mathsf{H}_0+\mathsf{V})t/\hbar\} - \exp\{-\mathrm{i}\mathsf{H}_0 t/\hbar\}]\, \mathrm{e}^{-st}\, \mathrm{d}t \qquad (15.21)$$

where $\mathscr{T}(s)$ is the Laplace transform of the T in equation (15.12).

However the form of $\mathscr{T}(s)$ given in (15.21) is not a useful one. It can be misinterpreted. Since H_0 and V do not usually commute the term in $\exp\{-(\mathrm{i}/\hbar)\mathsf{H}_0 t\}$ cannot be factored out. To get a more useful form, the properties of $\mathscr{T}(s)$ must be obtained from other considerations.

In scattering problems, as well as in some others, it is convenient to define another operator $\mathscr{T}_{\mathrm{I}}(s)$ by the relationship

$$\mathscr{T}(s) = -\mathscr{T}_0(s)\mathscr{T}_{\mathrm{I}}(s)\mathscr{T}_0(s), \qquad (15.22)$$

so that equation (15.20) becomes

$$\mathsf{T}_S(t) = \mathsf{T}_0(t) + (\hbar/2\pi)\int_\gamma \mathrm{e}^{st}\mathscr{T}_0(s)\mathscr{T}_{\mathrm{I}}(s)\mathscr{T}_0(s)\,\mathrm{d}s. \qquad (15.23)$$

The subscript on $\mathscr{T}_{\mathrm{I}}(s)$ stands for interaction and indicates that the operator $\mathscr{T}_{\mathrm{I}}(s)$ describes the interaction between the particle and the center of force, while the operators $\mathscr{T}_0(s)$ describe the motion of the particle under its own inertia. If one can learn the nature of this new operator, then the operator T_S can be evaluated. The problem becomes one of writing and solving an integral equation for $\mathscr{T}_{\mathrm{I}}(s)$ rather than for $\mathscr{T}(s)$ or for the differential equation for $\mathsf{T}(t)$.

The Laplace transform $\mathscr{T}(s)$ of $\mathsf{T}(t)$ is based on an initial condition such that $\mathsf{T}(t)$ is different from zero only for $t > 0$. The Schroedinger equation for $\mathsf{T}(t)$ is

$$\mathsf{H}_S\mathsf{T}_S = (\mathsf{H}_0+\mathsf{V})(\mathsf{T}_0+\mathsf{T}) = (-\hbar/\mathrm{i})(\dot{\mathsf{T}}_0+\dot{\mathsf{T}}). \qquad (15.24)$$

Since $\mathsf{H}_0\mathsf{T}_0 = (-\hbar/\mathrm{i})\dot{\mathsf{T}}_0$ it follows that

$$\mathsf{H}_0\mathsf{T} + \mathsf{V}\mathsf{T}_0 + \mathsf{V}\mathsf{T} = (-\hbar/\mathrm{i})\dot{\mathsf{T}}. \qquad (15.25)$$

In this equation both H_0 and V are independent of the time and $\mathsf{T}(t<0) = 0$. Hence (15.25) can be transformed to give

$$(\mathsf{H}_0+\mathsf{V})\int_0^\infty \mathsf{T}(t)\,\mathrm{e}^{-st}\,\mathrm{d}t + \mathsf{V}\int_0^\infty \mathsf{T}_0(t)\,\mathrm{e}^{-st}\,\mathrm{d}t =$$
$$(\mathsf{H}_0+\mathsf{V})\mathscr{T}(s) + \mathsf{V}\mathscr{T}_0(s) = (-\hbar/\mathrm{i})s\,\mathscr{T}(s). \qquad (15.26)$$

The last term in (15.26) is the Laplace transform of the time derivative of $\mathsf{T}(t)$ which follows the rule that for any function $\mathsf{F}(t)$

$$(\mathrm{i}/\hbar)\int_0^\infty \dot{\mathsf{F}}(t)\,\mathrm{e}^{-st}\,\mathrm{d}t = s\mathscr{F}(s)-(\mathrm{i}/\hbar)\mathsf{F}(t=0). \tag{15.27}$$

Using equation (15.22) to express $\mathscr{T}(s)$ in terms of $\mathscr{T}_{\mathrm{I}}(s)$ leads to

$$(\mathsf{H}_0+\mathsf{V})\mathscr{T}_0(s)\mathscr{T}_{\mathrm{I}}(s)\mathscr{T}_0(s)-\mathsf{V}\mathscr{T}_0(s) = -(\hbar/\mathrm{i})s\mathscr{T}_0(s)\mathscr{T}_{\mathrm{I}}(s)\mathscr{T}_0(s). \tag{15.28}$$

Since $\mathscr{T}_0(s)$ is a common factor on the right of each term it can be dropped, and since $\mathsf{H}_0\mathscr{T}_0(s) = (-\hbar/\mathrm{i})s\mathscr{T}_0(s)+1$ the equation becomes

$$\mathscr{T}_{\mathrm{I}}(s) = \mathsf{V}-\mathsf{V}\mathscr{T}_0(s)\mathscr{T}_{\mathrm{I}}(s). \tag{15.29}$$

This is known as the Lippmann-Schwinger equation. It is an integral equation for the determination of $\mathscr{T}_{\mathrm{I}}(s)$ in terms of the scattering potential V and the known free particle propagator $\mathscr{T}_0(s)$. A solution of (15.29) is equivalent to a solution of the Schroedinger equation with the specified initial conditions of the scattering problem at $t = 0$.

The Lippmann-Schwinger equation can also be obtained by noting that

$$\begin{aligned}\mathscr{T}_{\mathrm{S}}(s) &= \frac{1}{[\mathscr{T}_0(s)]^{-1}+\mathsf{V}} = \mathscr{T}_0(s)-\mathscr{T}_0(s)\mathsf{V}\mathscr{T}_0(s)+\mathscr{T}_0(s)\mathsf{V}\mathscr{T}_0(s)\mathsf{V}\mathscr{T}_0(s)+\dots\\ &= \mathscr{T}_0(s)-\mathscr{T}_0(s)\mathsf{V}\mathscr{T}_{\mathrm{S}}(s).\end{aligned} \tag{15.30}$$

Then using $\mathscr{T}_{\mathrm{S}} = \mathscr{T}_0+\mathscr{T}$ and equation (15.22), equation (15.29) follows.

PROBLEM 3. Show that $\mathsf{H}_0\mathscr{T}_0(s) = (-\hbar/\mathrm{i})s\mathscr{T}_0(s)+1$ and work out the details of deriving equation (15.29).

PROBLEM 4. Show that the series in (15.30) is the reciprocal of $[\mathscr{T}_0(s)]^{-1}+$ $+\mathsf{V}$ for both right-hand and left-hand multiplication.

PROBLEM 5. Show that a singularity of $\mathscr{T}(s)$, such as $R/(s+\alpha)$ must have $\mathrm{Re}(\alpha) \leq 0$ to conform to reality.

4. The Born series

An approximate solution of (15.29) may be obtained by iteration. The first approximation gives

$$\mathscr{T}_{\mathrm{I}}(s) = \mathsf{V}. \tag{15.31}$$

If this is inserted on the right of (15.29) it leads to

$$\mathscr{T}_{\mathrm{I}}(s) = \mathsf{V} - \mathsf{V}\mathscr{T}_0(s)\mathsf{V}. \tag{15.32}$$

A continuation of this process gives the Born series

$$\mathscr{T}_{\mathrm{I}}(s) = \mathsf{V} - \mathsf{V}\mathscr{T}_0(s)\mathsf{V} + \mathsf{V}\mathscr{T}_0(s)\mathsf{V}\mathscr{T}_0(s)\mathsf{V} - \dots. \tag{15.33}$$

This series may or may not converge depending on V. Since V appears in increasing powers it may be expected that the series will converge if V is sufficiently small. However if V is large, and if, for example, a resonance occurs (a pole for $\mathscr{T}_{\mathrm{I}}(s)$, or a zero denominator in a matrix multiplication) the series may not converge in any order.

When the series does converge the successive terms may be calculated; they are called the first, second, etc. Born approximation terms. Note that the terms in this series alternate in sign.

5. Feynman diagrams for the Born series

The matrix elements of $\mathscr{T}_{\mathrm{I}}(s)$ are the fundamental quantities needed for the scattering problem. The Born series as given above provides a way of calculating them when the potential is known and when the series converges. The successive Born series terms may be visualized in terms of diagrams. If the matrices are based on the ket vectors of a free particle, $|\,\bar{p}\rangle$, the diagrams represent the scattering from a state $|\,\bar{p}\rangle$ to a state $|\,\bar{p}'\rangle$.

Let the various terms in the Born series for the interaction operator be designated by $\mathscr{T}_{\mathrm{B}}{}^{(i)}(s)$, and the corresponding terms in the series for the propagator by $\mathsf{T}_{\mathrm{B}}{}^{(i)}(t)$. $\mathscr{T}_{\mathrm{B}}{}^{(1)}(s) = \mathsf{V}$ and hence

$$\mathsf{T}_{\mathrm{B}}^{(1)}(t) = (-\hbar/2\pi)\int_{\gamma} \mathsf{V}e^{st}\,ds = -i\hbar\mathsf{V}\delta(t), \tag{15.34}$$

since the inverse Laplace transform of a constant is a δ-function. If the propagator $\mathsf{T}_1{}^{\mathrm{B}}(t)$ is given a matrix representation in terms of plane waves (eigenkets of momentum $\bar{\boldsymbol{p}}$), the matrix elements are

$$\langle \bar{p}' \,|\, \mathsf{T}_{\mathrm{B}}^{(1)}(t) \,|\, \bar{p}\rangle = -i\hbar\,\langle \bar{p}' \,|\, \mathsf{V} \,|\, \bar{p}\rangle\,\delta(t). \tag{15.35}$$

The form of (15.35) suggests that at $t = 0$ there is a probability $|\,\langle \bar{p}' \,|\, \mathsf{V} \,|\, \bar{p}\rangle\,|^2$ that each ket vector $|\,\bar{p}\rangle$ becomes $|\,\bar{p}'\rangle$, i.e. the particle is scattered from the momentum $\bar{p}$ to the momentum $\bar{p}'$.

This process can be represented by the diagram

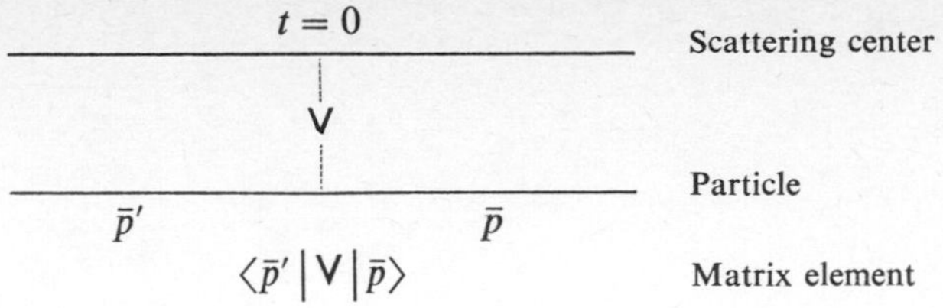

The diagram is usually only briefly sketched as above, but it has a detailed significance as follows:

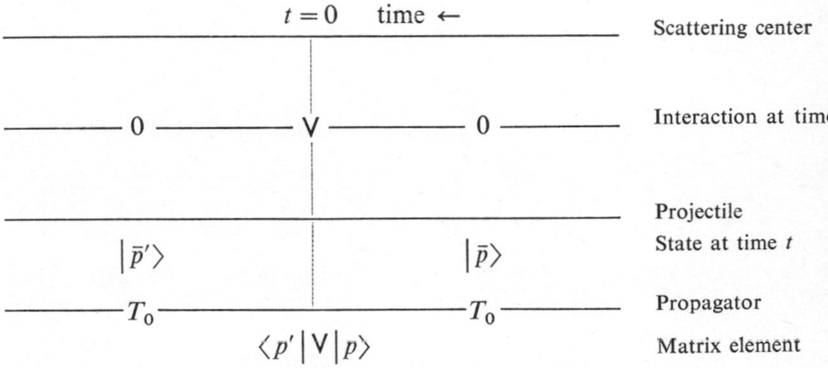

The diagram gives a simple schematic way of remembering the first Born term in terms of a simple classical analog: The particle in state $|\bar{p}\rangle$ scatters via the interaction V, at the time $t = 0$, to a state $|\bar{p}'\rangle$. It may seem strange that this scattering can take place at $t = 0$ when the particle is just about to enter the field of the sacttering center, but it must be remembered that the momentum eigenket $|\bar{p}\rangle$is a plane wave, extending over all space. In this first Born approximation the situation is treated as though the scattering center existed only at $t = 0$. No sum over states other than the initial or the final one is involved at this stage. This first Born term is obviously inadequate; it does not include any description of multiple scattering from the potential or of scattering extended over a period of time. These effects are included in the higher Born terms, and the higher order Feynman diagrams as discussed below.

To see the meaning of the second Born term $\mathscr{T}_B^{(2)}(s) = \mathsf{V}\mathscr{T}_0(s)\mathsf{V}$ one must use the convolution theorem of Laplace transformations. This second Born term is a product of three Laplace transforms, but the inverse transform is not the product of the three operators from which the transforms were derived. According to the convolution theorem the product of two Laplace transforms leads to

$$(-\hbar/2\pi)\int_{\gamma}\mathscr{F}(s)\,\mathscr{G}(s)\,e^{st}\,ds = (-\hbar/2\pi)\int_{0}^{t}\mathsf{F}(t-\tau)\,\mathsf{G}(\tau)\,d\tau$$
$$= (-\hbar/2\pi)\int_{0}^{t}\mathsf{F}(\tau)\,\mathsf{G}(t-\tau)\,d\tau\,. \tag{15.36}$$

This convolution process can be applied to the second term of (15.33) in two steps,

$$(-\hbar/2\pi)\int_{\gamma}\mathscr{T}_0(s)\,\mathsf{V}(s)\,e^{st}\,ds = (-\hbar/2\pi)\int_{0}^{t}\mathsf{T}_0(t-\tau)\,\mathsf{V}(\tau)\,d\tau\,. \tag{15.37}$$

Since the operator V is a constant with respect to s its inverse is a delta function with respect to the time. Hence

$$(-\hbar/2\pi)\int_{\gamma}\mathscr{T}_0(s)\,\mathsf{V}(s)\,e^{st}\,ds = \int_{0}^{t}\exp\{-i\mathsf{H}_0(t-\tau)/\hbar\}\,\mathsf{V}\,\delta(\tau)\,d\tau$$
$$= \exp\{-i\mathsf{H}_0 t/\hbar\}\,\mathsf{V}\,. \tag{15.38}$$

Both H_0 and V are operators. They do not necessarily commute and their order must be preserved. If now the process is repeated

$$(-\hbar/2\pi)\int_{\gamma}\mathsf{V}[\mathscr{T}_0(s\mathsf{V})]\,e^{st}\,ds =$$
$$= \int_{0}^{t}\mathsf{V}\delta(t-\tau')\int_{0}^{\tau'}\exp\{-i\mathsf{H}_0(\tau'-\tau)/\hbar\}\mathsf{V}\delta(\tau)\,d\tau' d\tau\,. \tag{15.39}$$

The integrations with respect to τ and τ' can be carried out to give the second Born term in the inverse transformation to be

$$\mathsf{V}\exp\{-(i/\hbar)\mathsf{H}_0 t\}\,\mathsf{V}\,. \tag{15.39a}$$

A matrix representation of the operator product in (15.39a) based on momentum eigenkets can be constructed by inserting two unit operators of the form $\mathbf{I} = \int d^3\boldsymbol{p}\,|\,\bar{p}\rangle\langle\bar{p}\,|$ which will be discussed in detail later,

$$\iint \langle\bar{p}'\,|\,\mathsf{V}\,|\,\bar{p}_2\rangle\,\langle\bar{p}_2\,|\,e^{-iE_1 t/\hbar}\,|\,\bar{p}_1\rangle\,\langle\bar{p}_1\,|\,\mathsf{V}\,|\,\bar{p}\rangle\,d^3\boldsymbol{p}_2\,d^3\boldsymbol{p}_1\,. \tag{15.39b}$$

However a physical picture which helps to an understanding of the scattering process is emphasized by using the expression (15.39) before carrying out the time integration, by forming the bracket between $\langle \bar{p}' |$ and $| \bar{p} \rangle$ and upon inserting one unit operator

$$\int d^3\boldsymbol{p}_1 \int_0^t d\tau' \langle \bar{p}' \,|\, \mathsf{V} \,|\, \bar{p}_1 \rangle \, \delta(t-\tau') \int_0^{\tau'} d\tau \, e^{-iE_1(\tau'-\tau)/\hbar} \langle \bar{p}_1 \,|\, \mathsf{V} \,|\, \bar{p} \;\rangle \delta(\tau) . \tag{15.39c}$$

This last expression can be described by saying that at $\tau = 0$ the ket vector $| \bar{p} \rangle$ is changed to $| \bar{p}_1 \rangle$ by the matrix element $\langle \bar{p}_1 \,|\, \mathsf{V} \,|\, \bar{p} \rangle$. This occurs for all values of $\bar{p}_1$ as is indicated by the integral $d^3\boldsymbol{p}_1$. Each of these ket vectors then develops with the propagator exp $[(-i/\hbar)E_1(\tau'-\tau)]$ until $\tau' = t$ when it is changed to $| \bar{p}' \rangle$ by the matrix element $\langle \bar{p}' \,|\, \mathsf{V} \,|\, \bar{p}_1 \rangle$.

The Feynman diagram for this second order term is:

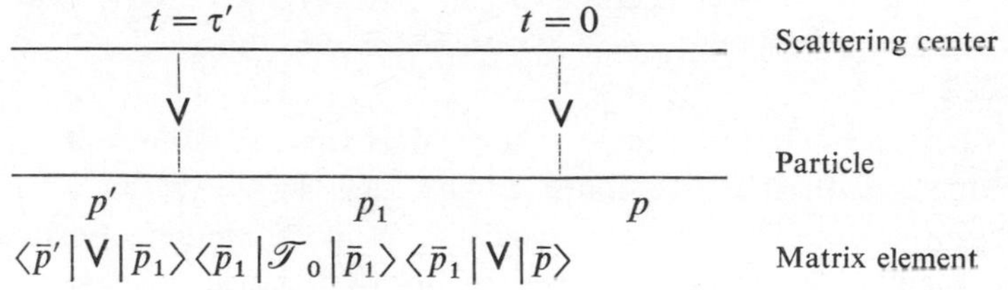

In more detail the diagram can be written:

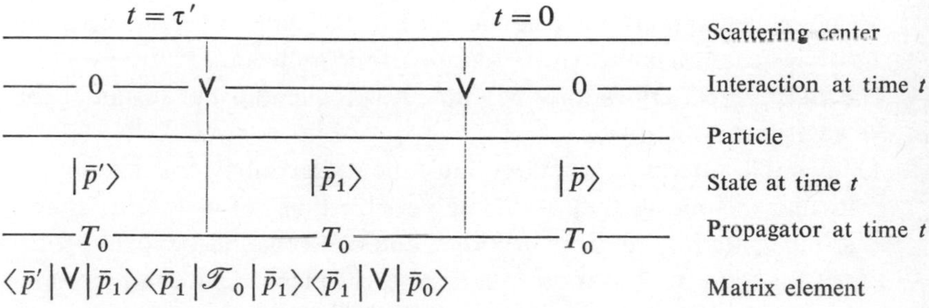

The propagator and the state ket are undefined and unimportant at the two instants of interaction, since the interaction as a function of the time is thought of as instantaneous. The intermediate state $| \bar{p}_1 \rangle$ is any state, and all states must be included in the integration over $d^3\boldsymbol{p}_1$. There is no question of conservation of energy or momentum between states such as $| \bar{p}_1 \rangle$ and $| \bar{p} \rangle$, or $| \bar{p}' \rangle$ although there is, as will be seen later, between $| \bar{p}' \rangle$ and $| p \rangle$, when the scattering is via a simple potential. This interpretation in terms of the propagators is helpful in visualizing and in thinking of the problem, but the more important matrix elements are those of the Laplace transforms themselves.

By successive application of these arguments one sees that such a term as the nth Born term $\mathscr{T}_{\mathrm{B}}^{(n)} = [(\mathsf{V}\mathscr{T}_0)]^{n-1}\mathsf{V}$ represents n successive scatterings between each of which the state vectors propagate as those of free particles. This n'th term can be represented by the Feynman diagram

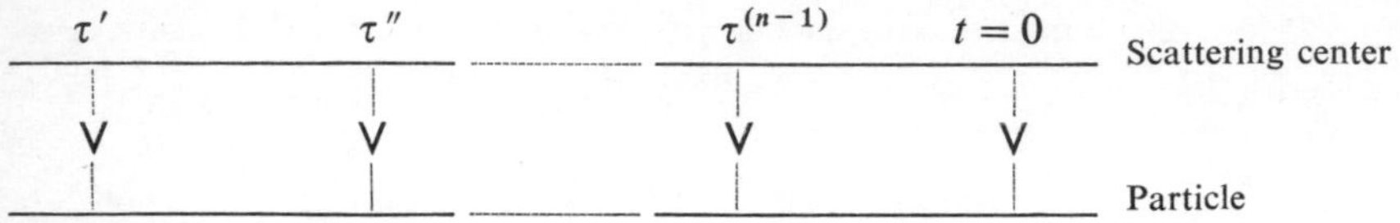

$\langle \bar{p}' | \mathsf{V} | \bar{p}'' \rangle \langle \bar{p}'' | \mathscr{T}_0 | \bar{p}'' \rangle \langle \bar{p}'' | \mathsf{V} | \bar{p}''' \rangle \ldots \langle \bar{p}^{(n-1)} | \mathsf{V} | \bar{p} \rangle$ Matrix element

It is seen that the following rules apply to these diagrams:

1. The diagrams have a one-to-one correspondence to the terms of the Born series.
2. Each diagram can be interpreted, in a classical sense, in terms of a number of multiple scatterings, sequential in the time, by the potential of the scattering center. This analogy seems to be both mathematically and conceptually exact. Whether or not the Born series converges is less important than that this view illuminates the physics of the scattering problem.
3. All possible diagrams must be included in the sum since this represents the basic quantum nature (wave nature) of the problem.
4. The multiple scatterings may be viewed as sequential but instantaneous for each scattering.
5. All possible intermediate states must be taken into consideration. In particular this means that a) all the possible times of a certain series of multiple scatterings must be included, and b) all the independent properties of the states such as momenta and other independent quantities must be considered in order to get all the states. Since for free particle states the vector momentum completely defines a state, and all values of the momentum must be taken into account regardless of the energy, this summation is referred to as scattering "off the energy shell".
6. This appeal to classical intuition has been based on time-dependent arguments. However it is not so much the matrix elements of the propagators T_{S}, T_0 and T that are useful as those of the Laplace transforms $\mathscr{T}_{\mathrm{S}}$, $\mathscr{T}_0$ and $\mathscr{T}$. This will become clearer below.

These rules are actually more general in their application than to the simple scattering problems considered here. They apply to a broad class of problems that includes, in addition, reactions and the production of new particles. These questions cannot be treated here.

A little consideration will show also that the process described here is only a slightly different form of the time dependent perturbation theory discussed in Chapter 7. The first Born term is the first order approximation, the second Born term is the second, etc. The formulation in terms of diagrams and Laplace transforms, however, is often helpful in finding approximate solutions by physical intuition rather than waiting for a complete process of logical development.

In summary, the Lippmann-Schwinger equation, with the Born series as a solution of it, allows a simple physical interpretation of the process of scattering in terms of Feynman diagrams. This interpretation is schematically very classical. A free particle comes in to the scattering center and is scattered by the interaction potential; it then propagates as a free particle "off the energy shell" to be scattered again to another state "off the energy shell", and then it propagates as a free particle to be scattered again, etc. Finally, after a number of scatterings it ends up as a free particle and if the scattering is due to a simple potential it ends up back on the energy shell again. However, all orders, or sequences, of such free particle propagators are coherent. They may interfere with all other propagators representing other scatterings. The sum of all these terms represents the scattering, and the Born series. The quantum mechanical view of the process is represented by the addition of the amplitudes, not the probabilities. The Feynman diagrams represent a schematic way of recalling the individual Born terms and provide a way to write down immediately the Born series. To visualize the problem is perhaps more simple than to visualize Newton's second law of motion. To carry it out may not be.

6. Momentum and coordinate bases for matrices and dyadic operators

The multiplication of the matrices representing operators is easily understood if the eigenkets are distinct and the eigenvalues are discrete. However the possible values of the vector $\bar{p}$ really form a continuum, and special attention must be given to replacing the sum over discrete values with an integral over the $\bar{p}$-space with the appropriate density factor. One satisfactory procedure is to approach the integral as the limit of a sum.

As was shown in Chapter 8 the transformation matrix between a representation based on coordinates, $\bar{r}$, and one based on momenta, $\bar{p}$, is just the exponential wavefunction of a free particle $\exp\{-(i/\hbar)\bar{\boldsymbol{p}}\cdot\bar{\boldsymbol{r}}\}$. In Chapter 8 its normalization was accomplished by the use of a convergence factor. Here, however, a slightly different procedure will be adopted.

For various purposes in the development of scattering theory it is also convenient to work in the spatial basis $|\,\bar{r}\rangle$ that continuously spans the three-space as well as working with the momentum base $|\,\bar{p}\rangle$ that summarily spans momentum three-space. It is also convenient to form dyadic operators from these bases. For example, it is convenient to use operators such as

$$\mathscr{P} \equiv \int \mathrm{d}^3\boldsymbol{p}\,|\,\bar{p}\rangle\langle\bar{p}\,| \tag{15.40a}$$

and

$$\mathscr{R} \equiv \int \mathrm{d}^3\boldsymbol{r}\,|\,\bar{r}\rangle\langle\bar{r}\,|. \tag{15.40b}$$

The properties of the various brackets and operators will now be discussed and their *relative* normalizations established.

Choose a presently arbitrary normalization

$$\langle\bar{p}\,|\,\bar{r}\rangle = N\exp\{-i\bar{\boldsymbol{p}}\cdot\bar{\boldsymbol{r}}/\hbar\}. \tag{15.41}$$

Also since brackets in either basis behave as Dirac delta functions set

$$\langle\bar{r}_2\,|\,\bar{r}_1\rangle = A\,\delta(\bar{r}_2-\bar{r}_1) \tag{15.42}$$

and

$$\langle\bar{p}_2\,|\,\bar{p}_1\rangle = B\,\delta(\bar{p}_2-\bar{p}_1). \tag{15.43}$$

The properties of Dirac delta functions are defined by

$$\int_\alpha^\beta \delta(t-a)f(t)\,\mathrm{d}t = f(a)$$

provided $\alpha \le a \le \beta$; otherwise the integral is zero. It is important to notice that the argument of the δ-function, and the function $f(t)$, is the same as the variable of integration. If a different argument occurs then a transformation of variables is necessary:

$$\int_\alpha^\beta \delta[g(t)-a]f(t)\,\mathrm{d}t = \int_{\alpha'}^{\beta'} \delta[g(t)-a]f[t(g)]\,\mathrm{d}g(t)/\mathrm{d}g'(t)$$
$$= \sum_{\text{roots of } g(t_0)=a} f(t_0)/g'(t_0).$$

This expression must be carefully considered if $g'(t_0)$ has zeros. Consider now:

$$\langle \bar{p} \,|\, \mathscr{R} \,|\, \bar{r} \rangle = \langle \bar{p} \,|\int \mathrm{d}^3\boldsymbol{r}_1 \,|\, \bar{r}_1 \rangle \langle \bar{r}_1 \,|\, \bar{r} \rangle$$

$$= \langle \bar{p} \,|\int \mathrm{d}^3\boldsymbol{r}_1 A\delta(\bar{r}_1 - \bar{r}) \,|\, \bar{r}_1 \rangle = A \langle \bar{p} \,|\, \bar{r} \rangle .$$

Thus $\mathscr{R}$ behaves as an identity operator

$$\mathsf{I} = A^{-1}\mathscr{R} = A^{-1} \int \mathrm{d}^3\boldsymbol{r} \,|\, \bar{r} \rangle \langle \bar{r} \,| . \tag{15.44}$$

In the same way $\mathscr{P}$ behaves as an identity operator

$$\mathsf{I} = B^{-1}\mathscr{P} = B^{-1} \int \mathrm{d}^3\boldsymbol{p} \,|\, \bar{p} \rangle \langle \bar{p} \,| . \tag{15.45}$$

Consider

$$\langle \bar{p}_2 \,|\, \mathscr{P} \,|\, \bar{p}_1 \rangle = \langle \bar{p}_2 \,|\int \mathrm{d}^3\boldsymbol{p}_3 \,|\, \bar{p}_3 \rangle \langle \bar{p}_3 \,|\, \bar{p}_1 \rangle$$

$$= B^2 \int \mathrm{d}^3\boldsymbol{p}_3 \, \delta(\bar{p}_2 - \bar{p}_3)\, \delta(\bar{p}_3 - \bar{p}_1)$$

$$= B \langle \bar{p}_2 \,|\, \mathsf{I} \,|\, \bar{p}_1 \rangle = B\, \delta(\bar{p}_2 - \bar{p}_1) .$$

Thus

$$\int \mathrm{d}^3\boldsymbol{p}_3 \, \delta(\bar{p}_2 - \bar{p}_3)\, \delta(\bar{p}_3 - \bar{p}_1) = B^{-1} \delta(\bar{p}_2 - \bar{p}_1) . \tag{15.46}$$

Also

$$\langle \bar{p}_2 \,|\, \mathscr{R} \,|\, \bar{p}_1 \rangle = A \langle \bar{p}_2 \,|\, \bar{p}_1 \rangle = AB\delta(\bar{p}_2 - \bar{p}_1)$$

$$= \int \mathrm{d}^3\boldsymbol{r} \, \langle \bar{p}_2 \,|\, \bar{r} \rangle \langle \bar{r} \,|\, \bar{p}_1 \rangle$$

$$= N^2 \int \mathrm{d}^3\boldsymbol{r} \exp \{\mathrm{i}(\bar{\boldsymbol{p}}_2 - \bar{\boldsymbol{p}}_1) \cdot \bar{\boldsymbol{r}}/\hbar\}$$

$$= \lim_{L \to \infty} N^2 \left\{ \int_{-\frac{1}{2}L}^{\frac{1}{2}L} \mathrm{d}x \exp \{\mathrm{i}(p_{2,x} - p_{1,x})x/\hbar\} \right\}^3$$

$$= N^2 \left\{ \lim_{L \to \infty} 2\hbar \frac{\sin [(L/2\hbar)\,(p_{2,x} - p_{1,x})]}{(p_{2,x} - p_{1,x})} \right\}^3$$

$$= N^2 \{2\pi\hbar \, \delta(p_{2,x} - p_{1,\,x})\}^3$$

$$= N^2 (2\pi\hbar)^3 \, \delta(\bar{p}_2 - \bar{p}_1) .$$

Thus

$$AB = N^2 (2\pi\hbar)^3 . \tag{15.47}$$

Consider

$$\langle \bar{r}_2 | \mathscr{R} | \bar{r}_1 \rangle = \int d^3\boldsymbol{r}_3 \langle \bar{r}_2 | \bar{r}_3 \rangle \langle \bar{r}_3 | \bar{r} \rangle$$

$$= A^2 \int d^3\boldsymbol{r}_3\, \delta(\bar{r}_2 - \bar{r}_3)\, \delta(\bar{r}_3 - \bar{r}_1)$$

$$= A \langle \bar{r}_2 | \mathsf{I} | \bar{r}_1 \rangle = A\, \delta(\bar{r}_2 - \bar{r}_1).$$

Thus

$$\int d^3\boldsymbol{r}_3\, \delta(\bar{r}_2 - \bar{r}_3)\, \delta(\bar{r}_3 - \bar{r}_1) = A^{-1} \delta(\bar{r}_2 - \bar{r}_1). \tag{15.48}$$

Also

$$\langle \bar{r}_2 | \mathscr{P} | \bar{r}_1 \rangle = AB\, \delta(F_2 - \bar{r}_1)$$

$$= \langle \bar{r}_2 | \int d^3\boldsymbol{p}\, | \bar{p} \rangle \langle \bar{p} | \bar{r}_1 \rangle$$

$$= N^2 \int d^3\boldsymbol{p} \exp\{i\bar{\boldsymbol{p}} \cdot (\bar{\boldsymbol{r}}_1 - \bar{\boldsymbol{r}}_2)/\hbar\}$$

$$= N^2 \left[\lim_{p \to \infty} \int_{-p}^{p} dp_x \exp\{ip_x(x_1 - x_2)/\hbar\} \right]^3$$

$$= N^2 (2\pi\hbar)^3\, \delta(\bar{r}_1 - \bar{r}_2). \tag{15.49}$$

Thus again $AB = (2\pi\hbar)^3 N^2$.
Thus any relative normalization that preserves $N^2 (2\pi\hbar)^3 = AB$ is allowed and gives the brackets and operators to be

$$\langle \bar{r}_2 | \bar{r}_1 \rangle = A\, \delta(\bar{r}_2 - \bar{r}_1), \tag{15.50a}$$

$$\langle \bar{p}_2 | \bar{p}_1 \rangle = B\, \delta(\bar{p}_2 - \bar{p}_1), \tag{15.50b}$$

$$\langle \bar{p} | \bar{r} \rangle = (AB)^{\frac{1}{2}} (2\pi\hbar)^{\frac{3}{2}} \exp\{i\bar{\boldsymbol{p}} \cdot \bar{\boldsymbol{r}}/\hbar\} \tag{15.50c}$$

$$= N \exp\{i\bar{\boldsymbol{p}} \cdot \bar{\boldsymbol{r}}/\hbar\},$$

$$\mathsf{I} = A^{-1} \int d^3\boldsymbol{r}\, | \bar{r} \rangle \langle \bar{r} |, \tag{15.50d}$$

$$\mathsf{I} = B^{-1} \int d^3\boldsymbol{p}\, | \bar{p} \rangle \langle \bar{p} |, \tag{15.50e}$$

$$\int d^3\boldsymbol{p} \exp\{i\bar{\boldsymbol{p}}\cdot(\bar{\boldsymbol{r}}_1-\bar{\boldsymbol{r}}_2)/\hbar\} = (2\pi\hbar)^3\,\delta(\bar{r}_1-\bar{r}_2), \quad (15.50\text{f})$$

$$\int d^3\boldsymbol{r} \exp\{i(\bar{\boldsymbol{p}}_2-\bar{\boldsymbol{p}}_1)\cdot\bar{\boldsymbol{r}}/\hbar\} = (2\pi\hbar)^3\,\delta(\bar{p}_1-\bar{p}_2). \quad (15.50\text{g})$$

On further setting

$$\int d^3\boldsymbol{r}\,\langle\bar{r}\mid\bar{p}_2\rangle\langle\bar{p}_1\mid\bar{r}\rangle = \delta(\bar{p}_2-\bar{p}_1)$$

$$= N^2(2\pi\hbar)^3\,\delta(\bar{p}_1-\bar{p}_2); \quad (15.50\text{h})$$

thus $N^2 = (2\pi\hbar)^{-3}$. (15.50i)

It is seen that a simple and convenient choice of A and B is to set

$$A = A^{-1} = B = B^{-1} = 1. \quad (15.50\text{j})$$

These results are summarized by the table below:

Symbol	Dimensions
$\langle\bar{r}_2\mid\bar{r}_1\rangle = \delta(\bar{r}_1-\bar{r}_2)$	L^{-3}
$\langle\bar{p}_2\mid\bar{p}_1\rangle = \delta(\bar{p}_1-\bar{p}_2)$	P^{-3}
$\langle\bar{p}\mid\bar{r}\rangle = (2\pi\hbar)^{-\frac{3}{2}}\exp(i\bar{\boldsymbol{p}}\cdot\bar{\boldsymbol{r}}/\hbar)$	$P^{-\frac{3}{2}}$
$\mathsf{I} = \int d^3\boldsymbol{r}\mid\bar{\boldsymbol{r}}\rangle\langle\bar{\boldsymbol{r}}\mid$	none
$\mathsf{I} = \int d^3\boldsymbol{p}\mid\bar{\boldsymbol{p}}\rangle\langle\bar{\boldsymbol{p}}\mid$	none
$\int d^3\boldsymbol{p}\exp\{i\bar{\boldsymbol{p}}\cdot(\bar{\boldsymbol{r}}_1-\bar{\boldsymbol{r}}_2)/\hbar\} = (2\pi\hbar)^3\delta(\bar{r}_1-\bar{r}_2)$	P^3L^{-3}
$\int d^3\boldsymbol{r}\exp\{i(\bar{\boldsymbol{p}}_2-\bar{\boldsymbol{p}}_1)\cdot\bar{\boldsymbol{r}}/h\} = (2\pi\hbar)^3\delta(p_1-p_2)$	none
$\int d^3\boldsymbol{p}_3\,\delta(\bar{p}_3-\bar{p}_2)\,\delta(\bar{p}_3-\bar{p}_1) = \delta(\bar{p}_1-\bar{p}_2)$	P^{-3}
$\int d^3\boldsymbol{r}_3\,\delta(\bar{r}_3-\bar{r}_2)\,\delta(\bar{r}_3-\bar{r}_1) = \delta(\bar{r}_1-\bar{r}_2)$	L^{-3}

This normalization, used in this and the following chapter, has the simplicities that brackets in a given basis integrate over their basis variables to a pure number, and that the insertion of either identity operator does not introduce any dimensions. Additional useful relationships are:

$$d^3\boldsymbol{p}_1 = p_1{}^2\,dp_1\,d\Omega_1 = \mu p_1\,dE_1\,d\Omega_1,\ d\Omega_1 = \sin\theta_1\,d\theta_1\,d\varphi_1, \quad (15.51)$$

$$\delta(\bar{p}_1-\bar{p}_2) = p_1{}^2\,\delta(p_1-p_2)\,\delta(\Omega_1-\Omega_2)$$
$$= p^2(\sin\theta)^{-1}\,\delta(p_1-p_2)\delta(\theta_1-\theta_2)\,\delta(\varphi_1-\varphi_2). \quad (15.52)$$

7. Time development of the scattering states of momentum

Consider momentum matrix elements of $T(t)$. Using (15.15) and (15.23),

$$\langle \bar{p}_2 \mid \mathsf{T}(t) \mid \bar{p}_1 \rangle =$$

$$= (\hbar/2\pi) \int_\gamma e^{st}\, ds \iint d^3\boldsymbol{p}_3\, d^3\boldsymbol{p}_4 \langle \bar{p}_2 \mid \frac{i/\hbar}{s+iE_3/\hbar} \mid \bar{p}_3 \rangle \langle \bar{p}_3 \mid \mathscr{T}_I(s) \mid \bar{p}_4 \rangle \times \langle \bar{p}_4 \mid \frac{i/\hbar}{s+iE_1/\hbar} \mid \bar{p}_1 \rangle$$

$$= (\hbar/2\pi) \int_\gamma e^{st}\, ds \iint d^3\boldsymbol{p}_3\, d^3\boldsymbol{p}_4 \frac{i/\hbar}{s+iE_3/\hbar}\, \delta(\bar{p}_3 - \bar{p}_2)\, \frac{i/\hbar}{s+iE_1/\hbar}\, \delta(\bar{p}_1 - \bar{p}_4) \times \langle \bar{p}_3 \mid \mathscr{T}_I(s) \mid \bar{p}_4 \rangle$$

$$= (\hbar/2\pi) \int e^{st}\, ds \frac{i/\hbar}{s+iE_2/\hbar}\, \frac{i/\hbar}{s+iE_1/\hbar} \langle \bar{p}_2 \mid \mathscr{T}_I(s) \mid \bar{p}_1 \rangle . \tag{15.53}$$

The singularities at $s = -iE_1/\hbar$ and $s = -iE_2/\hbar$ produce time solutions that have factors $\exp(-iE_1 t/\hbar)$ and $\exp(-iE_2 t/\hbar)$ respectively. The singularities of $\mathscr{T}_I(s)$ are all located at the left half plane at positions $s = -\alpha + i\beta$, $\alpha > 0$ and so behave as $e^{-\alpha t}$. These transient solutions decay to zero for large times (when observations are made) and can be neglected. Thus the Cauchy integral theorem (for the principal value) can be used to calculate the integral for large values of the times; it is just $2\pi i$ times the sum of the residues at the two poles located at $s = -iE_1/\hbar$ or $-iE_2/\hbar$. Note, however, that the effects of singularities of $\mathscr{T}_I(s)$ are still present in the residues of these two poles.

PROBLEM 6. Evaluate

$$\int_{-\infty}^{+\infty} dx\, f(x)\, \delta(x^2 - a^2) .$$

PROBLEM 7. Prove that

$$\lim_{t\to\infty} \left(\frac{\sin xt}{x} \right) = 2\pi \delta(x)$$

by considering the contour integral of

$$\lim_{t\to\infty} e^{ixt}/x .$$

8. The time propagator at large times

As the wave packet passes the scattering center it will be distorted and scattered into directions other than the incident direction. After a long enough time the packet will have moved away from the short range of the scattering forces and will move as a free particle. Its propagator must again become T_0. However, some of the wave packet will now be somewhat of a spherical shell wave packet, expanding in all directions with an amplitude and shape dependent on the direction. In equation (15.12) the first term, T_0, operating on the initial state gives the undeflected beam, while the second part leads to the scattering.

The second part is given by

$$\mathsf{T}(t) = (-\hbar/2\pi)\int_\gamma e^{st}\mathscr{T}(s)\,ds$$
$$= (\hbar/2\pi)\int_\gamma e^{st}\mathscr{T}_0(s)\mathscr{T}_I(s)\mathscr{T}_0(s)\,ds\,. \tag{15.54}$$

This may be evaluated using the Cauchy integral theorem, where for large values of the time,

$$\langle\bar{p}_2\,|\,\mathsf{T}(t\to\infty)\,|\,\bar{p}_1\rangle = -2\left\{\begin{matrix} e^{-iE_1t/\hbar}\,e^{i\varepsilon t/\hbar}\ \langle\bar{p}_2\,|\,\mathscr{T}_I(-iE_2/\hbar)\,|\,\bar{p}_1\rangle \\ -e^{-iE_2t/\hbar}\,e^{-i\varepsilon t/\hbar}\,\langle\bar{p}_2\,|\,\mathscr{T}_I(-iE_1/\hbar)\,|\,\bar{p}_1\rangle \end{matrix}\right\}, \tag{15.55}$$

where $\varepsilon = E_1 - E_2$.

Now

$$\lim_{t\to\infty}\int_{-\infty}^{\infty}\frac{\exp(\pm i\varepsilon t/\hbar)}{\varepsilon}\,d\varepsilon = \pm\,\pi i.$$

Thus we can treat $\exp(\pm i\varepsilon t/\hbar)/\varepsilon$ as equal $\pm\pi i\delta(\varepsilon)$ since

$$\lim_{t\to\infty}\int\frac{\exp(\pm i\varepsilon\, t/\hbar)}{\varepsilon}f(\varepsilon)d\varepsilon = \pm\pi i f(0)$$

for any function $f(\varepsilon)$ that is analytic in the upper half plane (+case), or in the lower half plane (− case). The first matrix element of $\mathscr{T}_I$ is not analytic for $\mathrm{Im}(\varepsilon) > 0$ but its integrand converges on a large semicircle in the upper half ε plane because of the factor $\exp\{-\mathrm{Im}(\varepsilon)t/\hbar)\}$ as $t\to\infty$. The second matrix element similarly is not analytic for $\mathrm{Im}(\varepsilon) < 0$, nevertheless the exponential $\exp(-\mathrm{i}\varepsilon t/\hbar)$ gives convergence for $\mathrm{Im}(\varepsilon) < 0$ as $t \to \infty$ on a large semicircle. Thus

$$\langle \bar{p}_2 \mid \mathsf{T}(t \to \infty) \mid \bar{p}_1 \rangle = -2\pi \mathrm{i} e^{-\mathrm{i}E_1 t/\hbar} \langle \bar{p}_2 \mid \mathscr{T}_\mathrm{I}(-\mathrm{i}E_1/\hbar) \mid \bar{p}_1 \rangle \, \delta(E_1 - E_2). \tag{15.56}$$

Note that the initial and final energy are the same (energy is conserved). However in general, the matrix elements of $\mathscr{T}_\mathrm{I}$ have three values of the energy inherent in them, one for each state vector and one for the energy corresponding to s in the operator $\mathscr{T}_\mathrm{I}$. If all are different the matrix element is said to be fully "off the energy shell", if two are the same and the third is different then it is "off the energy shell" "at one end". If all the same, it is fully "*on* the energy shell". If the two momenta are the same and their corresponding energy is the same as that of the operator $\mathscr{T}_\mathrm{I}$ then the element is said to be "on the momentum shell".

PROBLEM 8. Consider that $\langle \bar{p}_2 \mid \mathscr{T}_\mathrm{I}(s) \mid \bar{p}_1 \rangle = A/(s+\alpha+\mathrm{i}\beta)$ with A, α, β real constants, $\alpha > 0$ in equation (15.53). Calculate $\langle \bar{p}_2 \mid \mathsf{T}(t) \mid \bar{p}_1 \rangle$, for all times, and discuss the behavior for $t \to \infty$.

PROBLEM 9. Consider $\langle \bar{r}_2 \mid \mathscr{T}_0(-\mathrm{i}E_0/h) \mid \bar{r}_1 \rangle$ and discuss its properties.

9. Unitarity of the scattering state

As discussed above the scattering is treated by conceiving of a wave packet represented by a ket $\mid A \rangle$. This propagates in the time according to $\mid A, t \rangle = \mathsf{T}_\mathrm{S} \mid A, 0 \rangle = \mathsf{T}_\mathrm{S} \mid A \rangle$. By using the results derived above this time development can be calculated in terms of matrix elements of $\mathscr{T}_\mathrm{I}$.

If the initial state is normalized, $\langle A \mid A \rangle = 1$ and it represents the initial probability of there being a projectile incident upon a scattering center. As will be discussed later this normalization will be preserved for the case of elastic scattering that is being considered. The time dependent probability of the total scattering state is

$$\begin{aligned} P_\mathrm{t}(t) &= \langle A, t \mid A, t \rangle = \langle A \mid \mathsf{T}_\mathrm{S}^\dagger \mathsf{T}_\mathrm{S} \mid A \rangle \\ &= \langle A \mid (\mathsf{T}_0^\dagger + \mathsf{T}^\dagger)(\mathsf{T}_0 + \mathsf{T}) \mid A \rangle \\ &= \langle A \mid \mathsf{T}_0^\dagger \mathsf{T}_0 \mid A \rangle + 2\,\mathrm{Re}\{\langle A \mid \mathsf{T}_0^\dagger \mathsf{T} \mid A \rangle\} + \langle A \mid \mathsf{T}^\dagger \mathsf{T} \mid A \rangle. \end{aligned} \tag{15.57}$$

This is a form of the optical theorem that will be discussed below. Since the scattering process conserves probability (if only elastic scattering occurs) then $\langle A, t \mid A, t\rangle = 1$ for all times, and since $\mathsf{T}_0^\dagger \mathsf{T}_0 = 1$ (that is the operator is unitary), it follows that

$$\mathsf{T}_S^\dagger \mathsf{T}_S = 1, \tag{15.58}$$

also and that T_S is unitary. Thus

$$P_t(t) \equiv 1 = 1 + 2\,\mathrm{Re}\langle A \mid \mathsf{T}_0^\dagger \mathsf{T} \mid A\rangle + \langle A \mid \mathsf{T}^\dagger \mathsf{T} \mid A\rangle .$$

It follows that

$$\langle A \mid \mathsf{T}^\dagger \mathsf{T} \mid A\rangle = -2\,\mathrm{Re}\langle A \mid \mathsf{T}_0^\dagger \mathsf{T} \mid A\rangle . \tag{15.59}$$

This is also a form of the optical theorem. As shown above T is the propagator for that portion of the state that suffers scattering thus $\mathsf{T} \mid A\rangle$ is that scattered state and $\langle A \mid \mathsf{T}^\dagger \mathsf{T} \mid A\rangle$ is its probability. However $\langle A \mid \mathsf{T}_0{}^\dagger$ is the unscattered state and $\langle A \mid \mathsf{T}_0{}^\dagger \mathsf{T} \mid A\rangle$ represents the probability that the unscattered state is changed, or the probability, that the projectile is removed from the beam (scattered forward) by the scattering center. Since this scattering effect is to reduce the beam transmission in the forward direction, the negative sign is explained since $\langle A \mid \mathsf{T}^\dagger \mathsf{T} \mid A\rangle$ must be a positive number, or zero.

Thus the probability for scattering is

$$\begin{aligned} P_S &\equiv \langle A \mid \mathsf{T}^\dagger \mathsf{T} \mid A\rangle \\ &= \int \mathrm{d}^3 \boldsymbol{p}_2 \,\langle A \mid \mathsf{T}^\dagger \mid \bar{p}_2\rangle \langle \bar{p}_2 \mid \mathsf{T} \mid A\rangle \\ &= \int \mathrm{d}^3 \boldsymbol{p}_2 \,|\langle \bar{p}_2 \mid \mathsf{T} \mid A\rangle|^2 \\ &= \int \mathrm{d}^3 \boldsymbol{p}_2 \left| \int \mathrm{d}^3 \boldsymbol{p}_1 \,\langle \bar{p}_2 \mid \mathsf{T} \mid \bar{p}_1\rangle \langle \bar{p}_1 \mid A\rangle \right|^2 . \end{aligned} \tag{15.59a}$$

Using the results of equation (15.56),

$$P_S(t \to \infty) =$$

$$= \int \mathrm{d}^3 \boldsymbol{p}_2 \left| -2\pi \mathrm{i} \int \mathrm{d}^3 \boldsymbol{p}_1 \,\langle \bar{p}_1 \mid A\rangle\, \delta(E_1 - E_2) \langle \bar{p}_2 \mid \mathscr{T}_{\mathrm{I}}(-\mathrm{i}E_1/\hbar) \mid \bar{p}_1\rangle \right|^2 ,$$

where the time exponential has been squared to one. Noting that

$$\mathrm{d}^3 \boldsymbol{p}_1 = \mu p_1 \,\mathrm{d}E_1 \,\mathrm{d}\Omega_1, \quad s = -\mathrm{i}E_1/\hbar ,$$

$$P_S(t \to \infty) = 4\pi^2\mu^2 \int d^3\boldsymbol{p}_2 \left| \iint dE_1\, d\Omega_1\, \langle \bar{p}_1 \mid A \rangle\, p_1\, \delta(E_1 - E_2) \times \langle \bar{p}_2 \mid \mathscr{T}_{\mathrm{I}}(-\mathrm{i}E_1/\hbar) \mid \bar{p}_1 \rangle \right|^2$$

$$= 4\pi\mu^2 \int d^3\boldsymbol{p}_2\, p_2^2 \left| \int d\Omega_1\, \langle p_2 \hat{\imath}_1 \mid A \rangle \times \langle p_2 \hat{\imath}_2 \mid \mathscr{T}_{\mathrm{I}}(-\mathrm{i}E_2/\hbar) \mid p_2 \hat{\imath}_1 \rangle \right|^2.$$

Upon squaring

$$P_S(t \to \infty) = 4\pi^2\mu^2 \int d^3\boldsymbol{p}_1\, p_1^2 \mid \langle \bar{p}_1 \mid A \rangle \mid^2 \times \left| \int d\Omega_2 \langle p_1 \hat{\imath}_2 \mid \mathscr{T}_{\mathrm{I}}(-\mathrm{i}E_1/\hbar) \mid p_1 \hat{\imath}_1 \rangle \right|^2. \tag{15.60}$$

In the same way the probability of absorption of the beam in the forward direction is

$$P_a(t \to \infty) = -4\pi\mu \int d^3\boldsymbol{p}_1\, p_1 \mid \langle \bar{p}_1 \mid A \rangle \mid^2 \operatorname{Im} \{ \langle \bar{p}_1 \mid \mathscr{T}_{\mathrm{I}}(-\mathrm{i}E_1/\hbar) \mid \bar{p}_1 \rangle \}. \tag{15.61}$$

Thus unitarity requires that for each component $\mid \langle \bar{p}_1 \mid A \rangle \mid^2 d^3\boldsymbol{p}_1$ in the incident beam

$$P_S(t \to \infty) = 4\pi^2\mu^2 p_1^2 \int d\Omega_2 \mid \langle p_1 \hat{\imath}_2 \mid \mathscr{T}_{\mathrm{I}}(-\mathrm{i}E_1/\hbar) \mid \bar{p}_1 \rangle \mid^2$$
$$= -4\pi\mu p_1 \operatorname{Im} \{ \langle \bar{p}_1 \mid \mathscr{T}_{\mathrm{I}}(-\mathrm{i}E_1/\hbar) \mid \bar{p}_1 \rangle \} = P_a(t \to \infty). \tag{15.62}$$

This gives the relationship

$$\operatorname{Im} \{ \langle \bar{p}_1 \mid \mathscr{T}_{\mathrm{I}}(-\mathrm{i}E_1/\hbar) \mid \bar{p}_1 \rangle - \pi\mu p_1 \int d\Omega_2 \mid \langle p_1 \hat{\imath}_2 \mid \mathscr{T}_{\mathrm{I}}(-\mathrm{i}E_1/\hbar) \mid \bar{p}_1 \rangle \mid^2. \tag{15.63}$$

10. Scattering cross section

From the above development

$$P_t = \int d^3 p_1 \mid \langle \bar{p}_1 \mid A \rangle \mid^2 \tag{15.64}$$

is the total probability,

$$P_{\mathrm{S}}(t \to \infty) = \int \mathrm{d}^3\boldsymbol{p}_1 \,|\, \langle \bar{p}_1 \,|\, A \rangle \,|^2\, (2\pi\mu p_1)^2$$
$$\times \int \mathrm{d}\Omega_2 \,|\, \langle p_1 \hat{\imath}_2 \,|\, \mathscr{T}_{\mathrm{I}}(-\mathrm{i}E_1/\hbar) \,|\, \bar{p}_1 \rangle \,|^2 \tag{15.65}$$

is the scattering probability, while

$$P_{\mathrm{a}}(t \to \infty) = \int \mathrm{d}^3\boldsymbol{p}_1 \,|\, \langle \bar{p}_1 \,|\, A \rangle \,|^2\, (-4\pi\mu p_1)$$
$$\times \operatorname{Im} \{\langle \bar{p}_1 \,|\, \mathscr{T}_{\mathrm{I}}(-\mathrm{i}E_1/\hbar) \,|\, \bar{p}_1 \rangle\}, \tag{15.66}$$

is the probability of absorption of the beam.

The projectile has a probability of $p_1{}^2 \mathrm{d}p_1$ of having momenta along a z axis, with its momentum lying between p_1 and $p_1 + \mathrm{d}p_1$. The density per unit area of this plane wave of projectile beam is $(2\pi\hbar)^{-2}$. These factors must be divided into the incremental probability of scattering. Thus the normalized probability of scattering is

$$\sigma = (2\pi\mu)^2 (2\pi\hbar)^2 \int \mathrm{d}\Omega_2 \left| \langle p_1 \hat{\imath}_2 \,|\, \mathscr{T}_{\mathrm{I}}(-\mathrm{i}E_1/\hbar) \,|\, \bar{p}_1 \rangle \right|^2 . \tag{15.67}$$

This quantity has the dimensions of an area and is defined as *the total cross section, σ, for scattering*: The differential scattering cross section is defined as

$$\frac{\mathrm{d}\sigma}{\mathrm{d}\Omega_2} = (2\pi\mu)^2 (2\pi\hbar)^2 \left| \langle \bar{p}_1 \hat{\imath}_2 \,|\, \mathscr{T}_{\mathrm{I}}(-\mathrm{i}E_1/\hbar) \,|\, \bar{p}_1 \rangle \right|^2 \tag{15.68}$$
$$\equiv |\, f(\theta_2, \varphi_2) \,|^2$$

where the scattering amplitude, with dimensions of a length, is defined as

$$f(\theta_2, \varphi_2) \equiv (2\pi\mu)\,(2\pi\hbar)\, \langle p_1 \hat{\imath}_2 \,|\, \mathscr{T}_{\mathrm{I}}(-\mathrm{i}E_1/\hbar) \,|\, \bar{p}_1 \rangle . \tag{15.69}$$

11. Schroedinger representation of scattering

Since $\langle \bar{r} \,|\, \bar{p} \rangle = (2\pi h)^{-\frac{3}{2}} \exp(+\mathrm{i}\bar{p} \cdot \bar{r}/\hbar)$,

$$f(\theta_2, \varphi_2) = (2\pi\mu)\,(2\pi\hbar) \iint \mathrm{d}^3\boldsymbol{r}_1\, \mathrm{d}^3\boldsymbol{r}_2\, \langle p_1 \hat{\imath}_2 \,|\, \bar{r}_2 \rangle \langle \bar{r}_2 \,|\, \mathscr{T}_{\mathrm{I}} \,|\, \bar{r}_1 \rangle \langle \bar{r}_1 \,|\, \bar{p}_1 \rangle$$
$$= (2\pi\mu)\,(2\pi\hbar)^{-2} \iint \mathrm{d}^3\boldsymbol{r}_1\, \mathrm{d}^3\boldsymbol{r}_2$$
$$\times \exp(-\mathrm{i}p_1 \hat{\boldsymbol{\imath}}_2 \cdot \bar{\boldsymbol{r}}_2/\hbar)\, \langle \bar{r}_2 \,|\, \mathscr{T}_{\mathrm{I}} \,|\, \bar{r}_1 \rangle \exp(\mathrm{i}\bar{\boldsymbol{p}}_1 \cdot \bar{\boldsymbol{r}}_1/\hbar). \tag{15.70}$$

12. The first Born approximation for local, central potentials

When the potential is local and central it obeys

$$\langle \bar{r}_2 \,|\, \mathsf{V} \,|\, \bar{r}_1 \rangle = V(r_1)\,\delta(\bar{r}_2 - \bar{r}_1). \tag{15.71}$$

The first Born term is then

$$\langle \bar{r}_2 \,|\, \mathscr{T}_{\mathrm{I}} \,|\, \bar{r}_1 \rangle = V(r_1)\,\delta(\bar{r}_2 - \bar{r}_1).$$

Thus

$$f_{\mathrm{B}}^{(1)}(\theta_2, \varphi_2) = (2\pi\mu)(2\pi\hbar)^{-2} \int \mathrm{d}^3\boldsymbol{r}_1 \exp\{\mathrm{i}p_1(\hat{\boldsymbol{\imath}}_1 - \hat{\boldsymbol{\imath}}_2) \cdot \bar{\boldsymbol{r}}_1/\hbar\} V(r_1), \tag{15.71a}$$

and the first Born approximation to the scattering amplitude is

$$\begin{aligned} f_{\mathrm{B}}^{(1)}(\theta_2) &= (2\pi\hbar^2)^{-2}\mu^2 \int \mathrm{d}^3\boldsymbol{r}_1 \exp\{\mathrm{i}p_1(\hat{\boldsymbol{\imath}}_1 - \hat{\boldsymbol{\imath}}_2) \cdot \bar{\boldsymbol{r}}_1/\hbar\}\, V(r_1) \\ &= \frac{1}{2\pi}(\mu/\hbar^2)^2 2\pi \int_0^\infty r_1^2\,\mathrm{d}r_1 \int_0^\pi \sin\theta\,\mathrm{d}\theta\, \mathrm{e}^{\mathrm{i}qr\cos\theta} V(r_1), \end{aligned}$$

where $\bar{q} = p_1(\hat{\imath}_1 - \hat{\imath}_2)$ and θ is the angle between $\bar{\boldsymbol{r}}_1$ and $\bar{\boldsymbol{q}}$. Thus

$$f_{\mathrm{B}}^{(1)}(\theta_2) = [2(\mu/\hbar^2)^2/q] \int_0^\infty r_1\,\mathrm{d}r_1\, V(r_1) \sin qr_1 .$$

Now $q = 2p_1 \sin\frac{1}{2}\theta_2$.

Thus

$$f_{\mathrm{B}}^{(1)}(\theta_2) = [\mu^2/\hbar^4 p_1 \sin\tfrac{1}{2}\theta_2] \int_0^\infty r_1\,\mathrm{d}r_1 V(r_1) \sin qr_1 . \tag{15.72}$$

This amplitude is explicitly real if $V(r)$ is real. This point will be discussed below.

PROBLEM 10. Work out the form of the second Born term for a central-local potential.

PROBLEM 11. Discuss the first Born approximation for the Coulomb potential $1/r$.

13. The optical theorem

The differentials of P_{S} and P_{a} must sum to zero. Thus by (15.65) and (15.66)

$$(2\pi\mu p_1)^2 \int d\Omega_2 \,|\langle \bar{p}_2 | \mathscr{T}_1 | \bar{p}_1 \rangle|^2 = 2(2\pi\mu p_1)\,\mathrm{Im}\,\{\langle \bar{p}_1 | \mathscr{T}_1 | \bar{p}_1 \rangle\}$$

or $$(2\pi\mu)p_1(2\pi\hbar)^{-2} \int |f(\theta_2)|^2 \, d\Omega_2 = 2(2\pi\hbar)^{-1}\,\mathrm{Im}\,\{f(0)\}\,.$$

This gives

$$\sigma = -\frac{4\pi}{k}\,\mathrm{Im}\,\{f(0)\}\,, \qquad (15.73)$$

where $k = p_1/\hbar$. This equation states the optical theorem for elastic scattering. It relates the imaginary part of the *forward* scattering amplitude to the total scattering cross section. As has been discussed above, this theorem results from the physical fact that the total scattering probability is just equal to the forward scattering probability, or to the absorption of the incident flux.

Note that if $f(0)$ has no imaginary part, then the total cross section is zero. In the preceding section the first Born amplitude was calculated for real, central, local potentials and was found to be real. This result predicts a zero total cross section, using the optical theorem — a result that is nonsense. The error, of course, is in neglecting all the higher order Born terms, or Feynman diagrams.

14. Unitarity, the R- and S-matrices, and phase shifts

The unitarity of a free particle or of the scattering state are expressed respectively by

$$\mathsf{T}_0^\dagger \mathsf{T}_0 = 1\,,$$

and

$$\mathsf{T}_S^\dagger \mathsf{T}_S = 1\,,$$

from which it follows that

$$\mathsf{T}^\dagger \mathsf{T} = -2\,\mathrm{Re}\,\{\mathsf{T}_0^\dagger \mathsf{T}\}$$

where T is defined so that

$$\mathsf{T}_S = \mathsf{T}_0 + \mathsf{T}\,,$$

and T is an additional operator to relate T_0 to T_S.

Another viewpoint is to define a multiplicative operator R by

$$\mathsf{T}_S \equiv \mathsf{T}_0 \mathsf{R}\,. \qquad (15.74)$$

Thus

$$\mathsf{T} = \mathsf{T}_0(\mathsf{R}-1)$$

and R obeys the unitarity equation

$$(\mathsf{R}^\dagger-1)(\mathsf{R}-1) = -2 \text{ Re } \{(\mathsf{R}-1)\}$$

or

$$\mathsf{R}^\dagger\mathsf{R} = 1. \tag{15.75}$$

Thus R is a unitary operator. The fact that T_S is represented by the product of T_0 on R implies that the initial state is first changed by the unitary operator R to a different state, that is then propagated as a free particle *via* the operator T_0.

A second operator, the S operator, may be defined by

$$\mathsf{S} \equiv \mathsf{R}-1 \tag{15.76}$$

and unitarity requires that

$$\mathsf{S}^\dagger\mathsf{S} = -2 \text{ Re}\, \mathsf{S}. \tag{15.77}$$

The S-operator is related to the interaction propagator by

$$\mathsf{T} = \mathsf{T}_0\mathsf{S}. \tag{15.78}$$

Thus the relationship between matrix elements of T, $\mathscr{T}_\text{I}$ and S are given by

$$\begin{aligned}\langle \bar{p}_2 \,|\, \mathsf{T} \,|\, \bar{p}_1\rangle &= \mathrm{e}^{-\mathrm{i}E_2t/\hbar}\langle \bar{p}_2 \,|\, \mathsf{S} \,|\, \bar{p}_1\rangle \\ &= -2\pi\mathrm{i}\, \langle \bar{p}_2 \,|\, \mathscr{T}_\text{I}(-\mathrm{i}E_1/\hbar) \,|\, \bar{p}_1\rangle\, \delta(E_1-E_2)\, \mathrm{e}^{-\mathrm{i}E_1t/\hbar}\end{aligned} \tag{15.79}$$

so that upon integrating over the energy E_2,

$$\langle p_1\hat{\imath}_2 \,|\, \mathscr{T}_\text{I}(-\mathrm{i}E_1/\hbar) \,|\, p_1\hat{\imath}_1\rangle = -\frac{1}{2\pi\mathrm{i}}\langle p_1\hat{\imath}_2 \,|\, \mathsf{S} \,|\, p_1\hat{\imath}_1\rangle. \tag{15.80}$$

Thus, except for a factor $2\pi\mathrm{i}$ the S and $\mathscr{T}_\text{I}$ matrix elements are the same. Thus the scattering amplitude is given by

$$f(\theta_2, \varphi_2) = 2\pi\mathrm{i}\mu\hbar\, \langle p_1\hat{\imath}_2 \,|\, \mathsf{S} \,|\, p_1\hat{\imath}_1\rangle. \tag{15.81}$$

Referring to the unitarity equations for $\mathsf{S} = -2\pi\mathrm{i}\mathscr{T}_\text{I}$ and R it is seen that a general parameterization of the matrix elements is

$$\langle p_1\hat{\imath}_2 \,|\, \mathsf{S} \,|\, p_1\hat{\imath}_1\rangle = \mathrm{e}^{\,2\mathrm{i}\delta} - | = 2\mathrm{i}\mathrm{e}^{\mathrm{i}\delta}\sin\delta|, \tag{15.82}$$

where δ = real and is termed the phase shift of the S matrix (or scattering amplitude). Thus

$$\langle p_1\hat{\imath}_2 \,|\, \mathsf{R} \,|\, p_1\hat{\imath}_1\rangle = \mathrm{e}^{2\mathrm{i}\delta}. \tag{15.83}$$

It is noted that R is explicitly unitary.

The description of the scattering amplitude in terms of the phase shift is an important one that allows simple physical interpretations. It is seen that

$$\begin{aligned} f(\theta_2, \varphi_2) &= -4\pi\mu\hbar\, e^{i\delta} \sin\delta \\ &\equiv |f(\theta_2, \varphi_2)| \exp\{i \arg f(\theta_2, \varphi_2)\}. \end{aligned} \tag{15.84}$$

Thus

$$|f(\theta_2, \varphi_2)| = -4\pi\mu\hbar \sin\delta \tag{15.85a}$$

and

$$\arg f(\theta_2, \varphi_2) = \delta, \tag{15.85b}$$

so that a knowledge of the phase shift is equivalent to knowing $f(\theta_2, \varphi_2)$.

15. Relationship of the phase shift to causality, time delay, and resonances

Matrix elements of the scattering propagator are given by

$$\begin{aligned} \langle \bar{p}_2 | \mathsf{T} | \bar{p}_1 \rangle &= \langle \bar{p}_2 | \mathsf{T}_0 | \bar{p}_1 \rangle + \langle \bar{p}_2 | \mathsf{T} | \bar{p}_1 \rangle \\ &= \langle \bar{p}_2 | \mathsf{T}_0 \mathsf{R} | \bar{p}_1 \rangle \\ &= \delta(\bar{p}_2 - \bar{p}_1)\, e^{-iE_1 t/\hbar} + e^{-iE_2 t/\hbar} \langle \bar{p}_2 | \mathsf{S} | \bar{p}_1 \rangle \\ &= e^{-iE_2 t/\hbar}\, e^{2i\delta}. \end{aligned}$$

This expression holds for all energy conserving scatterings, and in particular for forward scattering. For comparison, when the strength of the scattering center's force vanishes, the matrix element is ($\mathsf{T}_S \to \mathsf{T}_0$)

$$\langle \bar{p}_1 | \mathsf{T}_0 | \bar{p} \rangle = e^{-iE_1 t/\hbar}.$$

Thus the scattering center effectively introduces a phase time difference in the propagator of amount

$$\gamma = (2\hbar/E_1)\delta, \tag{15.86}$$

so that

$$\langle \bar{p}_1 | \mathsf{T}_S | \bar{p}_1 \rangle = e^{-i(E_1/\hbar)(t-\gamma)}. \tag{15.87}$$

Upon Taylor expansion of δ about $E_1 = 0$

$$\delta = \delta(E = 0) + (d\delta/dE_1)_{E=0}(E_1) + \dots,$$

gives

$$\gamma = (2\hbar\delta\,(E_1 = 0)/E_1) + 2\hbar(d\delta/dE_1)_{E_1=0} + \dots.$$

Levenson's theorem (to be discussed later) requires that $\delta(E_1 = 0) = m\pi$,

m an integer. Thus the leading term of the expansion is not physically observable by scattering alone. Thus the scattering center introduces an effective time delay due to the second term that is given

$$\tau = 2\hbar(\mathrm{d}\delta/\mathrm{d}E_1)_{E_1=0} \tag{15.88}$$

or more generally at any energy E_1 near E_0

$$\tau = 2\hbar(\mathrm{d}\delta/\mathrm{d}E_1)_{E_1=E_0}\,. \tag{15.89}$$

Causality corresponds to the physical statement that no observable effects can propagate faster than the particles can propagate, whatever the nature of the scattering center of force, and always subject to relativity. Since τ and $\mathrm{d}\delta/\mathrm{d}E_1$ are linearly related, a positive derivative for δ corresponds to a positive time delay and *vice-versa*. Thus, for example, if the scattering center's forces are negligible outside some size b, the largest time *advance* would be

$$\tau \geq -b/c \tag{15.90}$$

where c is the velocity of light. This would place a limit on the negativeness of $\mathrm{d}\delta/\mathrm{d}E$:

$$\mathrm{d}\delta/\mathrm{d}E \geq -(b/c)/2\hbar\,. \tag{15.91}$$

Other similar causality limits can be placed upon the negativeness of phase shift derivatives and were first discussed by Wigner. Note, however, that no similar limit, for elastic scattering, can be placed upon how positive $\mathrm{d}\delta/\mathrm{d}E$ can be, or how steep the function δ *versus* E can rise, since such cases correspond to positive time delay.

When the phase shift has a positive excursion in a given region of energy (i.e. the quantity $\mathrm{d}\delta/\mathrm{d}E$ is larger in the region than to either side) the scattering is characterized by a time delay. A particular case of interest occurs when the matrix for $\mathscr{T}_\mathrm{I}$ has a simple pole for E_1 near E_0. Let

$$\langle p_1\hat{\imath}_2 \mid \mathscr{T}_\mathrm{I}(s_1) \mid p_1\hat{\imath}_1\rangle = -A\,\mathrm{e}^{\mathrm{i}\alpha}/[(E_1-E_0)+\tfrac{1}{2}\mathrm{i}\Gamma] \tag{15.92}$$

where A, α, E_0 and Γ are real and constant. Note that the singularity for $\mathscr{T}_\mathrm{I}$ must be chosen in the lower half E_1 plane (left hand s_1 plane) otherwise the scattering will diverge for $t \to \infty$. This behavior of the $\mathscr{T}_\mathrm{I}$ matrix gives a positive time delay

$$\tau = \frac{\tfrac{1}{2}\hbar\Gamma}{(E_1-E_0)^2+\tfrac{1}{4}\Gamma^2} \tag{15.93}$$

which attains a maximum when $E_1 = E_0$. Such a behavior of the $\mathscr{T}_\mathrm{I}$ matrix is termed a *resonance*. At resonance the cross section attains its largest value when the phase shift passes through an odd integral multiple of $\frac{1}{2}\pi$.

There are two conditions when the phase shift may pass through $\frac{1}{2}(2n+1)\pi$: with positive or with negative slopes. Only the former case corresponds to a positive time delay, can be associated with the attractiveness of the scattering force, and is a resonance; the case of passage through $\frac{1}{2}(2n+1)\pi$ with a negative slope is often termed an "echo" of resonances at lower energies, does not correspond to an enhanced time delay, and is not a resonance in the present sense.

It is important to note, however, that the phase shift need not pass through $\frac{1}{2}(2n+1)\pi$ to correspond to manifestations of attractive forces that appear only as positive excursions of the phase shift.

16. Partial wave representations

Since L^2 and L_z (the square of the total angular momentum and its z component) commute with the Hamiltonians H_A and H_0 it is possible to have simultaneous eigenkets of energy (or $k = \sqrt{(2\mu E/\hbar^2)}$) and L^2 and L_z. These are especially useful in some problems where only a few low values of angular momentum are possible. Such cases are common and arise when the range of the scattering center of force is short. Suppose the force has an appreciable effect only for a spatial region of a, then incident particles whose impact parameter is Δ, have an angular momentum of about $L = \mu V\Delta = \hbar k\Delta$. But if the scattering occurs strongly only for $\Delta \leq a$ then, approximately,

$$\sqrt{\{L(L+1)\}} \leq \hbar ka$$

and there is an upper limit to the values of L that are important. This suggests expanding the $\mathscr{T}_\mathrm{I}$ matrix in simultaneous eigenkets of L, L_z.

Thus it is often convenient to expand the kets $|\,\bar{p}\rangle$ into kets that are simultaneous eigenkets of scalar momentum or energy (eigenvalues p or $E = p^2/2\mu$), of the square of angular momentum (eigenvalue $L(L+1)\hbar^2$), and z projection of angular momentum (eigenvalue $M\hbar$). Such kets will be denoted by $|\,E, L, M\rangle$. They obey

$$\begin{aligned} \mathsf{H}_0\,|\,E, L, M\rangle &= E\,|\,E, L, M\rangle \\ L_z\,|\,E, L, M\rangle &= M\hbar\,|\,E, L, M\rangle \\ L^2\,|\,E, L, M\rangle &= L(L+1)\hbar^2\,|\,E, L, M\rangle \end{aligned} \tag{15.94}$$

where H_0 is the free particle Hamiltonian.

These kets and bras are complete and orthonormal and it is convenient to set

$$\langle E_2, L_2, M_2 \mid E_1, L_1, M_1\rangle = \delta(E_2 - E_1)\, \delta_{L_2, L_1}\, \delta_{M_2, M_1}\,. \tag{15.95}$$

With this normalization it follows that

$$\mathsf{I} = \int \mathrm{d}E \sum_{L=0}^{\infty} \sum_{M=-L}^{L} \mid E, L, M\rangle \langle E, L, M \mid \tag{15.96}$$

is a unit or identity operator. The brackets $\langle E_2, L, M \mid \bar{p}_1\rangle$ are required and can be calculated as follows. Consider that by (15.96)

$$\sum_{L, M} \int \mathrm{d}E_3 \langle \bar{p}_2 \mid E_3, L, M\rangle \langle E_3, L, M\rangle = \delta(\bar{p}_2 - \bar{p}_1)\,. \tag{15.97}$$

Each bracket $\langle E_3 LM \mid \bar{p}_1\rangle$ is a function of the scalar variables displayed in the symbol and is thus a function only of E_1, E_3, L, M, and $\cos\theta$, where θ is the angle between the z direction and $\bar{p}_1$: $\cos\theta = (\bar{p}_1/p_1)\cdot(\bar{L}_z/M\hbar)$. Thus

$$\langle E_3, L, M \mid \bar{p}_1\rangle = F(E_1, E_3)\, \Theta_L^M(\cos\theta)\,. \tag{15.98}$$

If $\bar{p}_1$ is taken as defining the z-axis then $L_z = M\hbar = 0$, or $M = 0$. This expression, by (15.94), must be a simultaneous eigenfunction of L^2, L_z, and H_0. Thus the angular dependence is given by

$$\Theta_L^M(\cos\theta) = (2\pi)^{-\frac{1}{2}}\, \Pi_L^0(\cos\theta)\,, \tag{15.99}$$

where Π_L^0 is a normalized associated Legendre function and is related to the Legendre polynomials P_L by

$$\Pi_L^0(\cos\theta) = \sqrt{2L+\tfrac{1}{2}}\, P_L(\cos\theta)\,, \tag{15.100}$$

$$P_L(1) = 1\,, \tag{15.100a}$$

where

$$\int \mathrm{d}\Omega\, P_L(\cos\theta)\, P_{L'}(\cos\theta) = 2\pi \left(\frac{2}{2L+1}\right) \delta_{L', L}\,. \tag{15.100b}$$

Similarly, the energy dependence is given by

$$F(E_1, E_3) = F(E_1)\, \delta(E_1 - E_3)\,. \tag{15.101}$$

Combining (15.98), (15.99) and (15.101)

$$\langle E_3, L, 0 \mid \bar{p}_1\rangle = \frac{1}{\sqrt{2\pi}} F(E_1)\, \Pi_L^0(\cos\theta)\, \delta(E_1 - E_3)\,. \tag{15.102}$$

Using (15.97), and choosing z along $\bar{p}_1$, the sum over M can be dropped, and

$$1 = \int d^3p_2 \sum_L \int_0^\infty dE_3 \frac{1}{2\pi} F^*(E_2)\, F(E_1)\, \Pi_L^{0*}(\cos\theta_2)\, \Pi_L^0(1) \times$$

$$\times\ \delta(E_1 - E_3)\, \delta(E_2 - E_3)$$

$$= \frac{1}{2\pi} \int_0^\infty \mu p_2\, dE_2 \int_0^\infty dE_3 \sum_L \int_0^{2\pi} d\varphi_2 \int_0^\pi \sin\theta_2\, d\theta_2\, \Pi_L^0(1)\, \Pi_0^0(\cos\theta)\, \Pi_L^0(\cos\theta)$$

$$\times \sqrt{2}\, F^*(E_2)\, F(E_1)\, \delta(E_1 - E_3)\, \delta(E_2 - E_3)$$

$$= \sqrt{2}\mu p_1\, |\, F(E_1)\, |^2 \sum_L \sqrt{\tfrac{1}{2}(2L+1)}\, \delta_{L,0}$$

since

$$\Pi_L^0(1) = \sqrt{\tfrac{1}{2}(2L+1)}\,.$$

Thus

$$|\, F(p_1, p_1)\, |^2 = 1/\mu p_1\,. \tag{15.103}$$

Taking the phase of F real, (15.102) and (15.103) give

$$\langle E_3, L, 0\, |\, \bar{p}_1\rangle = \frac{1}{\sqrt{2\pi}} \frac{1}{\sqrt{\mu p_1}} \Pi_L^0(\cos\theta_2)\, \delta(E_1 - E_3)\,. \tag{15.104}$$

Consider a partial wave decomposition of the scattering amplitude:

$$f(\theta_2, E_1) = (2\pi\mu)\,(2\pi\hbar)\, \langle p_1 \hat{\imath}_2\, |\, \mathscr{T}_1(-iE_1/\hbar)\, |\, p_1 \hat{\imath}_1\rangle$$

$$= (2\pi\mu)\,(2\pi\hbar) \sum_{L,L'} \iint dE_3\, dE_4\, \langle p_1 \hat{\imath}_2\, |\, E_4, L', 0\rangle \times$$

$$\times\ \langle E_4, L', 0\, |\, \mathscr{T}_1(-iE_1/\hbar)\, |\, E_3, L, 0\rangle\, \langle E_3, L, 0\, |\, p_1 \hat{\imath}_1\rangle$$

$$= (2\pi\hbar) \sum_{L,L'} \iint dE_3\, dE_4 \frac{1}{\sqrt{p_1 p_2}} \Pi_L^0(1)\, \Pi_{L'}^0(\cos\theta_2)\, \delta(E_3 - E_1) \times$$

$$\times\ \delta(E_4 - E_1)\, \langle E_4, L', 0\, |\, \mathscr{T}_1(-iE_1/\hbar)\, |\, E_3, L, 0\rangle\,. \tag{15.105}$$

Now the integrations over the energies E_3 and E_4 can be carried out to give

$$f(\theta_2, E_1) = (2\pi\hbar) \sum_{L,L'} \frac{1}{p_1} \Pi_L^0(1)\, \Pi_{L'}^0(\cos\theta)\, \langle E_1, L', 0\, |\, \mathscr{T}_1(-iE_1/\hbar)\, |\, E_1, L, 0\rangle.$$

Thus the only scattering is between states $|\, E_1, L, 0\rangle$ and $|\, E_1, L', 0\rangle$. Since the operator L^2 commutes with H_0 it can have simultaneous eigenvalues. Thus one may get

$$\langle E_1, L', 0 \,|\, \mathscr{T}_{\mathrm{I}}(-\mathrm{i}E_1/\hbar) \,|\, E_1, L, 0\rangle = \delta_{L', L} \langle E_1, L, 0 \,|\, \mathscr{T}_{\mathrm{I}}(-\mathrm{i}E_1/\hbar) \,|\, E_1, L, 0\rangle$$
$$\equiv \delta_{L', L} \mathscr{T}_{\mathrm{I}}^{(L)}(E_1) \qquad (15.106)$$

to obtain

$$f(\theta_2, E_1) = (\pi/k_1) \sum_L (2L+1) P_L(\cos\theta_2) \mathscr{T}_{\mathrm{I}}^{(L)}(E_1) . \qquad (15.107)$$

This is the desired expansion of $f(\theta_2, E_1)$ in a series, each term of which corresponds to scattering of the particle with precise angular momentum quantum numbers, as well as a sharp value of the energy. Note that the matrix elements $\mathscr{T}_{\mathrm{I}}^{(L)}(E)$ are dimensionless.

17. Partial wave phase shifts, unitarity, and dispersion relationships

By the optical theorem, unitarity requires that

$$\sigma = \int \mathrm{d}\Omega_2 \,|\, f(\theta_2, E_1) \,|^2 = (-4\pi/k_1) \operatorname{Im} \{f(0, E_1)\} .$$

Using (15.107) this becomes

$$\frac{\pi^2}{k_1^2} \int \mathrm{d}\Omega_2 \sum_{L', L} (2L'+1)(2L+1) P_{L'}(\cos\theta_2) P_L(\cos\theta_2) \mathscr{T}_{\mathrm{I}}^{*(L')}(E_1) \mathscr{T}_{\mathrm{I}}^{(L)}(E_1) =$$
$$= -\frac{4\pi^2}{k_1^2} \sum_L (2L+1) P_L(1) \operatorname{Im} \{\mathscr{T}_{\mathrm{I}}^{(L)}(E_1)\} .$$

Since the P_L are orthogonal

$$\sigma = (4\pi^3/k_1^2) \sum_L (2L+1) \,|\, \mathscr{T}_{\mathrm{I}}^{(L)}(E_1) \,|^2$$
$$= -(4\pi^2/k_1^2) \sum_L (2L+1) \operatorname{Im} \{\mathscr{T}_{\mathrm{I}}^{(L)}(E_1)\} . \qquad (15.108)$$

Thus, term by term,

$$|\, \mathscr{T}_{\mathrm{I}}^{(L)}(E_1) \,|^2 = -(1/\pi) \operatorname{Im} \{\mathscr{T}_{\mathrm{I}}^{(L)}(E_1)\} . \qquad (15.109)$$

This has a solution

$$\mathscr{T}_{\mathrm{I}}^{(L)}(E_1) = (-1/\pi) \sin \delta_L \, \mathrm{e}^{\mathrm{i}\delta_L} \qquad (15.110)$$

valid for all real phase angles δ_L. δ_L is called the partial wave phase shift. Thus by (15.110) and (15.107)

$$f(\theta_2, E_1) = -(1/k_1) \sum_L (2L+1) P_L(\cos\theta_2) \sin \delta_L \, \mathrm{e}^{\mathrm{i}\delta_L} , \qquad (15.111)$$

gives the partial wave expansion of the scattering amplitude.

Since the partial wave amplitude $f(\theta_2, E_1)$, or equivalently the partial wave matrix elements, $\mathscr{T}_{\mathrm{I}}^{(L)}(E_1)$, are analytic in the upper half plane of E_1 they have imaginary and real parts that are related by the Cauchy integral theorem. Thus

$$\mathrm{Re}\,\{\mathscr{T}_{\mathrm{I}}^{(L)}(E_1')\} = \frac{1}{\pi}\int_{-\infty}^{\infty} \frac{\mathrm{Im}\,\{\mathscr{T}_{\mathrm{I}}^{(L)}(E_1)\}}{E_1 - E_1'}\,\mathrm{d}E_1\,, \tag{15.112}$$

and

$$\mathrm{Im}\,\{\mathscr{T}_{\mathrm{I}}^{(L)}(E_1')\} = -\frac{1}{\pi}\int \frac{\mathrm{Re}\,\{\mathscr{T}_{\mathrm{I}}^{(L)}(E_1)\}}{E_1 - E_1'}\,\mathrm{d}E_1\,. \tag{15.113}$$

If there are any singularities of $\mathscr{T}_{\mathrm{I}}^{(L)}(E_1)$ on the real, negative E_1 axis, the contour must be deflected to pass above these singularities.

Using unitarity, or the optical theorem, statement of (15.109) one finds the following relationships:

$$\mathscr{T}_{\mathrm{I}}^{(L)}(E_1') = \frac{1}{\pi}\int_{-\infty}^{\infty} \frac{\mathrm{Im}\,\{\mathscr{F}_{\mathrm{I}}^{(L)}(E_1)\}}{E_1 - E_1'}\,\mathrm{d}E_1 + \mathrm{i}\,\mathrm{Im}\,\{\mathscr{T}_{\mathrm{I}}^{(L)}(E_1)\}\,, \tag{15.114}$$

$$= -\int_{-\infty}^{\infty} \frac{|\mathscr{T}_{\mathrm{I}}^{(L)}(E_1)|^2}{E_1 - E_1'}\,\mathrm{d}E_1 - \pi\mathrm{i}\,|\mathscr{T}_{\mathrm{I}}^{(L)}(E_1')|^2\,. \tag{15.115}$$

Such formulae are termed *Dispersion relationships* and are seen to relate the (measurable) scattering amplitude to integrals over *the (measurable) partial wave cross sections*.

Earlier it was seen that all singularities of $\mathscr{T}_{\mathrm{I}}$ are restricted to a half plane; these dispersion relationships further restrict the nature of the singularities in that half plane.

PROBLEM 12. Consider

$$\mathscr{T}_{\mathrm{I}}(s) = \iint \mathrm{d}^3\boldsymbol{p}_1\,\mathrm{d}^3\boldsymbol{p}_2\,|\,\bar{p}_2\rangle\,\frac{g}{s+\alpha}\,\langle\bar{p}_1\,|$$

with α, real > 0 and with $\langle\bar{p}_1\,|\,\bar{p}_2\rangle = \delta(\bar{p}_1 - \bar{p}_2)$. Calculate a) the scattering amplitude $f(\theta)$, $\cos\theta = \bar{p}_2\bar{p}_1/p^2$ and the scattering cross section. b) What is the energy and angular dependence of the scattering cross section? c) Calculate the time domain scattering propagator $\mathsf{T}(t)$ for *all* times t and discuss its behavior for $t \to \infty$. d) Consider this problem in relation to the optical theorem as given in equation (15.56) and show that unitarity of T_{S} requires a particular value for g.

18. Deduction of phase shifts from cross sections

If the interaction has a "short range", as discussed earlier, one expects the terms to converge rapidly to zero as L increases. This is indeed the case in many scattering problems observed in atomic, nuclear, and high energy processes. This results in a simplification for real, physical problems, and is the reason for the importance of expression (15.111).

It is seen that unitarity requires that each partial wave phase shift be real, just as the overall phase, discussed in 15.14, was required to be real.

The differential scattering cross section is thus given by

$$\frac{d\sigma}{d\Omega_2} = \frac{1}{k^2} \sum_{L,\,L'} (2L'+1)(2L+1)\, P_{L'}(\cos\theta_2)\, P_L(\cos\theta_2) \times$$

$$\times \sin\delta_{L'} \exp\{i(\delta_L - \delta_{L'})\} \sin\delta_{L'}. \quad (15.116)$$

This expression may be further simplified by using the fact it is real, and the fact that products of the Legendre polynomials are sums of other Legendre polynomials.

The simpler angular dependence of the cross sections, in terms of the orthogonal Legendre polynomials, allows experimental deductions of the scattering phase shifts. In terms of the phase shifts, the total cross section becomes

$$\sigma = (4\pi/k^2) \sum_L (2L+1) \sin^2\delta_L. \quad (15.117)$$

In practical problems for elastic scattering a number of properties of the scattering amplitude need to be employed. A list of these includes the following:

1. The phase shifts are assumed to be continuous functions of the energy.
2. A phase shift with a positive energy derivative in an energy region implies an unusually long time delay for the system at that energy; such regions represent resonant phenomena.
3. A phase shift, to be physical and not disobey causality, must have lower limit placed on how negative its energy derivative can be.
4. Each phase shift, as $E_1 \to 0$, must tend to $n_L\pi$ where n_L is the number of bound states of angular momentum $L\hbar$. This is Levinson's theorem.
5. Each phase shift, as $E_1 \to \infty$, must tend to zero at least as fast as $1/k$, for forces of short range.

PROBLEM 13. Consider S-wave scattering by a "hard-sphere" ($V \to +\infty$ for $0 \leq r \leq a$). Show that there is a limit upon the phase shift: $\delta \geq -ka$.

PROBLEM 14. By considering the number of loops and nodes of the wavefunction for S-wave scattering, show that Levinson's theorem is true.

19. The relationship between the time dependent treatment and the wavefunction treatment

The scattering state has been discussed in terms of the time development of an arbitrary initial state $|A\rangle$ that at any later time is $\mathsf{T}|A\rangle = |A, t\rangle$, and in particular for the time development of an initial eigenket of momentum $|\bar{p}\rangle$ that becomes $\mathsf{T}|\bar{p}\rangle$. Actually, only the long-time behavior was determined, since this is all that can be observed. However, if the time behavior of T is known, or equivalently the energy dependence of $\mathscr{T}_{\mathrm{I}}$ is known, the scattering state can be determined for *all* times.

The more familiar viewpoint is to consider the incident state to be a plane wavefunction $\langle\bar{p}|\bar{r}\rangle = (2\pi\hbar)^{-\frac{3}{2}} \exp(\mathrm{i}\bar{p}\cdot\bar{r}/\hbar)$, that scatters as waves into all directions. If the Hamiltonian is known then the solution of the Schroedinger equation, with the incident wave boundary conditions, gives the wavefunction at all values of $\bar{r}$. The assumed knowledge of the Hamiltonian for all $\bar{r}$ is equivalent to a knowledge of T for all time or of $\mathscr{T}_{\mathrm{I}}$ for all energy. This viewpoint will be discussed in Chapter 16.

However it should be noted that the time dependent treatment is more fundamental since, in fact, one usually does not know the Hamiltonian for all $\bar{r}$ while one can, in some cases, measure the energy dependence of $\mathscr{T}_{\mathrm{I}}$ for all important on-shell energies and use analyticity to construct it elsewhere. Furthermore, in many cases it is not true that the Hamiltonian is a simple function of $\bar{r}$, for example when the interaction is non-local, or equivalently when the interaction is velocity dependent.

Nevertheless the use of Schroedinger representatives, and solution of the Schroedinger equation upon assuming shapes of potentials is very important since it allows simple calculations and useful models of scattering phenomena to be constructed. This viewpoint will be discussed in Chapter 16.

CHAPTER 16

SOME SPECIAL CASES OF SCATTERING

1. Use of spherical coordinates

In the previous chapter the general theory of scattering was developed in terms of operators, the time development operators or propagators and their Laplace transforms. When specific matrix representations were used they were based on the eigenkets of linear momentum. However, since most scattering problems have a spherical symmetry it would seem that spherical coordinates would be appropriate. The operators involved in the scattering, $\mathsf{H}_S, \mathsf{H}_0, \mathsf{V}, \mathsf{T}_S, \mathsf{T}_0, \mathsf{T}$, as well as $\mathscr{T}_S, \mathscr{T}_0, \mathscr{T}_I$ all commute with the operators for angular momentum and so the matrix representations for any one of them can be divided into blocks referring to an eigenvalue of the angular momentum.

The relationship between the kets of linear momentum and those of angular momentum are obtained by expressing a plane wave in terms of spherical waves. The equation for the energy eigenfunctions of a free particle in spherical coordinates is that of equation (6.4) if the term in $V(r)$ is omitted. A solution can then be expressed as

$$U(r, \vartheta, \varphi) = \Pi_l^m \, e^{im\varphi} f_l(r)/(2\pi)^{\frac{1}{2}}. \tag{16.1}$$

Since we shall be concerned first with expressing a plane wave in this form we may take the polar axis in the direction of the wave and reduce the expression to

$$e^{(i/\hbar)pz} = e^{(i/\hbar)pr\cos\vartheta} = \sum_{l=0}^{\infty} A_l \, \Pi_l^0 (\cos\vartheta) f_l(r), \tag{16.1a}$$

where the coefficients A_l remain to be evaluated.

The evaluation of A_l is carried out in the usual way by multiplying both sides of the equation by Π_l^0 (cos ϑ) sin ϑ dϑ dφ and integrating over the solid angle,

$$A_l f_l(r) = \int_{-1}^{1} e^{(i/\hbar)prz}\, \Pi_l(z)\, dz \tag{16.2}$$

$$= \frac{\hbar}{ipr}\,[e^{(i/\hbar)r}\,\Pi_l(1) - e^{-(i/\hbar)r}\,\Pi_l(-1)] - \frac{\hbar}{ipr}\int e^{(i/\hbar)prz}\,\frac{d}{dz}\,\Pi_l(z)\,dz\,.$$

Since $\Pi_l(z)$ is a finite polynomial, a continuation of this integration by parts will lead to a finite polynomial in $(\hbar/ipr)$ whose coefficients are sines and cosines of $(pr/\hbar)$. For large values of r, only the first terms are important. Hence, asymptotically

$$A_l f_l(r) \to \frac{\hbar}{ipr}\,[e^{(i/\hbar)pr}\,\Pi_l(1) - e^{-(i/\hbar)pr}\,\Pi_l(-1)]\,. \tag{16.3}$$

Since $\Pi_l(1) = \frac{1}{2}(2l+1)$ and $\Pi_l(-1) = \frac{1}{2}(2l+1)(-1)^l$ it follows that

$$A_l f_l(r) \to \frac{(2l+1)^{\frac{1}{2}}\hbar}{pr}\, i^l \sin\left[\frac{pr}{\hbar} - \tfrac{1}{2}l\pi\right]. \tag{16.4}$$

If the asymptotic form of $f_l(r)$ is known, the constant A_l is given.

The differential equation for $f_l(r)$ is

$$\frac{d^2 f_l}{dr^3} + \frac{2}{r}\frac{df_l}{dr} + \left[\left(\frac{p}{\hbar}\right)^2 - \frac{l(l+1)}{r^2}\right] f_l = 0\,, \tag{16.5}$$

and the solutions are

$$f_l = \left(\frac{\pi\hbar}{2pr}\right)^{\frac{1}{2}} J_{l+\frac{1}{2}}(pr/\hbar) = j_l(pr/\hbar) \tag{16.6}$$

where j_l is a spherical Bessel function. The asymptotic form is then

$$f_l \to (\hbar/pr)\sin\,[(pr/\hbar) - \tfrac{1}{2}l\pi]$$

and

$$A_l = (2l+1)^{\frac{1}{2}}\, i^l\,.$$

The values of the A_l have been determined from the asymptotic forms, but the expansion (16.1a) is valid everywhere,

$$e^{(i/\hbar)pz} = \sum_{l=0}^{\infty} (4l+2)^{\frac{1}{2}}\, i^l\, \Pi_l^0(\cos\vartheta)\, f_l(r)\,, \tag{16.7}$$

where $f_l(r)$ is given by (16.6).

Each term in the above expansion is an eigenfunction of the square of the angular momentum with the eigenvalue $l(l+1)\hbar^2$. Each term can be pictured classically as representing those particles which approach the origin to a certain minimum distance. If a particle with mass μ and velocity v is directed so as to pass the scattering center at a distance Δ, if undeflected, its angular momentum is $mv\Delta$. If the square of this quantity is given a quantized value, $(\mu v \Delta_l)^2 = l(l+1)\hbar^2$, the kinetic energy can be expressed in terms of l:

$$\tfrac{1}{2}\mu v^2 = \frac{\hbar^2}{2\mu}\frac{l(l+1)}{\Delta_l^2}. \qquad (16.8)$$

In equation (16.5) only kinetic energy is considered, and it is equal to $p^2/2\mu$. Accordingly the coefficient of $f_l(r)$ in the differential equation can be written

$$l(l+1)\,[(1/\Delta_l^2)-(1/r^2)].$$

When $r < \Delta_l$ this expression is negative and $f_l(r)$ does not oscillate; and the corresponding term in (16.7) approaches zero. Each term in (16.7) may then be interpreted as representing those particles whose distance of closest approach to the origin is approximately Δ_l.

PROBLEM 1. Consider a uniform stream of classical particles with velocity v moving in the z-direction. How many of them will have an angular momentum relative to the origin whose square lies between $l(l+1)\hbar^2$ and $(l+1)(l+2)\hbar^2$?

2. A central scattering potential

In the previous chapter many of the integrations involved in operator or matrix multiplication were simplified if the scattering potential is taken to be spherically symmetrical. If V is a function of the radius only, the matrices for V can be evaluated in spherical coordinates,

$$\langle p'' \mid \mathrm{V} \mid p' \rangle = (1/2\pi\hbar)^3 \int e^{(i/\hbar)(\boldsymbol{p}'-\boldsymbol{p}'')\cdot\boldsymbol{r}}\, V(r) r^2\, \mathrm{d}r \sin\vartheta\, \mathrm{d}\vartheta\, \mathrm{d}\varphi$$

$$= \sum_{l',l''} 2[(2l''+1)(2l'+1)]^{\frac{1}{2}}\, i^{(l'+l'')} \Pi_{l''}(\cos\vartheta'')\Pi_{l'}(\cos\vartheta') \sin\vartheta\, \mathrm{d}\vartheta\, \mathrm{d}\varphi \times$$

$$\times \int_0^\infty f_{l''}^*(p''r/\hbar)\, V(r)\, f_{l'}(p'r/\hbar)\, V^2\, \mathrm{d}r. \qquad (16.9)$$

In this equation the direction of $\boldsymbol{r}$ is given by (ϑ, φ). ϑ'' is the angle between $\boldsymbol{r}$ and $\boldsymbol{p}''$, and ϑ' is the angle between $\boldsymbol{r}$ and $\boldsymbol{p}'$. When the integration over the angles is carried out, only the cases for $l' = l''$ remain and the result is

$$\langle p'' \mid \mathsf{V} \mid p' \rangle = [4\pi/(2\pi\hbar)^3]\, \Sigma\, (2l+1)^{\frac{1}{2}}(-1)^l \Pi_l\, (\cos\vartheta)\, V_l(p'', p'), \qquad (16.10)$$

where θ is the angle between $\boldsymbol{p}'$ and $\boldsymbol{p}''$, and

$$V_l(p'', p') = \int_0^\infty j_l(p''r/\hbar)V(r)j_l(p'r/\hbar)r^2\, \mathrm{d}r\,. \qquad (16.11)$$

The fact that the sum over l occurs in the matrix element, the amplitude, shows that the different angular momenta may interfere with each other. The presence of $l = 1$ scattering may reduce the scattering in some directions below its value if only $l = 0$ scattering is present.

As an illustration consider the potential

$$V = W \mathrm{e}^{-\alpha' r}/r\,. \qquad (16.12)$$

The $V_0(p, p)$ can be evaluated at once

$$W \int_0^\infty j_0(pr/\hbar)\, \mathrm{e}^{-\alpha' r} j_0(pr/\hbar) r\, \mathrm{d}r = (\hbar/p)^2\, W \int_0^\infty \mathrm{e}^{-\alpha x} \frac{\sin^2 x}{x} \mathrm{d}x$$

$$= W(\hbar/p)^2\, \tfrac{1}{4} \log(1+4/\alpha^2) \rightarrow W(h/p\alpha)^2[1-2/\alpha^2] \qquad (16.13)$$

where $\alpha = \alpha'\hbar/p$. The integral for V_1 is more complicated but if α is large the integrand can be expanded and leads to

$$V(p, p) = W(\hbar/p)^2(2/3\alpha^4)\,. \qquad (16.14)$$

The first two terms of the matrix element are then

$$\langle p' \mid \mathsf{V} \mid p \rangle = (1/2\pi^2\hbar^3)W(h/p)^2[(1/2^{\frac{1}{2}}\alpha^2)+(\tfrac{2}{3})^{\frac{1}{2}}(1/\alpha^4)\,(\cos\theta)]$$

$$= (1/2\pi^2\hbar^3)W(h/\alpha p)^2\, 2^{\frac{1}{2}}\, [\tfrac{1}{2}-(1/\alpha^2)\,(1-\cos\theta)] \qquad (16.15)$$

when p' and p have the same magnitude but different directions. The square of this expression contains a term in $\cos\theta$ which is negative for values of $\theta > \frac{1}{2}\pi$.

PROBLEM 2. For the scattering potential in equation (16.12) examine the scattering as a function of angle for various energies as far as the results can be obtained from the first two angular momenta.

PROBLEM 3. For the scattering potential in equation (16.12) work out the scattering as a function of angle and energy by evaluating the matrix $\langle p' \mid \mathsf{V} \mid p \rangle$ directly. Show how this approaches the Rutherford scattering as $\alpha' \rightarrow 0$.

PROBLEM 4. Work out the first term of the Born series for the case of a rectangular well potential, i.e. $V = W$ for $0 < r < R$ and $V = 0$ for $r > R$.

3. Scattering of identical particles

As has been shown in previous chapters the wave function or the state vector representing a pair of particles must be either symmetrical or antisymmetrical in the coordinates of the particles. Such conditions must be satisfied if α-particles are scattered by α-particles, or protons are scattered by protons. In the case of protons scattered by protons, the spin must be taken into account. For the "singlet" state, an eigenstate of σ^2 with the eigenvalue $\frac{3}{4}\hbar^2$, the functions of the coordinates will be symmetrical. For the "triplet" states, the coordinate functions must be antisymmetrical.

The operator representing an interchange of the particle coordinates commutes with the scattering operators and so the antisymmetric and the symmetric cases are independent, the symmetry is preserved in the scattering.

Consider first the matrix elements based on eigenkets of momentum. To provide the symmetry these kets must be

$$(\mathrm{e}^{\mathrm{i}(p/\hbar)r} \pm \mathrm{e}^{-\mathrm{i}(p/\hbar)r})/2^{\frac{1}{2}}$$

and the matrix elements are then

$$\langle p' \mid \mathscr{T}_{\mathrm{I}}(s) \mid p\rangle = \tfrac{1}{2}\int (\mathrm{e}^{-\mathrm{i}(p'/\hbar)r} \pm \mathrm{e}^{\mathrm{i}(p'/\hbar)r})\mathscr{T}_{\mathrm{I}}(s)\,(\mathrm{e}^{\mathrm{i}(p/\hbar)r} \pm \mathrm{e}^{-\mathrm{i}(p/\hbar)r})\,\mathrm{d}^3 r. \tag{16.16}$$

If the scattering is central and local $\mathscr{T}_{\mathrm{I}}(s)$ is a function of r only and the matrix element can be written

$$\begin{aligned}\langle p' \mid \mathscr{T}_{\mathrm{I}}(s) \mid p\rangle = \pi \int_0^\infty \mathscr{T}_{\mathrm{I}}(s, r) r^2\,\mathrm{d}r \int_0^\pi &[\exp\{\mathrm{i}\,(p'-p)\frac{r}{\hbar}\cos\vartheta\} \\ &+\exp\{-\mathrm{i}\,(p'-p)\frac{r}{\hbar}\cos\vartheta\} \pm \exp\{\mathrm{i}\,(p'+p)\frac{r}{\hbar}\cos\vartheta\} \\ &\pm\exp\{-\mathrm{i}\,(p'+p)\frac{r}{\hbar}\cos\vartheta\}]\,\sin\vartheta\,\mathrm{d}\vartheta \\ = 4\pi \int_0^\infty \mathscr{T}_{\mathrm{I}}(s)\, r\mathrm{d}r &\left[\frac{\sin(p'-p)(r/\hbar)}{(p'-p)/\hbar} \pm \frac{\sin(p'+p)(r/\hbar)}{(p'+p)/\hbar}\right].\end{aligned} \tag{16.17}$$

Since the scattering conserves energy and the magnitude of the momentum, the magnitudes of the vector difference and sum are

$$p' - p = 2p \sin\tfrac{1}{2}\theta \quad \text{and} \quad p' + p = 2p \cos\tfrac{1}{2}\theta. \tag{16.18}$$

Using as an illustration the potential of (16.12) and the first Born approximation to $\mathscr{T}(s, r)$

$$\begin{aligned}\langle p' \mid \mathsf{V} \mid p \rangle &= 8\pi\hbar^2 \frac{1}{\hbar^2\alpha'^2 + (2p \sin\frac{1}{2}\theta)^2} \pm \frac{1}{\hbar^2\alpha'^2 + (2p\cos\frac{1}{2}\theta)^2} \\ &= 16\pi(\hbar/p)^2 \frac{(\hbar\alpha'/p)^2 - 2(\cos^2\frac{1}{2}\theta - \sin^2\frac{1}{2}\theta)}{(\hbar\alpha'/p)^4 + 2(\hbar\alpha/p)^2 + 4\sin^2\frac{1}{2}\theta\cos^2\frac{1}{2}\theta}.\end{aligned} \tag{16.19}$$

PROBLEM 5. Illustrate the significance of equation (16.19) by sketching the dependence of the scattering on θ for various values of $(\hbar\alpha'/p)$ to show the difference between the symmetric and the antisymmetric cases.

PROBLEM 6. Work out the difference between the symmetric and the antisymmetric cases of a central scatterer using the formulation in polar coordinates.

PROBLEM 7. What is the cross section as a function of angle for protons scattered by protons when nothing is known about the polarization of either the target or the incident particle?

CHAPTER 17

QUANTUM STATISTICAL MECHANICS

In most problems of physical interest, such as the behavior of electrons and nuclei in a solid, or a stream of particles with unpolarized spins, it is impossible to know the state of the system precisely. This is true in classical as well as in quantum mechanics. There is usually a lack of precise information over and above the fundamental indetermination of which quantum mechanics takes account. This lack is due in part to the extreme complexity of a system of many particles, and also in part to the fact that such systems cannot really be completely isolated from their surroundings. In other cases, it is due merely to the fact that the apparatus used does not precisely fix all the commuting quantities. Ordinary classical mechanics, and quantum mechanics of isolated systems, represent limiting cases which only approximately correspond to real situations. Often the approximation by such a limiting case is not adequate, and statistical mechanics must be called upon to obtain useful results.

The standard problem in classical mechanics concerns the motion of a physical system from some initial state at the time $t = t_0$ to a final state at a later time t. The initial state is given by the values, at $t = t_0$, of all the coordinates and all the momenta. The equations of motion then permit the calculation, for the later time t, of all the values of these changing quantities, provided there is no interference with the motion from sources not included in the formulation of the problem. In the case of electrons in a solid there are some 10^{22} particles per cubic centimeter, if only the valence electrons are considered. Obviously it is out of the question to devise an experiment to fix precisely the values of so many coordinates and momenta. Furthermore, a piece of a solid cannot be completely isolated from its surroundings, and these surroundings influence the behavior of the particles. Nevertheless, in such cases, useful information about the behavior of the system can be obtained from the techniques of statistical mechanics.

In statistical mechanics one studies the average behavior of a large number of identical systems in all the different states consistent with the limited information available about the system of interest. Such an imagined collection of identical systems was called by Gibbs an "ensemble".

It is easy to confuse the results of quantum mechanics with those of statistical mechanics, but it is important to distinguish them clearly. Both in quantum and in statistical mechanics it is necessary to consider average values. In quantum statistical mechanics it is necessary to consider double averages. One must average first for quantum-mechanical reasons and then average these averages for the purposes of statistical mechanics.

If, for example, a system is in a particular state described by the wave function $\psi_n(q, t)$ the quantum-mechanical average of a quantity Q is given by

$$\langle Q \rangle_n(t) = \int \psi^*_n(q, t)\, Q\, \psi_n(q, t)\, dq\,. \tag{17.1}$$

This average is denoted by the bra and ket symbols around the Q. If, however, it is not known that the system is in the state ψ_n, but merely that it is in some one of the possible states ψ_s, where s lies between s_0 and s_1, then the mean value for consideration is obtained by averaging with equal weights over these possible states,

$$\overline{\langle Q \rangle} = \frac{1}{s_1 - s_0} \sum_{s=s_0}^{s_1} \langle Q \rangle_s(t) = \frac{1}{N} \sum_s \int \psi_s^* Q \psi_s\, dq\,. \tag{17.2}$$

N is the number of different states over which the sum is taken. This latter average will be designated by a bar over the symbol for the quantum mechanical average. Equal weights are given to all of these states since there is no reason for believing that one state is more probable than another. Obviously both $\langle \overline{Q} \rangle$ and $\langle Q \rangle_s$ will, in general be functions of the time.

In quantum statistical mechanics, as in the corresponding classical theory, the first problem is that of finding the appropriate ensemble to represent the probable occurrence of the various possible states. The ensemble must correspond to the physical situation at hand, and it is at this point that the physicist must exercise his intuition and his judgment. He must devise suitable means of evaluating the desired averages over the ensemble. This chapter will be devoted to illustrating some of such methods.

1. The phase space in classical mechanics

If the coordinates and the momenta of all the N particles in a system are taken as the coordinates of a $6N$ dimensional space, the motion of the system can be described as the motion of a single point in this phase space. The point may describe a closed orbit, as in some simple cases, or it may never return to its original position as in the case of a free particle. Since the Hamiltonian equations are of the first order, the orbit will never cross itself, nor will it cross an orbit corresponding to different initial conditions. Because of the many dimensions of the phase space, a graphical representation is difficult. It is difficult even to imagine such a multidimensional space with any great degree of accuracy. Nevertheless one-dimensional cases, which have only a two-dimensional phase space, can often be used as illustrations.

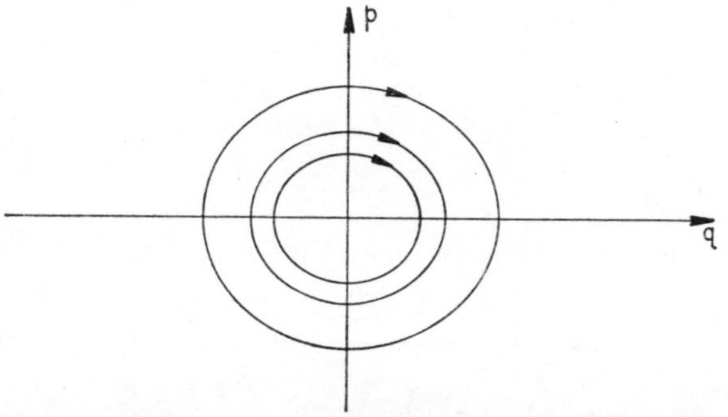

Fig. 17.1. Some orbits in the phase space of the representative point of a simple harmonic oscillator.

In the case of a one-dimensional simple harmonic oscillator, the representative point in the phase space describes a closed orbit in the form of an ellipse. Figure 17.1 illustrates this situation with several ellipses showing the motion of the oscillator with different sets of initial conditions. Since all the orbits represent the same oscillator they will be described with the same period. It is also clear that the initial conditions are not uniquely determined by specifying an orbit, since any point on the ellipse could serve as the starting point.

An oscillator moving on a given ellipse in the phase space has an energy characteristic of the ellipse. If nothing is known about the oscillator except its energy, any point on the ellipse, at $t = 0$, is as good as any other point to represent its behavior. The appropriate "ensemble" is then a set of points uniformly distributed along the elliptical orbit. Average and probable values can then be determined from this distribution. The average coordinate is zero, as is the average momentum. The most probable positions are the extreme positions and the most probable momenta are the maximum values in two directions.

If nothing whatever is known about the state of the oscillator at $t = t_0$, the representative point might lie anywhere in the phase plane, and the appropriate ensemble would be a uniform distribution of points over the whole plane. According to Liouville's theorem, such a uniform distribution will remain uniformly dense as each point pursues its independent motion in the plane. If we know nothing whatever about the state of the oscillator at $t = t_0$, we continue to know nothing whatever about it at later times, in spite of having a general solution of the equation of motion.

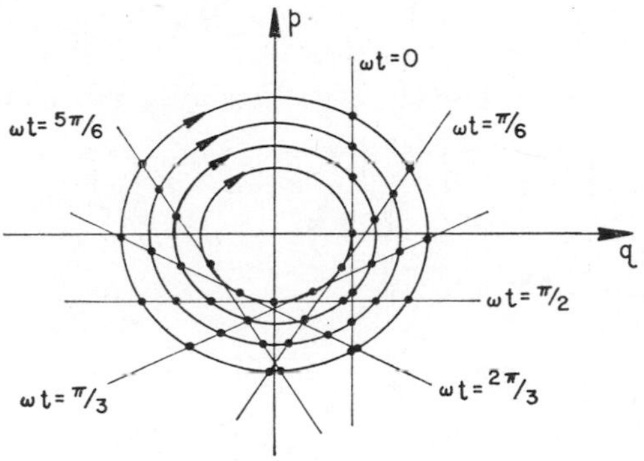

Fig. 17.2. Changing positions in the phase space of the representative points of a simple harmonic oscillator when at $t = 0$ the points were distributed along a straight line at constant q.

On the other hand, if one knows a little about the state of the oscillator at $t = t_0$, he can predict a little something about it at a later time. Figure 17.2 indicates the behavior of the points in phase space that represent an oscillator at $q = q_0$ when $t = 0$, but with unknown momentum. The points are distributed uniformly along the line $q = q_0$ parallel to the p-axis. At later times this line of points takes other positions as shown in the figure, and one can say that at time t the point representing the state of the oscillator will lie somewhere along the appropriate line. It also follows that the representative point never gets inside the inner ellipse.

If the information available about a mechanical system is such that the appropriate distribution of points in the phase space is a function of constants of the motion only, the distribution will be invariant in time. For a simple oscillator the only constant is the energy, and an ensemble given as a function of the energy would be a distribution over all ellipses such as in fig. 17.1, but uniform on each ellipse.

The familiar canonical distribution

$$\rho = A\mathrm{e}^{-E/kT} = \mathrm{e}^{(F-E)/kT} \tag{17.3}$$

is usually used to represent a complex system in equilibrium with a constant temperature bath at temperature T. Clearly this does not represent a system with a known energy, but it can be shown that the average energy, as well as the energy of maximum probability, is related to T. Extensive discussions of the justification for the canonical ensemble can be found in books devoted to statistical mechanics.

It is usually convenient to normalize a phase space density, such as given in equation (17.3), so that the total probability, the integral of the density over the whole of the phase space, is set equal to unity. Thus if

$$\int \rho\, \mathrm{d}p\mathrm{d}q = A \int \mathrm{e}^{-E/kT}\, \mathrm{d}p\, \mathrm{d}q = 1\,, \tag{17.4}$$

A can be determined. The quantity

$$F = kT \log A \tag{17.5}$$

can be shown to be analogous to the free energy of a thermodynamic system.

PROBLEM 1. If it is known that at $t = 0$ a classical oscillator has an energy less than W and a positive value of the coordinate, sketch the appropriate distribution of points in the phase space. What will be the statistical average coordinate and momentum? Find the same quantities at $t = \pi/2\omega$.

PROBLEM 2. At $t = 0$ a classical free particle in one dimension is known to be at $x = 0$, but its momentum is known only to lie between $-p_0$ and $+p_0$. Find the phase space distribution at a later time t. What can be said at time t about the position and velocity of the particle?

PROBLEM 3. By evaluating integrals such as in equation (17.4) find the statistical mean value of the energy and the square of the energy of a simple harmonic oscillator in thermal equilibrium with a temperature bath.

PROBLEM 4. Sketch the orbits in the phase space for a particle moving vertically in a gravitational field. If at $t = 0$ a number of particles are projected upward from $z = 0$ with velocities between v_1 and v_2, what can be said about their positions and momenta at a later time? If they are restricted to $z > 0$ what is their distribution in z?

2. Pure states and mixed states in quantum mechanics

In classical mechanics the term "pure state", although rarely used, could refer to a set of precise values for all the coordinates and all the momenta of the system. Classical mechanics treats of pure states, when possible, arbitrary pure states. In quantum mechanics such a specification is impossible; a pure state is a state defined ônly as sharply as is consistent with the principle of indetermination. Thus in quantum mechanics a pure state is a *simultaneous eigenstate* of a physically complete set of commuting operators. For a system of particles such a set of commuting operators will consist of as many independent operators as there are independent coordinates. There are just half as many precisely specified quantities in the quantum as in the classical case.

In a one-dimensional system there is just one independent operator in a physically complete set of commuting operators. This may be taken to be the coordinate x or the momentum p. But other operators may be taken instead, such as

$$O_1 = ap^2+b(x-x_0)^2. \tag{17.6}$$

Eigenfunctions of this operator will form a mathematically complete set of orthogonal functions centered about the point $x = x_0$. Each of such functions can be expanded in terms of other complete sets of eigenfunctions, such for example as the energy eigenfunctions of an oscillator, or the eigenfunctions of momentum.

Thus a pure state is represented by a well defined single valued function of a complete set of commuting variables. It can be designated as well by eigenvalues of a set of independent commuting operators. For a three-dimensional harmonic oscillator there must be three eigenvalues. These could be the energy values along the three coordinate axes. They could also be two quantum numbers giving the angular momentum and a third referring to the total energy. If a system is described by a pure state wave function, or by a set of eigenvalues, it is specified as precisely as is possible, and the development of the state in time is given by the equation of motion. This development is completely determinate.

If the condition of an oscillator is less precisely known than possible, it may be said to be in a "mixed state", and may be represented by an ensemble. If one knows only that an oscillator is, with equal probability, to be found in one of its three lowest energy states, at least something is known about it, and it is said to be in a mixed state. In this case, the energy distribution is known, and an energy distribution is known to be constant in time. The mean energy is $\frac{3}{2}\hbar\omega$. It is constant in time because each individual pure state has a constant energy and the distribution over these states is constant. Also the mean square of the energy is $\frac{35}{12}(\hbar\omega)^2$. The mean displacement is zero, since it is zero for each of the pure states of which the mixed state is composed and the mean square displacement is $\frac{3}{4}\,\hbar\omega/a$, where a is the force constant. Note that such a "mixed" state cannot be described by a wave function, nor by a linear combination of the three pure states involved.

Thus in a mixed state there are two kinds of averages involved. There is the (quantum mechanical) average for each individual pure state, and there is the (statistical) average over these averages. In computing expected experimental results, these two kinds of average are usually combined into a single statistical prediction. It is important to distinguish them, however. The first is irreducible by any experimental arrangement. The second average represents only the limitation of the experimental arrangement, and such a limitation can, in principle, be reduced.

PROBLEM 5. Consider the three lowest energy states of a simple oscillator. These may be combined to make three other orthogonal states. Let

$$|A\rangle = \tfrac{1}{2}|0\rangle + \frac{1}{\sqrt{2}}|1\rangle + \tfrac{1}{2}|2\rangle$$

$$|B\rangle = \frac{1}{\sqrt{2}}|0\rangle \qquad\qquad - \frac{1}{\sqrt{2}}|2\rangle$$

$$|C\rangle = \tfrac{1}{2}|0\rangle - \frac{1}{\sqrt{2}}|1\rangle + \tfrac{1}{2}|2\rangle.$$

For each of these states evaluate the quantum mechanical mean energy, the mean displacement, and the mean of the displacement squared.

Show that if the states $|A\rangle$, $|B\rangle$ and $C\rangle$ are regarded as equally probable, the quantum statistical mean values of the above quantities are the same as if the equally probable states had been taken as $|0\rangle$, $|1\rangle$ and $|2\rangle$. Note that this result applies only to the case of equal probabilities for the different states involved.

PROBLEM 6. If it is known only that a hydrogen atom (without spin) is in a state with $n = 2$, it may be assumed that all such states are equally probable. Under these circumstances compute the mean value of the coordinate x^2 and show how it differs from the mean value of x^2 computed for a pure state made up as a definite linear combination of the four states with $n = 2$.

3. The density operator or density matrix

In classical statistical mechanics the knowledge, or lack of knowledge, of the state of the mechanical system is expressed by a density function in the phase space. In quantum statistical mechanics it is expressed by a density operator, or by its representation as a density matrix. The properties of the density operator are most easily illustrated on a system with a small number of eigenstates so we may consider a particle with spin $\frac{1}{2}$ and ignore any other coordinates it may have. Then take a representation in which s_z is diagonal. The representations of the other spin matrices are given in Chapter 9.

If nothing whatever is known about the spin of the particle it may be in either spin state, and the probability of being in one state is equal to the probability of being in the other. These states may be designated by the ket vectors $|+\frac{1}{2}\rangle$ and $|-\frac{1}{2}\rangle$, the eigenvectors of the spin along the z-axis. The statistical average of any quantity R will then be one-half the quantum mechanical mean value in $|+\frac{1}{2}\rangle$ plus the corresponding mean value in $|-\frac{1}{2}\rangle$,

$$\overline{\langle R\rangle} = \tfrac{1}{2}\{\langle+\tfrac{1}{2}|R|+\tfrac{1}{2}\rangle+\langle-\tfrac{1}{2}|R|-\tfrac{1}{2}\rangle\}. \tag{17.7}$$

If the statistical probability that the particle is in the $|+\frac{1}{2}\rangle$ state is twice as great as that it is in the $|-\frac{1}{2}\rangle$ state, the combined mean value is given by

$$\overline{\langle R\rangle} = \tfrac{1}{3}\{2\langle+\tfrac{1}{2}|R|+\tfrac{1}{2}\rangle+\langle-\tfrac{1}{2}|R|-\tfrac{1}{2}\rangle\}. \tag{17.8}$$

With the two eigenvectors as basic vectors the density may be represented as a diagonal matrix. In the first case

$$\rho = \begin{pmatrix} \frac{1}{2} & 0 \\ 0 & \frac{1}{2} \end{pmatrix} \tag{17.9}$$

and in the second case

$$\rho = \begin{pmatrix} \frac{2}{3} & 0 \\ 0 & \frac{1}{3} \end{pmatrix}. \tag{17.10}$$

Since the bracket expressions in equations (17.7) and (17.8) are the diagonal elements of the matrix for R, it is clear that in this representation

$$\overline{\langle R\rangle} = \mathrm{Tr}\,(\rho\, R) \tag{17.11}$$

and also that

$$\mathrm{Tr}\,(\rho) = 1. \tag{17.12}$$

The trace of a matrix, written $\mathrm{Tr}(M)$, is the sum of the diagonal terms. Equation (17.12) is simply the normalization of the density matrix to give a total probability equal to unity.

It is obvious in this case that equation (17.11) gives the statistical mean value of R, and since the trace of a matrix is invariant to a unitary transformation, equation (17.11) may be taken as the rule for determining the statistical mean value of any quantity in any representation.

The "quantum statistical" mean values obtained from equations (17.7) or (17.8), or from equation (17.11) using the density matrices (17.9) or (17.10) are different from the "quantum" mean values to be obtained from a pure state composed with equal probabilities of the two basic states, $|\frac{1}{2}\rangle$ and $|-\frac{1}{2}\rangle$. The latter mean value would have the form

$$\left[\frac{1}{\sqrt{2}}\langle\tfrac{1}{2}| + \frac{e^{-i\alpha}}{\sqrt{2}}\langle-\tfrac{1}{2}|\right] R \left[\frac{1}{\sqrt{2}}|\tfrac{1}{2}\rangle + \frac{e^{+i\alpha}}{\sqrt{2}}|-\tfrac{1}{2}\rangle\right]$$

$$= \tfrac{1}{2}\langle\tfrac{1}{2}|R|\tfrac{1}{2}\rangle + \tfrac{1}{2}\langle-\tfrac{1}{2}|R|-\tfrac{1}{2}\rangle + \tfrac{1}{2}[\langle-\tfrac{1}{2}|R|\tfrac{1}{2}\rangle e^{-i\alpha} + \langle\tfrac{1}{2}|R|-\tfrac{1}{2}\rangle e^{i\alpha}], \tag{17.13a}$$

which includes an arbitrary phase constant α. The mean value depends on α since the single pure state represents a definite phase relationship between $|\frac{1}{2}\rangle$ and $|-\frac{1}{2}\rangle$, but the value of α is not given by the statement of the problem.

When there are just two basic states, as in this case, any state has just one other state orthogonal to it. In the above illustration this orthogonal state is

$$(1/\sqrt{2})\,[\,|\tfrac{1}{2}\rangle + e^{i(\alpha+\pi)}|-\tfrac{1}{2}\rangle],$$

and the "quantum" mean value of R in this state is

$$\langle R\rangle = \tfrac{1}{2}\langle\tfrac{1}{2}|R|\tfrac{1}{2}\rangle + \tfrac{1}{2}\langle-\tfrac{1}{2}|R|-\tfrac{1}{2}\rangle - \tfrac{1}{2}[\langle\tfrac{1}{2}|R|\tfrac{1}{2}\rangle e^{-i\alpha} + \langle\tfrac{1}{2}|R|-\tfrac{1}{2}\rangle e^{i\alpha}]. \tag{17.13b}$$

When the results (17.13a) and (17.13b) are averaged with equal weights the result is just that of equation (17.7). They must be given equal weights since both states satisfy the specification that $|\frac{1}{2}\rangle$ and $|-\frac{1}{2}\rangle$ are equally probable.

It is apparent that in this case the same result could have been obtained by considering all values of α as equally probable and averaging over α in equation (17.13a). Hence the use of the "quantum statistical" mean value is sometimes referred to as the use of random, or equally probable phases. The averaging is usually deferred until the final step in the calculation. This procedure, of course, is possible only when all of the orthogonal pure states represented by the relative phases are equally probable. In the more general case in which some states are regarded as more probable than others, averages of the form of equation (17.11) are necessary.

The statistical density is often most useful in matrix form, but it is also expressible in operator notation. The matrix (17.9) is a representation of the identity or unit operator so we have for that case

$$\rho = \tfrac{1}{2} I . \tag{17.9a}$$

For the second case

$$\rho = \tfrac{1}{2} I + \frac{1}{3\hbar} s_z . \tag{17.10a}$$

The coefficients in equations (17.9a) and (17.10a) are selected to make $\mathrm{Tr}\,(\rho) = 1$. Since all elements on the diagonal of the unit matrix are the same, the coefficient of the unit operator I is just the reciprocal of the number of basic states. For the important case in which the number of bacis states is infinite, the coefficient of the unit operator approaches zero.

PROBLEM 7. Show that the trace of a matrix is invariant to a unitary transformation, and that the trace of the product of two matrices is independent of the order in which they are multiplied.

PROBLEM 8. Transform the density matrix (17.10) to express it in terms of

$$A = (\tfrac{2}{3})^{\frac{1}{2}} \mid \tfrac{1}{2}\rangle + \mathrm{e}^{\mathrm{i}\alpha} (\tfrac{1}{3})^{\frac{1}{2}} \mid -\tfrac{1}{2}\rangle$$

and the state orthogonal to A as the basic states of the matrix.

PROBLEM 9. Show that the density matrix which represents a system in a pure state satisfies the equation

$$\rho_{\mathrm{p}}^2 = \rho_{\mathrm{p}} . \tag{17.13}$$

PROBLEM 10. Find the statistical mean values of s_x and s_y for the two densities given in (17.9) and (17.10).

Any pure state may be represented in terms of a complete set of orthogonal states $\mid n\rangle$, where n represents the eigenvalues of a maximum set of independent operators. Thus an arbitrary state $\mid A_s\rangle$ may be written

$$\mid A_s\rangle = \sum_n a_{n,s} \mid n\rangle \tag{17.14}$$

with $\Sigma_n a_{ns}{}^2 = 1$. The quantum mechanical mean value of any quantity represented by an operator Q is then

$$\langle A_s \mid \mathrm{Q} \mid A_s\rangle = \sum_{n, n'} \langle n' \mid \mathrm{Q} \mid n\rangle a_{n',s}^* \mathrm{a}_{ns} , \tag{17.15}$$

where n and n' take on all possible values. If there are a number of different states $\mid A_s\rangle$, and if P_s designates the probability that the system is in $\mid A_s\rangle$, the quantum statistical mean value of Q, $\langle \overline{Q} \rangle$, is

$$\overline{\langle Q \rangle} = \sum_s P_s \langle A_s \mid \mathrm{Q} \mid A_s \rangle = \sum_{s,\, n,\, n'} P_s\, a^*_{n',\, s}\, a_{ns}\, \langle n' \mid \mathrm{Q} \mid n \rangle . \tag{17.16}$$

If the states s are orthogonal, a_{ns} is a unitary matrix, and $a^{-1}_{n',s} = a^*_{sn'}$ is the inverse. The density operator may then be defined by its matrix representation

$$\rho_{n,\, n'} = \sum_s P_s\, a_{ns}\, a^{-1}_{sn'} = \langle n \mid \rho \mid n' \rangle . \tag{17.17}$$

With this form of the density matrix

$$\overline{\langle Q \rangle} = \sum_{n,\, n'} \langle n \mid \rho \mid n' \rangle \langle n' \mid \mathrm{Q} \mid n \rangle$$

$$= \sum_n \langle n \mid \rho \mathrm{Q} \mid n \rangle = \mathrm{Tr}\,(\rho Q) = \mathrm{Tr}\,(Q\rho) . \tag{17.18}$$

The expression (17.18) is invariant to the choice of basic states in terms of which ρ and Q are expressed. It is only required that they be orthogonal. The expression "different states" used above implies "orthogonal states", for only when states are orthogonal can they be considered as really and completely different.

The expression (17.17) may be taken as the general definition of the density matrix. When the desired density can be expressed in terms of operators, the terms of the matrix can be evaluated according to the second equality in (17.17). When it is expressed as a probability of various states, the first equality is useful.

Equation (17.17) gives the components of the density matrix expressed in terms of the eigenvectors $\mid n \rangle$. In operator form this becomes

$$\rho = \sum_{n,\, n'} \mid n \rangle \langle n \mid \rho \mid n' \rangle \langle n' \mid \tag{17.17a}$$

as indicated in equations (8.9) and (8.10). The eigenvector $\mid n \rangle$ is equivalent to the wave function $\psi_n(x)$ so the operator becomes

$$\rho = \sum_{n,\, n'} \langle n \mid \rho \mid n' \rangle\, \psi_n(x)\, \psi^*_{n'}(x') \tag{17.17b}$$

where x and x' each represent all the coordinates of the system. If ρ represents a pure state

$$\rho^{(\mathrm{p})} = \psi_n(x)\psi^*_n(x') . \tag{17.17c}$$

For a distribution over pure states

$$\rho = \sum_n P_n \psi_n(x)\psi^*_n(x') , \tag{17.17d}$$

and a canonical distribution expressed in terms of energy eigenstates gives

$$\langle x \mid \rho \mid x' \rangle = \sum_n \exp\left\{\frac{F-E_n}{kT}\right\} \psi_n(x)\, \psi_n^*(x'). \qquad (17.17e)$$

PROBLEM 11. Find the statistical mean value of the three components of spin when $\rho = \frac{1}{2}(I + s_x/\hbar)$.

PROBLEM 12. Use as basis states the eigenstates of s_z, and write a wave function for a pure state such as to represent a quantum mechanical distribution of probabilities for s_x equal to the statistical distribution given by equation (17.10a). Show that such a function is not unique, and that various quantum mechanical distributions of s_z and s_y can be so represented.

PROBLEM 13. Evaluate the coefficients in equation (17.14) so that $\mid A_s \rangle$ represents an angular momentum $\hbar$ in the (ϑ, φ) direction. Evaluate also the coefficients for two other states orthogonal to the first. Then using arbitrary relative probabilities of these three states construct the density matrix according to the prescription in equation (17.17).

PROBLEM 14. With the density matrix of the above problem evaluate the mean values of J_x and J_y.

4. The canonical density operator

If the density operator is a function of constants of the motion only, the operator and its matrix representation will be independent of the time, i.e. will commute with the Hamiltonian. Also the various mean values will be independent of the time, as in the corresponding classical case. In particular, the density operator useful in the statistical interpretation of the thermodynamics of systems in equilibrium is that of the canonical distribution:

$$\rho_c = e^{(F-H)/kT} = e^{F/kT} \sum_{s=0} \frac{1}{s!}\left(\frac{-1}{kT}\right)^s H^s. \qquad (17.19)$$

In the matrix representation based on energy eigenfunctions, ρ_c is diagonal and represents the presence of each energy eigenstate with the probability $\exp\{(F-H)/kT\}$. The constant F is the statistical mechanical analog of the free energy. It permits the normalization condition (17.12) to be satisfied.

The exponential form in (17.19) is not easy to use except when applying the operator to energy eigenfunctions. In other cases the series expansion must be used, or the transformation to the desired basic functions can be carried out after the operator has been applied.

In the energy eigenfunction representation, the trace of the density matrix is simply related to the "Zustandsumme", or the sum over states. For discrete energy values

$$\mathrm{Tr}\,(\rho) = \sum_n \mathrm{e}^{(F-E_n)/kT} = \mathrm{e}^{F/kT} \sum_n \mathrm{e}^{-E_n/kT} = 1\,.$$

Then

$$\mathrm{e}^{-F/kT} = Z = \sum_n \mathrm{e}^{-E_n/kT}, \qquad (17.20)$$

where Z is the sum over states. It is important to remember that the E_n in equation (17.20) are the energies of the whole system, not merely of one particle. On account of the high degeneracy usually associated with systems of many coordinates, the same value of E_n may occur many times, and each independent state must be included in the sum.

To determine what states are independent one must remember that the sum is a sum over the diagonal terms of a matrix. The states are then the basic states on which the matrix is built and they are orthogonal to each other. They constitute a complete orthogonal set.

To illustrate the use of a density matrix, consider the very simple case of a system with a magnetic moment proportional to its angular momentum and placed in a magnetic field. The energy may be taken as $-\mu J \cdot B$. The direct treatment is to take the z-axis along the magnetic field. Then the energy is $-\mu BJ_z$ and the density operator for a canonical ensemble is $\exp\{(F-\mu BJ_z)/kT\}$. Since the energy is diagonal, the evaluation of mean values involves simply the sum over the different energy states. However, this simple problem may also be treated by other methods to illustrate their nature. Such apparently more complicated methods may be the most useful in less simple problems.

For a specific case let $J = 2\hbar$. The matrices are then 5×5 and the trace of the density matrix gives

$$\mathrm{e}^{-F/kT} = 1+2\cosh\frac{\mu\hbar B}{kT} + 2\cosh\frac{2\mu\hbar B}{kT}\,.$$

The mean value of the energy is the trace of the product of the density matrix and the energy matrix. Both of these are diagonal and the product is a diagonal matrix. If all energy states were equally probable the statistical mean energy would be zero. For the canonical distribution however, it is

$$\langle \bar{E} \rangle = \frac{2\sinh(\mu\hbar B/kT)+4\sinh(2\mu\hbar B/kT)}{1+2\cosh(\mu\hbar B/kT)+2\cosh(2\mu\hbar B/kT)}(-\mu\hbar B)\,. \qquad (17.21)$$

This approaches zero as $\mu B/kT \to 0$ and approaches $-2\mu\hbar B$ as $(B/kT) \to \infty$. The first limit is that of a uniform distribution over the five states. The second indicates the dominant probability of the state of lowest energy.

In this simple case the evaluation of the various mean values is trivial, but this example can be used to illustrate what might be done in other situations. Consider for example a case in which for some reason it is not convenient to put the z-axis along the magnetic field. The energy term in the exponent of the density operator will then have the form

$$\begin{aligned} H &= -\mu(J_x B_x + J_y B_y + J_z B_z) \\ &= -\mu B(\alpha J_x + \beta J_y + \gamma J_z) \end{aligned} \tag{17.22}$$

where α, β, γ are the direction cosines of the magnetic field.

There are several ways in which to deal with the density matrix in such a case. If $\mu\hbar B/kT$ is very small, a series expansion of the exponential can be used since only a few terms would be needed. The first three terms would give

$$\begin{aligned} \rho = \mathrm{e}^{F/kT} \Big\{ 1 + \frac{\mu B}{kT}(\alpha J_x + \beta J_y + \gamma J_z) + \tfrac{1}{2}\left(\frac{\mu B}{kT}\right)^2 [\alpha^2 J_x^2 + \alpha\beta(J_x J_y + J_y J_x) + \\ + \alpha\gamma(J_x J_z + J_z J_x) + \beta^2 J_y^2 + \beta\gamma(J_y J_z + J_z J_y) + \gamma^2 J_z^2] + \ldots \Big\}. \end{aligned} \tag{17.23}$$

For this special case, with $J = 2$, the matrix for ρ expressed on a basis of the eigenkets of J_z is given in table 17.1. It is clear that further development of even this simple matrix function to higher powers would be laborious.

If, as in this case, one is concerned with only a finite matrix, the canonical density operator can be exactly expressed in terms of a few of the angular momentum operators as in equation (17.10a). If the z-axis is along the direction of the magnetic field the operator becomes

$$\begin{aligned} \rho = \mathrm{e}^{F/kT} \Big\{ 1 &+ \left[\tfrac{4}{3}\sinh\left(\frac{\mu\hbar B}{kT}\right) - \tfrac{1}{6}\sinh\left(\frac{2\mu\hbar B}{kT}\right)\right]\frac{J_z}{\hbar} \\ &- \left[\tfrac{5}{4} - \tfrac{4}{3}\cosh\left(\frac{\mu\hbar B}{kT}\right) + \tfrac{1}{12}\cosh\left(\frac{2\mu\hbar B}{kT}\right)\right]\left(\frac{J_z}{\hbar}\right)^2 \\ &- \left[\tfrac{1}{3}\sinh\left(\frac{\mu\hbar B}{kT}\right) - \tfrac{1}{6}\sinh\left(\frac{2\mu\hbar B}{kT}\right)\right]\left(\frac{J_z}{\hbar}\right)^3 \\ &+ \left[\tfrac{1}{4} - \tfrac{1}{3}\cosh\left(\frac{\mu\hbar B}{kT}\right) + \tfrac{1}{12}\cosh\left(\frac{2\mu\hbar B}{kT}\right)\right]\left(\frac{J_z}{\hbar}\right)^4 \Big\}. \end{aligned} \tag{17.24}$$

This is not an approximation, but is valid for all values of $(\mu\hbar B/kT)$. It is obtained by evaluating the five coefficients of the five operators (I, J_z, J_z^2, J_z^3, J_z^4) to give the five diagonal terms of the matrix. If then it is so desired, a transformation to other axes can be carried out on the operators.

The trace of ρ in equation (17.24) is the linear combination of the traces of its component operators. Similarly

$$\begin{aligned}\mathrm{Tr}\,(\rho J_z) &= \mathrm{e}^{F/kT}\left\{2\sinh\left(\frac{\mu\hbar B}{kT}\right) + 4\sinh\left(\frac{2\mu\hbar B}{kT}\right)\right\}\hbar \\ &= \frac{2\sinh(\mu\hbar B/kT)+4\sinh(2\mu\hbar B/kT)}{1+2\cosh(\mu\hbar B/kT)+2\cosh(2\mu\hbar B/kT)}\,\hbar. \qquad (17.25)\end{aligned}$$

Since the trace of each odd power of J_z is zero, this expression reduces to two terms, in agreement with equation (17.21).

PROBLEM 15. Evaluate F in the above situation.

The operators in equation (17.24) can be given a matrix representation in terms of the eigenkets of J_z and then transformed to the direction (α, β, γ). Or the original density matrix (equation (17.21)) in terms of these ket vectors can be transformed to a basis made up of eigenvectors of $J_{z'}$ where z' is in the direction (α, β, γ). The transformation matrix is that of Chapter 9 where (ϑ, φ) are the polar angles corresponding to the direction cosines (α, β, γ).

TABLE 17.1

Expansion of $\exp(-\mu \boldsymbol{J}\cdot\boldsymbol{B}/kT)$ to second power in x

J_z' / J_z''	$2\hbar$	$\hbar$	0	$-\hbar$	$-2\hbar$
$2\hbar$	$1+2\gamma x+$ $+(1+3\gamma^2)x^2/2$	$(\alpha-\mathrm{i}\beta)x+$ $+\frac{3}{2}\gamma(\alpha-\mathrm{i}\beta)x^2$	$\frac{3}{2}(\alpha-\mathrm{i}\beta)^2x^2/2$	0	0
$\hbar$	$(\alpha+\mathrm{i}\beta)x+$ $+\frac{3}{2}\gamma(\alpha+\mathrm{i}\beta)x^2$	$1+\gamma x+$ $+\frac{5}{2}(1-\frac{3}{5}\gamma^2)x^2/2$	$\frac{3}{2}(\alpha-\mathrm{i}\beta)x+$ $+\frac{3}{2}\gamma(\alpha-\mathrm{i}\beta)x^2/2$	$\frac{3}{2}(\alpha-\mathrm{i}\beta)^2x^2/2$	0
0	$\frac{3}{2}(\alpha+\mathrm{i}\beta)^2x^2/2$	$\frac{3}{2}(\alpha+\mathrm{i}\beta)x+$ $+\frac{3}{2}\gamma(\alpha+\mathrm{i}\beta)x^2/2$	$1+3(1-\gamma^2)x^2/2$	$\frac{3}{2}(\alpha-\mathrm{i}\beta)x$ $-\frac{3}{2}\gamma(\alpha+\mathrm{i}\beta)x^2/2$	$\frac{3}{2}(\alpha-\mathrm{i}\beta)^2x^2/2$
$-\hbar$	0	$\frac{3}{2}(\alpha+\mathrm{i}\beta)^2x^2/2$	$\frac{3}{2}(\alpha+\mathrm{i}\beta)x$ $-\frac{3}{2}\gamma(\alpha-\mathrm{i}\beta)x^2/2$	$1-\gamma x+$ $+\frac{5}{2}(1-\frac{3}{5}\gamma^2)x^2/2$	$(\alpha-\mathrm{i}\beta)x$ $-3\gamma(\alpha-\mathrm{i}\beta)x^2/2$
$-2\hbar$	0	0	$\frac{3}{2}(\alpha+\mathrm{i}\beta)^2x^2/2$	$(\alpha+\mathrm{i}\beta)x$ $-3\gamma(\alpha+\mathrm{i}\beta)x^2/2$	$1-2\gamma x+$ $+(1+3\gamma^2)x^2/2$

$x = \mu B/kT$; α, β, γ are the direction cosines of $\boldsymbol{B}$.

Finally the density matrix may be expressed as in equation (17.17). The coefficients of expansion $a_{s,n}$ are the elements of the transformation matrix from the eigenvectors of J_z to those of $J_{z'}$. The s represent the eigenvalues of J_z, and the n those of $J_{z'}$. Thus

$$\rho_{n,n'} = \sum_{s=-2}^{s=2} a_{n,s} a_{s,n'}^{\dagger} \, \mathrm{e}^{-s\mu\hbar B/kT} . \tag{17.26}$$

Problem 16. Show that the density matrix of equation (17.24) represents a completely unpolarized system when $B/kT \to 0$. This implies that the expectation value of each component of the angular momentum is zero and the expectation values of the squares of all components are identical.

Problem 17. Evaluate two components of the density matrix when the magnetic field has the direction cosines (α, β, γ). Do this by using equation (17.26) and then show that the corresponding terms in the matrix of table 17.1 are the first terms in an expansion in powers of $\mu\hbar B/kT$.

5. Use of the sum over states

The quantity F in equation (17.20) is analogous to the free energy of thermodynamics, so that when the sum over states can be evaluated

$$F = -kT \log Z . \tag{17.27}$$

With the canonical distribution, the statistical mean energy is given in terms of Z or F by

$$H = \sum_n E_n \, \mathrm{e}^{(F-E_n)/kT} = -\frac{1}{Z} \frac{\partial Z}{\partial(1/kT)} = \frac{\partial(F/kT)}{\partial(1/kT)} . \tag{17.28}$$

Similarly the mean square of the energy is

$$\overline{\langle H^2 \rangle} = \sum_n E_n^2 \, \mathrm{e}^{(F-E_n)/kT} = \frac{1}{Z} \frac{\partial^2 Z}{\partial(1/kT)^2} . \tag{17.29}$$

Knowledge of the mean square of the energy permits an estimate of the accuracy with which the mean energy represents the probable energy.

Equation (17.27), which defines the sum over states, appears so simple as to be almost trivial. Two difficulties, however, may appear in evaluating it. In the first place it is necessary to know the energy eigenvalues and to know all of them. In some simple cases these can be computed, but in most cases of physical interest the determination of the energy eigenstates for a many-particle system can be made only approximately.

A second problem is estimating the degree of degeneracy and carrying out the sum. In some cases the sum can be approximated by an integral, but care must always be taken to see that this approximation does not conceal some observable characteristically quantum-mechanical effects.

PROBLEM 18. Consider a three-dimensional isotropic harmonic oscillator. Find the energy eigenvalues and the corresponding functions in polar coordinates.

PROBLEM 19. Consider a canonical distribution of three-dimensional harmonic oscillators corresponding to the temperature T and also such that $\rho \sim \exp(-L_z/\Delta)$. Evaluate the sum over states.

6. Systems of harmonic oscillators

6.1. Consider a single harmonic oscillator of frequency $\omega/2\pi$. The sum of states is given by

$$Z = \sum_{n=0}^{\infty} \mathrm{e}^{-(n+\frac{1}{2})\hbar\omega/kT} = \mathrm{e}^{-\hbar\omega/2kT} \sum_{n=0}^{\infty} \mathrm{e}^{-n\hbar\omega/kT} = \frac{\mathrm{e}^{-\hbar\omega/2kT}}{1-\mathrm{e}^{-\hbar\omega/kT}}. \tag{17.30}$$

From equation (17.28) follows the well-known expression for the average energy:

$$\overline{\langle H\rangle} = \hbar\omega\left[\tfrac{1}{2} + \frac{1}{\mathrm{e}^{\hbar\omega/kT}-1}\right]. \tag{17.31}$$

Such a system, of a single one-dimensional harmonic oscillator, is hardly one to which methods of statistical mechanics would normally be applied. It is treated here because it is a simple illustration of the general methods outlined above and because such oscillators are frequently useful as approximations to portions of more complex systems.

PROBLEM 20. Evaluate the average x^4 of a harmonic oscillator at the temperature T.

PROBLEM 21. Evaluate the average of the energy squared for a harmonic oscillator at the temperature T.

6.2. Consider a three-dimensional oscillator with three different frequencies along perpendicular axes. Let the angular frequencies be ω_1, ω_2, ω_3. The sum over states is then

$$Z = \frac{e^{\hbar(\omega_1+\omega_2+\omega_3)/2kT}}{(e^{\hbar\omega_1/kT}-1)(e^{\hbar\omega_2/kT}-1)(e^{\hbar\omega_3/kT}-1)}. \tag{17.32}$$

The energy of each state is given by the sum of three terms, and each possible state is counted once by summing over all values of the individual quantum numbers n_1, n_2, n_3. This is a possibility peculiar to this situation because the three coordinates of the oscillator are entirely independent.

PROBLEM 22. Evaluate the average energy for the above problem, and also the average square of the energy.

6.3. An N-dimensional oscillator can be treated as a simple extension of the above three-dimensional case. When all of the frequencies are different

$$Z = \prod_{i=1}^{N} \frac{e^{\hbar\omega_i/2kT}}{e^{\hbar\omega_i/kT}-1}. \tag{17.33}$$

In the special case of an N-dimensional isotropic oscillator

$$Z = \frac{e^{N\hbar\omega/2kT}}{(e^{\hbar\omega/kT}-1)^N}. \tag{17.33a}$$

As the number of dimensions is increased, the average energy becomes more and more representative of the ensemble, and it becomes significant, in the limit, to speak of the energy of the ensemble. For a three-dimensional isotropic oscillator there is, of course, a single ground state with the energy $\frac{3}{2}\hbar\omega$. The next states have energies of $\frac{5}{2}\hbar\omega$, and there are three of them depending upon the coordinate with which the additional energy is associated. The state with the energy $\frac{7}{2}\hbar\omega$ occurs in six different ways. The number of statesi n N dimensions with the energy $(n+\frac{1}{2}N)\hbar\omega$ is $(n+N-1)!/n!(n-1)!$. For large N, the function $\{(n+N-1)!/n!(N-1)!\} \exp\{-(n+\frac{1}{2}N)\,\hbar\omega/kT\}$ has a sharp maximum near the average energy, and the sum over states is composed largely of terms in the neighborhood of the maximum.

PROBLEM 23. Evaluate the average energy and the average square of the energy for an oscillator of N dimensions. Show that

$$\frac{\overline{\langle H^2\rangle} - \langle \overline{H}\rangle^2}{\langle \overline{H}\rangle^2} \to 0$$

as N increases.

7. Systems of identical particles without interactions

In the case of a group of electrons moving in a potential field the electrostatic interaction between them may sometimes be lumped with the attractive potential to give an average in which each particle moves independently of the others, as a first approximation. This gives a moderately good approximation for electrons in a crystal. In such a case, each electron may be in any one of an infinite number of states, and the wave function representing the system is a product, or a linear combination of products, of the wave functions for the individual particles. Two cases must be distinguished. When the particles are subject to the Pauli exclusion principle, the state of the whole system must be antisymmetric in the coordinates and spin of each pair of particles. The requirement of antisymmetry limits the products of functions to those in which no two factors are alike. This restriction gives rise to the Fermi-Dirac statistics. If, however, the particles under consideration are not electrons, but are, for example, alpha particles, such a restriction does not apply. The requirement of symmetric functions gives rise to the Einstein-Bose statistics.

In dealing with this approximation, numerous expressions are used which contain the energies of the individual electron states. Such expressions often show a marked similarity to expressions like equation (17.27) for the sum over states. However, the distinction must be clearly made between general expressions, such as equation (17.27), and derived expressions for particular cases such as will be considered in this section.

7.1. *The Fermi statistics*

Each electron of the system is assumed to be moving under the influence of given forces due to the atoms of the crystal or the walls of a hypothetical box in which the system is confined. When any remaining electron interaction is neglected, the Schroedinger equation can be separated into equations for each individual electron and these separate equations are identical in form. One may then determine states and energies for the individual electrons. Let each electron state be designated by the subscript i, and let ε_i be the energy of a single electron in the ith state. A state of the whole system can be specified by giving the number of electrons n_i in each of the electron states. Because of the Pauli exclusion principle, n_i can have only the values 0 or 1. The numbers n_i are also clearly subject to the restriction

$$\sum_i n_i = N. \tag{17.34}$$

N is the total number of electrons in the system.

The sum over states is

$$Z = \sum_n \exp\{-E_n/kT\} = \sum_n \exp\{-\sum_i n_i\varepsilon_i/kT\}. \tag{17.35}$$

The quantity E_n is the total energy of the system in the state n, and the first sum is over all of the possible states of the whole system. With each state is associated the appropriate value of E_n. The second sum is expressed in terms of the n_i and the ε_i and the sum is over all combinations of the n_i that satisfy equation (17.34).

T evaluate Z let $x_i = \exp\{-\varepsilon_i/kT\}$, and consider the infinite product

$$P(z) = \prod_{i=1}^{\infty} (1+x_i z). \tag{17.36}$$

This product contains all powers of z, no term in it contains a given x_i more than once, and the coefficient of z^N is just the desired sum of states Z. It contains all the products of the x_i, taken N at a time. This coefficient can be evaluated by a means of a contour integral in the complex plane about the origin,

$$Z = \frac{1}{2\pi i} \oint \frac{\prod_{i=1}^{\infty} (1+x_i z)}{z^{N+1}} dz. \tag{17.37}$$

Equation (17.37) is an exact expression for the sum of states, but the evaluation of the integral presents some difficulty. A good approximation can be obtained by integrating through a saddle point on the real axis.

Along the negative real axis is a series of zeros, and at no place does the integrand become very large. For positive real values of z the integrand is everywhere positive. It has a pole of order $(N+1)$ at the origin and becomes infinite as $z \to \infty$. Along the real axis is a minimum, which is sharp for large values of N.

Since the integrand is an analytic function of z, the presence of a sharp minimum on the real axis indicates that there is a correspondingly sharp maximum, at the same point, along the direction parallel to the imaginary axis. This point is a saddle point. Let $z = A$ be the point at which the minimum occurs on the real axis. If the integral is taken around a contour in the form of a circle with radius A, the principal contribution will come from the saddle point. If the integral can be evaluated in this neighborhood, the result will be an approximation to the whole integral. The procedure is then to locate the minimum at $z = A$, to expand the integrand in a Taylor series about this point, and to integrate the expansion.

For purposes of the expansion it is easier to deal with the logarithm of the integrand. Let

$$\log \frac{P(z)}{z^{N+1}} = G(z) = \sum_i \log(1+x_i z)-(N+1)\log z. \tag{17.38}$$

The derivative of G is

$$\frac{\mathrm{d}G}{\mathrm{d}z} = \sum_i \frac{x_i}{1+x_i z} - \frac{N+1}{z}, \tag{17.40}$$

so that the value of A is given by

$$N+1 = \sum_i \frac{x_i A}{1+x_i A}. \tag{17.39}$$

The second derivative along the real axis does not vanish at $z = A$, but is positive, and is large if the states are close enough together in energy,

$$\frac{\mathrm{d}^2 G}{\mathrm{d}z^2} = \frac{N+1}{z^2} - \sum_i \frac{x_i^2}{(1+x_i z)^2} \tag{17.40}$$

$$\left(\frac{\mathrm{d}^2 G}{\mathrm{d}z^2}\right)_A = \frac{1}{A^2}\sum_i \frac{x_i A}{(1+x_i A)^2} \gg 0. \tag{17.40a}$$

The higher derivatives will be neglected.

The approximate value of G in the neighborhood of A is given by

$$G(z) = \sum_i \log(1+x_i A)-(N+1)\log A+\tfrac{1}{2}\left(\frac{z-A}{A}\right)^2 \sum_i \frac{x_i A}{(1+x_i A)^2}. \tag{17.41}$$

To approximate the contour integral, let $(z-A) = \mathrm{i}y$, and integrate over y from $-\infty$ to ∞. This is, of course, not around the circle but is justified as an approximation,

$$Z = \frac{1}{2\pi}\exp\{\sum_i \log(1+x_i A)-(N+1)\log A\}$$

$$\times \int_{-\infty}^{\infty} \exp\left\{-\left[\frac{1}{2A^2}\sum_i \frac{x_i A}{(1+x_i A)^2}\right] y^2\right\} \mathrm{d}y$$

$$= \frac{1}{2\pi}\exp\{\sum_i \log(1+x_i A)-(N+1)\log A\}\left\{2\pi A^2 \Big/ \sum_i \frac{x_i A}{(1+x_i A)^2}\right\}^{\frac{1}{2}} \tag{17.42}$$

$$\log Z = \sum_i \log(1+x_iA) - N\log A - \tfrac{1}{2}\log \sum_i \frac{2\pi x_i A}{(1+x_iA)^2}. \tag{17.43}$$

It is customary to let $A = e^{\zeta/kT}$ where ζ is called the Fermi energy. Then

$$\log Z = \sum_i \log\left(1+\exp\frac{\zeta-\varepsilon_i}{kT}\right) - \frac{N}{kT} - \tfrac{1}{2}\log\sum_i \frac{2\pi}{(1+\exp\{(\zeta-\varepsilon_i)/kT\})(\exp\{(\varepsilon_i-\zeta)/kT\}+1)}. \tag{17.43a}$$

The average number of particles in the state j can be obtained from equation (17.35) by differentiation:

$$\overline{n_j} = -\frac{\partial(\log Z)}{\partial(\varepsilon_j/kT)} = \frac{1}{\exp\{(\varepsilon_j-\zeta)/kT\}+1} \tag{17.44}$$

when the last term in $\log Z$ is neglected. Similarly

$$\overline{\langle H\rangle} = \sum_i \frac{\varepsilon_i}{\exp\{(\varepsilon_i-\zeta)/kT\}+1}. \tag{17.45}$$

Equation (17.45) for the statistical mean energy is just what would be obtained by multiplying the average number of particles in an electron state by the energy of such an electron and adding over all the states. To evaluate these quantities further, it is necessary to know something about the distribution of the energies ε_i. Only then is the particular mechanical system characterized.

7.2. *The Einstein-Bose statistics*

When the restriction imposed by the Pauli exclusion principle is removed, the sum over states can be obtained from the contour integral

$$Z = \frac{1}{2\pi i}\oint \prod_{i=1}^{\infty} \frac{(1+x_iz+x_i^2z^2+x_i^3z^3+\ldots)}{z^{N+1}}\,dz. \tag{17.46}$$

This gives the coefficient of z^N, but each x_i may appear up to N times in the coefficient. Since all x_i are less than unity, if z is restricted to the unit circle around the origin,

$$Z = \frac{1}{2\pi i}\oint \frac{1}{z^{N+1}} \prod_{i=1}^{\infty} \frac{dz}{(1-x_iz)}. \tag{17.46a}$$

The integrand has a pole at the origin and a series of poles at $z = 1/x_i$ along the real axis. The first, however, is at a value of z greater than unity, and between 0 and 1 is a sharp minimum which constitutes a saddle point. The integral over the saddle point may be taken as an approximation to the contour integral.

The logarithm of the integrand is

$$G(z) = -\sum_i \log(1-x_i z) - (N+1)\log z\,. \tag{17.47}$$

The minimum can be determined from the derivative, which leads to

$$\sum_i \frac{x_i A}{1-x_i A} = N+1\,. \tag{17.48}$$

By using the second derivative

$$G = -\sum_i \log(1-x_i A) - (N+1)\log A + \tfrac{1}{2}\left(\frac{z-A}{A}\right)^2 \sum_i \frac{x_i A}{(1-x_i A)^2} + \dots \tag{17.49}$$

and

$$Z = \frac{1}{(2\pi)^{\frac{1}{2}} A^N} \prod_{i=1}^{\infty} (1-x_i A)^{-1} \left\{\sum_i \frac{x_i A}{(1-x_i A)^2}\right\}^{-\frac{1}{2}}. \tag{17.50}$$

The value of $\log Z$ from equation (17.50) gives

$$n_j = \frac{x_j A}{1-x_j A} = \frac{1}{(1/A)\exp(\varepsilon_j/kT)-1} \tag{17.51}$$

which is the well-known Einstein-Bose distribution function. This differs from the expression for the mean energy of an oscillator in that it contains A. Equation (17.51) represents the distribution of a given number, N, of particles. In the case of an oscillator, or of black body radiation, the number of quanta is indeterminate.

CHAPTER 18

ELECTRONS IN SOLIDS

The quantum theory of crystalline solids is inherently more complicated than that of the simple atomic systems we considered in Chapters 13 and 14. Indeed, a macroscopic piece of solid on which measurements are performed consists of some 10^{23} atomic nuclei vibrating about their regular positions at lattice sites. Around each nucleus a number of *core electrons* are moving more or less the same way as they do in the free atoms, since their distance from other nuclei in the solid is much larger than from the nucleus "they belong to". This is not the case for the *valence electrons*, which cannot be asssigned to individual atoms, their properties being determined by the lattice of ions (nuclei + core electrons) as a whole. In particular, their interaction with each other is significant and completely different from the mutual interaction of valence electrons in a free atom or ion.

In this wilderness of ion–ion, ion–electron, and electron–electron interactions we cannot hope to find our way and apply quantum mechanical methods without discerning some simplifying features. Such features were indeed realized in the early days of quantum mechanics, and their exploitation made it possible to calculate rather accurately some electrical, optical, and thermodynamic properties by relatively elementary methods.

The first one of these features is the independence of ionic and electronic motion, which is stated by the *Born–Oppenheimer theorem*. The essence of this theorem is that, due to their smaller masses, the electrons always move as if the lattice of ions were rigid. We can thus calculate their basic properties assuming that they are moving in a rigid, perfectly periodic potential, and their interaction with lattice vibrations (phonons) can be studied by taking "snapshots" of the vibrating lattice at different times and calculating the properties of electrons moving in this distorted lattice. Similarly, the lattice vibrations are in first approximation independent of the state of the electrons, the electron–phonon interaction being a small perturbation.

Another simplifying feature is the applicability of the *molecular field method* (Hartree, or Hartree–Fock approximation), which enables us to avoid attacking the problem of 10^{23} interacting valence electrons moving in the potential of the ions by replacing it by the simpler one of a single electron moving in the potential of the ions and the average potential of all the other electrons. The validity of this approximation is less obvious and less well established than that of the Born–Oppenheimer theorem, and indeed, there are several cases where it cannot be applied (the most notable ones being magnetism and superconductivity in solids). However, a whole branch of solid state physics, band theory, is devoted to the determination of single-electron eigenfunctions and energy eigenvalues appropriate to various known solids, and many experiments can be successfully interpreted within the framework of this approximation.

In the present chapter we will treat the general problem of eigenstates of an electron moving in a periodic potential. In Section 1 we will derive a few general properties of the eigenstates, first for a one-dimensional model and then for a potential representing a real three-dimensional crystal. In Section 2 we discuss the general properties of the energy spectrum, and in Section 3 we present two of the methods by which the eigenstates can be determined. In Section 4 we will apply our results to derive some macroscopic properties determined by the nature of these electronic states.

1. General form of the wave functions

The Hamiltonian of an electron moving in a solid consists of three parts: the kinetic energy, the electron–ion interaction energy, and the electron–electron interaction energy. If we adopt the one-electron approximation outlined in the introduction, the latter two terms can be represented by a potential energy: the sum of the attractive potential of the ions forming a rigid lattice and the repulsive potential arising from the average interaction of the electron under consideration with all the other electrons. To find the energy eigenstates we have to solve the equation

$$\nabla^2 U + \frac{2m}{\hbar}[E - V(r)]U = 0. \tag{18.1}$$

In this section we will assume that $V(r)$ is given and study those aspects of the eigenstates which do not depend on the specific form of $V(r)$, but only on the fact that $V(r)$ has the same symmetry properties as the crystal lattice.

1.1. *One-dimensional case*

In a hypothetical infinite one-dimensional "crystal" composed of identical "atoms" equation (18.1) becomes

$$\frac{d^2U}{dx^2} + \frac{2m}{\hbar^2}[E - V(x)]U = 0, \tag{18.2}$$

where the potential $V(x)$ is periodic, which means that for any integer n, $V(x+nd) = V(x)$, where d is the interatomic distance. This periodicity can be expressed by saying that $V(x)$ is invariant under the transformations $x \to x+nd$ corresponding to translations of the coordinate system by $-nd$ (or to that of the function $V(x)$ by $+nd$). Let us introduce the translation operators $\mathsf{T}(nd)$ by the definition

$$\mathsf{T}(nd)f(x) = f(x+nd).$$

It we furthermore define the product of two translation operators as the operator corresponding to that translation which arises by carrying out the two translations subsequently, we have from the above definition

$$\mathsf{T}(n_1 d)\mathsf{T}(n_2 d) = \mathsf{T}(n_1 d + n_2 d). \tag{18.3}$$

Because in our eigenvalue problem the potential is periodic, any observable properties associated with the eigenfunctions U must be also periodic. This does not necessarily mean that the eigenfunctions themselves have to be invariant under the translations $\mathsf{T}(nd)$, $U(x)$ and $\mathsf{T}(nd)U(x)$ may differ in a phase factor $e^{i\varphi(nd)}$ without influencing the expectation value of any observable:

$$\mathsf{T}(nd)U(x) = U(x+nd) = e^{i\varphi(nd)}U(x).$$

The phase factors corresponding to various translations are not unrelated: to satisfy equation (18.3) we must have

$$\varphi(n_1 d) + \varphi(n_2 d) = \varphi(n_1 d + n_2 d),$$

which means that $\varphi(nd)$ must be proportional to nd, $\varphi(nd) = knd$. The proportionality factor k, called the wave number, serves as an index (quantum number) classifying the eigenfunctions according to their transformation properties,

$$\mathsf{T}(nd)U_k(x) = e^{iknd}U_k(x). \tag{18.4}$$

This property of the functions $U_k(x)$ can be used to define a new set of functions $u_k(x) = e^{-ikx}U_k(x)$, which are invariant under the translations $\mathsf{T}(nd)$:

$$\mathsf{T}(nd)u_k(x) = \mathrm{e}^{-ik(x+nd)}\mathsf{T}(nd)U_k(x) = \mathrm{e}^{-ikx}U_k(x) = u_k(x).$$

We have shown thus that the eigenfunctions satisfying the one-dimensional Schroedinger equation involving a periodic potential $V(x)$ can be always written in the form

$$U_k(x) = \mathrm{e}^{ikx}u_k(x), \tag{18.5}$$

where $u_k(x)$ has the same periodicity as $V(x)$. (This is called Floquet's theorem, which was first applied to the problem of electrons in a periodic potential by F. Bloch; the wave functions of the form (18.5) are often referred to as Bloch functions.)

Substitution of the form (18.5) into equation (18.2) gives an equation for the periodic function $u_k(x)$:

$$\frac{\mathrm{d}^2 u_k}{\mathrm{d}x^2} + 2\mathrm{i}k\frac{\mathrm{d}u_k}{\mathrm{d}x} + \frac{2m}{\hbar^2}\left[E_k - \frac{\hbar^2 k^2}{2m} - V(x)\right]u_k = 0. \tag{18.6}$$

Several conclusions can be drawn directly from equation (18.6) without more precise knowledge of $V(x)$.

1. It is clear from the imaginary coefficient in the equation that $u_k(x)$ cannot be real unless $k = 0$, or unless $u_k(x)$ is a constant. The latter occurs for free electrons, when $V(x)$ is a constant.

2. Since a change in sign of k is equivalent to a change in sign of the imaginary term, it follows that $u_{-k}(x) = u_k^*(x)$. The eigenfunctions u_k and u_{-k} are associated with the same value of the energy, so that E_k is an *even* function of k.

3. Since E_k is also an *analytic* function of k, it follows from 2 that $\mathrm{d}E_k/\mathrm{d}k = 0$ when $k = 0$, hence $E_k = E_0 + \omega k^2$, where ω is a constant. This property can be used to define an *effective mass* $m^* = \hbar^2/2\omega$, in terms of which

$$E_k - E_0 = \frac{\hbar^2 k^2}{2m^*}, \tag{18.7}$$

in obvious analogy with the kinetic energy of a free particle of mass m^*.

4. The wave number k is defined only modulo $2\pi/d$. This can be seen from the fact that for any integer s

$$U_k = u_k \mathrm{e}^{ikx} = u'_{k+(2\pi s/d)}\mathrm{e}^{\mathrm{i}[k+(2\pi s/d)]},$$

where

$$u'_{k+(2\pi s/d)} = u_k \mathrm{e}^{-\mathrm{i}(2\pi s/d)x},$$

and $u'_{k+2\pi s/d}$ is still periodic with the period d.

This property leads to the division of the values of k into regions, known as Brillouin zones. Although the boundaries and the numbering of the zones are somewhat arbitrary, it is convenient to take them symmetrically about the point $k = 0$. Thus the values of k between $-\pi/d$ and $+\pi/d$ constitute the first Brillouin zone. The next zone may be divided into two parts. It includes the values of k between $-2\pi/d$ and $-\pi/d$, and also those between $+\pi/d$ and $+2\pi/d$. Similarly, each of the higher zones includes a region of length π/d of positive values of k and a corresponding region of negative values.

5. The first Brillouin zone, the region between $-\pi/d$ and $+\pi/d$, is the extent of the range of k necessary to include all solutions of the eigenvalue equation. For a fixed k, equation (18.6) has an infinity of solutions with the corresponding energy values which may be denoted by $E_{n,k}$. For a fixed n, however, $E_{n,k}$ is a periodic function of k with the period $2\pi/d$. This, combined with the fact that $E_{n,k}$ is an even function of k and an analytic function, leads to the property

$$\mathrm{d}E_{n,k}/\mathrm{d}k = 0 \tag{18.8}$$

at the boundaries of a Brillouin zone. It is clear from this property of $E_{n,k}$ that the validity of the effective mass approximation (18.7) is limited to the center of the first Brillouin zone, i.e., to $k \ll 2\pi/d$.

Figure 18.1a shows a possible set of values of $E_{n,k}$ as functions of k. Each value of E appears in each Brillouin zone and so such a diagram is unnecessarily repetitious. There are two principal ways in which the energy eigenvalues may be depicted without such repetition. On the one hand all energies may be referred to the wave numbers in the first zone. Thus the energies corresponding to a given n constitute one continuous curve, a *band*, and the different bands are denoted by different n's. This is called the reduced zone scheme. On the other hand, the lowest band may be considered to be in the first zone, the next band to be in the second zone, and so on. This leads to the extended zone scheme, represented by the heavy line in figure 18. 1a, and to the designation of each energy, and the corresponding Bloch function, uniquely by a value of k. The energy is then generally discontinuous at the zone boundaries.

6. The set of functions, for a given k, that satisfy equation (18.6) form an orthonormal set, and any function periodic with the period d can be expressed in terms of such a set. If the various values of k are also included, one has a complete set, suitable for the expansion of any function, after the manner of a Fourier integral. The designation of the function by a single index k, which runs from $-\infty$ to $+\infty$ is convenient for this purpose.

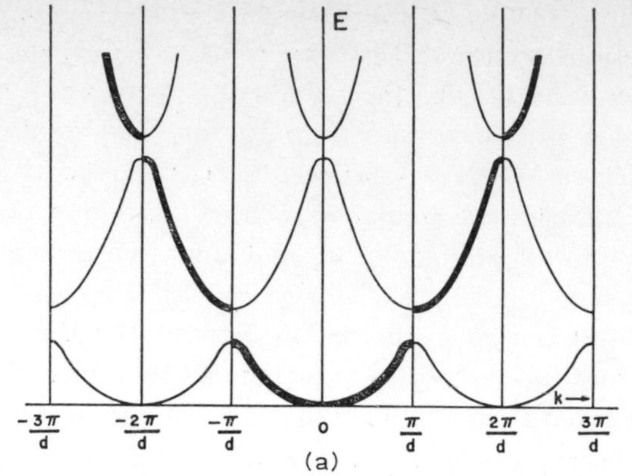

(a)

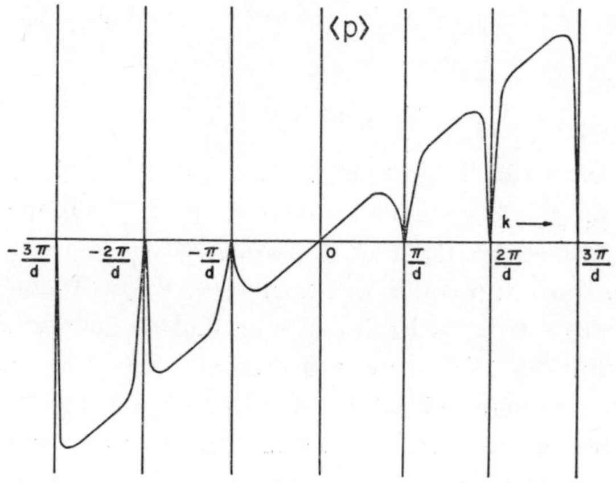

(b)

Fig. 18.1. The upper figure represents the energy as a function of k for the one-dimensional case. The heavy line represents a single valued function in which the different energy bands are assigned to different Brillouin zones. The lower figure represents the quantum mechanical mean momentum as a function of k corresponding to the heavy line in the top figure.

7. The quantity $\hbar k$ is often called the "crystal momentum", but it is not a true momentum. The Bloch functions are not eigenfunctions of the momentum operator, but the expectation value of the momentum for such a state can be evaluated if the energy is known as a function of k.

From the general principles of quantum mechanics,

$$\langle p \rangle = \frac{\hbar}{\mathrm{i}} \int u_k^* \left(\mathrm{i}k u_k + \frac{\partial u_k}{\partial x} \right) \mathrm{d}x = \hbar k + \frac{\hbar}{\mathrm{i}} \int u_k^* \frac{\partial u_k}{\partial x} \, \mathrm{d}x \,. \qquad (18.9)$$

To evaluate this integral, differentiate equation (18.6) with respect to k, multiply by u_k^*, and integrate over one period. If the function u_k is normalized so that the integral of the square of the absolute value over one period is unity, the result is

$$\int u_k^* \left(\frac{\partial^2}{\partial x^2} + 2ik\frac{\partial}{\partial x} - k^2 - \frac{2m}{\hbar^2}V\right)\frac{\partial u_k}{\partial k}\,\mathrm{d}x + 2i\int u_k^* \frac{\partial u_k}{\partial x}\,\mathrm{d}x +$$

$$+\frac{2m}{\hbar^2}E_k\int u_k^* \frac{u_k}{k}\,\mathrm{d}x + \frac{2m}{\hbar^2}\frac{\mathrm{d}E_k}{\mathrm{d}k} - 2\mathrm{k} = 0\,.$$

By partial integration, the first integral can be transformed so that the operators operate on u_k^*, and from the complex conjugate of equation (18.6) the result is $-(2m/\hbar^2)E_k$. The whole expression then leaves

$$\int u_k^* \frac{\partial u_k}{\partial k}\,\mathrm{d}x = \frac{im}{\hbar^2}\frac{\mathrm{d}E_k}{\mathrm{d}k} - ik\,,$$

and equation (18.9) gives

$$\langle p\rangle = \frac{m}{\hbar}\frac{\mathrm{d}E_k}{\mathrm{d}k}. \qquad (18.10)$$

For free electrons, $E = (\hbar^2/2m)k^2$, and $p = \hbar k$. In the general case, $\langle p\rangle = (m/m^*)\,\hbar k$ for very small values of k ($k \ll 2\pi/d$), and $\langle p\rangle = 0$ at the boundaries of the Brillouin zones [cf. equations (18.7) and (18.8)].

Figure 18.1b gives the expectation values of the momentum as a function of the crystal momentum for the states represented by the heavy curve in fig. 18.1a. It is seen that $\langle p\rangle$ indeed goes to zero at each zone boundary. It is just at the wavelengths corresponding to the zone boundaries that no propagation is possible.

PROBLEM 1. Construct a wave packet from states in the neighbourhood of k_0 and show that the maximum of the packet moves with the velocity $\hbar^{-1}(\mathrm{d}E_k/\mathrm{d}k)$. A convenient form is given by taking the amplitude of the constituent functions to be $\exp\,[-\alpha(k-k_0)^2]$ and assuming that $u_k(x)$ varies slowly with k.

1.2. *Three-dimensional case*

In a real crystal the electrons move in a three-dimensional potential having the full symmetry of the crystal lattice. This includes translational and rotational symmetry.

The translational symmetry imposes the following condition on the potential:

$$V(\boldsymbol{r}) = V(\boldsymbol{r}+n_1\boldsymbol{a}_1+n_2\boldsymbol{a}_2+n_3\boldsymbol{a}_3),$$

where n_1, n_2 and n_3 are integers and $\boldsymbol{a}_1$, $\boldsymbol{a}_2$ and $\boldsymbol{a}_3$ are three non-coplanar basis vectors which specify the crystal lattice. The parallelepiped defined by the basis vectors is the unit cell of the crystal. It can be shown by repeating the arguments leading up to equation (18.5) that the general form of the energy eigenfunctions in this case is

$$U_{\boldsymbol{k}}(\boldsymbol{r}) = u_{\boldsymbol{k}}(\boldsymbol{r})\, \mathrm{e}^{\mathrm{i}\boldsymbol{k}\cdot\boldsymbol{r}},$$

where $u_{\boldsymbol{k}}(\boldsymbol{r})$ satisfies the same periodicity condition as the potential $V(\boldsymbol{r})$. The differential equation satisfied by $u_{\boldsymbol{k}}$ is

$$\nabla^2 u_{\boldsymbol{k}} + 2\mathrm{i}\boldsymbol{k}\cdot\nabla u_{\boldsymbol{k}} + \frac{2m}{\hbar^2}\left[E_{\boldsymbol{k}} - \frac{\hbar^2 k^2}{2m} - V(\boldsymbol{r})\right] u_{\boldsymbol{k}} = 0. \qquad (18.11)$$

The vector $\boldsymbol{k}$ is called the wave vector.

The formal similarity of equations (18.6) and (18.11) makes it possible to generalize our conclusions regarding the one-dimensional case to the present three-dimensional case:

1–3. These conclusions carry over rather directly, including the definition of the effective mass (equation (18.7)), if we keep in mind that the wave vector $\boldsymbol{k}$, unlike the wave number k, is not a scalar. Consequently, in the general case, the effective mass becomes the inverse of a tensor, in terms of which $E_{\boldsymbol{k}}$ is approximated as

$$E_{\boldsymbol{k}} - E_0 = \tfrac{1}{2}\hbar^2[(m^*)^{-1}]_{\alpha\beta} k_\alpha k_\beta, \qquad (18.12)$$

where α and b refer to cartesian coordinates.

4. In order to define the Brillouin zones, it is convenient to introduce the reciprocal lattice. The basis vectors of the reciprocal lattice ($\boldsymbol{b}_1$, $\boldsymbol{b}_2$, $\boldsymbol{b}_3$) are defined by the relations

$$\boldsymbol{a}_i \cdot \boldsymbol{b}_j = \delta(i,j); \qquad \boldsymbol{b}_i = \frac{\boldsymbol{a}_j \times \boldsymbol{a}_k}{\boldsymbol{a}_1 \cdot (\boldsymbol{a}_2 \times \boldsymbol{a}_3)}. \qquad (18.13)$$

The components of $\boldsymbol{k}$ in the directions of the $\boldsymbol{b}_i$ vectors (which are non-coplanar, but not necessarily mutually orthogonal) are defined only modulo $2\pi\boldsymbol{b}_i$, because for any set m_1, m_2, m_3 of integers

$$u_{\boldsymbol{k}}\, \mathrm{e}^{\mathrm{i}\boldsymbol{k}\cdot\boldsymbol{r}} = u'_{\boldsymbol{k}+2\pi(m_1\boldsymbol{b}_1+m_2\boldsymbol{b}_2+m_3\boldsymbol{b}_3)} \exp\{\mathrm{i}[\boldsymbol{k}+2\pi(m_1\boldsymbol{b}_1+m_2\boldsymbol{b}_2+m_3\boldsymbol{b}_3)]\cdot\boldsymbol{r}\},$$

where

$$u'_{\boldsymbol{k}+2\pi(m_1\boldsymbol{b}_1+m_2\boldsymbol{b}_2+m_3\boldsymbol{b}_3)} = u_{\boldsymbol{k}} \exp\{-2\pi i(m_1\boldsymbol{b}_1+m_2\boldsymbol{b}_2+m_3\boldsymbol{b}_3)\cdot\boldsymbol{r}\}, \quad (18.14)$$

which has the required periodicity.

The region of $\boldsymbol{k}$-space, the reciprocal space, surrounding the origin and such that no two vectors, except those ending on the surface, differ by a vector of the reciprocal lattice, constitutes the first Brillouin zone. In the reduced zone scheme we introduce a band index n and refer all functions to the first Brillouin zone ascribing to them wave vectors lying within it. It is often more convenient, however, to use the extended zone scheme, regarding the energy values and the eigenfunctions as designated by $\boldsymbol{k}$ only. In that case the lowest energy band will be represented by $\boldsymbol{k}$ vectors in the first zone, the second band by those in the second zone, etc.

As it can be verified from equation (18.13), in the case of a simple cubic

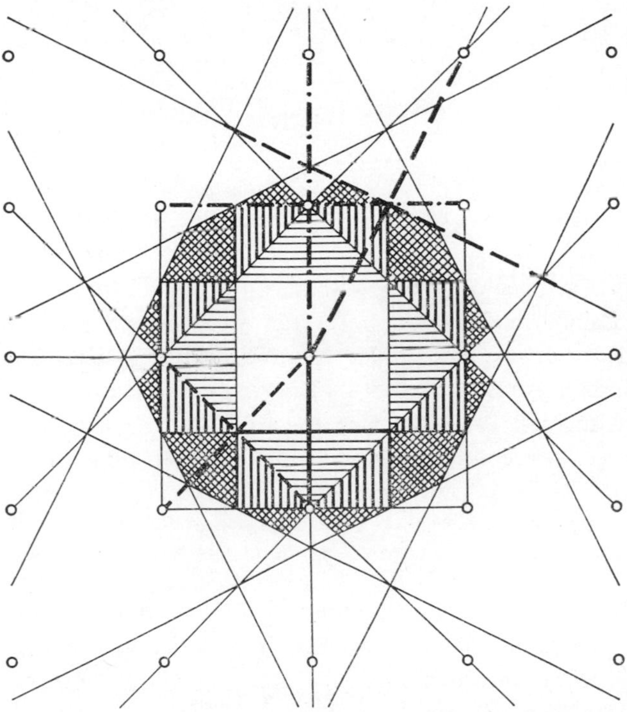

Fig. 18.2. Brillouin zones for a two-dimensional square lattice. The shading indicates the portions that belong together to form a single zone. The heavy lines represent vectors in the reciprocal lattice whose bisectors (also heavily drawn) form zone boundaries.

lattice the reciprocal vectors are parallel to the lattice vectors, and all have the same length $1/a$. The reciprocal lattice is thus also simple cubic, with a lattice constant of $1/a$. If one point of this lattice is taken as the origin, the zones are conveniently constructed by connecting this point with every other lattice point and bisecting the connecting lines by perpendicular planes.

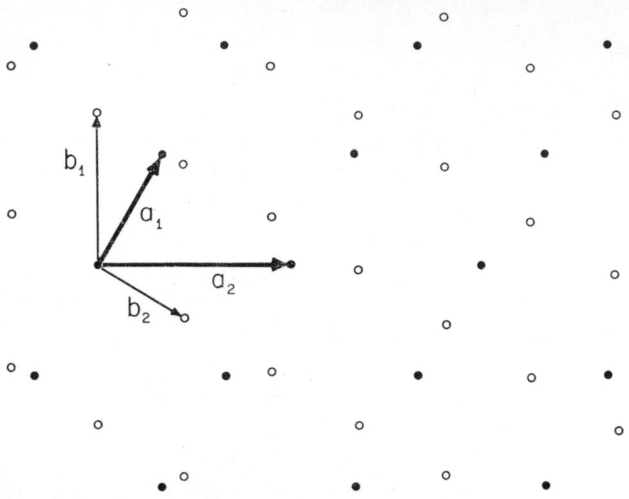

Fig. 18.3. A non-orthogonal two-dimensional lattice designated by heavy dots, and the corresponding reciprocal lattice designated by open circles.

These planes will enclose the zones. This process is difficult to illustrate in three dimensions, but fig. 18.2 shows it in two dimensions. The correspondingly shaded segments belong together to constitute a zone. All zones have the same area. Figures 18.3 and 18.4 show the construction of the reciprocal lattice and the first Brillouin zone for a two-dimensional lattice in which the lattice vectors are not orthogonal and are of different lengths.

PROBLEM 2. Show that $u'_{\boldsymbol{k}+2\pi(m_1\boldsymbol{b}_1+m_2\boldsymbol{b}_2+m_3\boldsymbol{b}_3)}$ in equation (18.14) has the periodicity of the lattice spanned by the basis vectors $\boldsymbol{a}_1$, $\boldsymbol{a}_2$ and $\boldsymbol{a}_3$.

5. It follows from the above definition of the Brillouin zones in terms of perpendicular bisecting planes of reciprocal lattice vectors, and from the periodicity, parity, and analiticity of $E_{n,\boldsymbol{k}}$ that on the Brillouin zone boundaries the normal component of $\text{grad}_{\boldsymbol{k}} E_{n,\boldsymbol{k}}$ vanishes.

6,7. Our remarks concerning the completeness of the set of one-dimensional Bloch functions and the meaning of the crystal momentum hold for

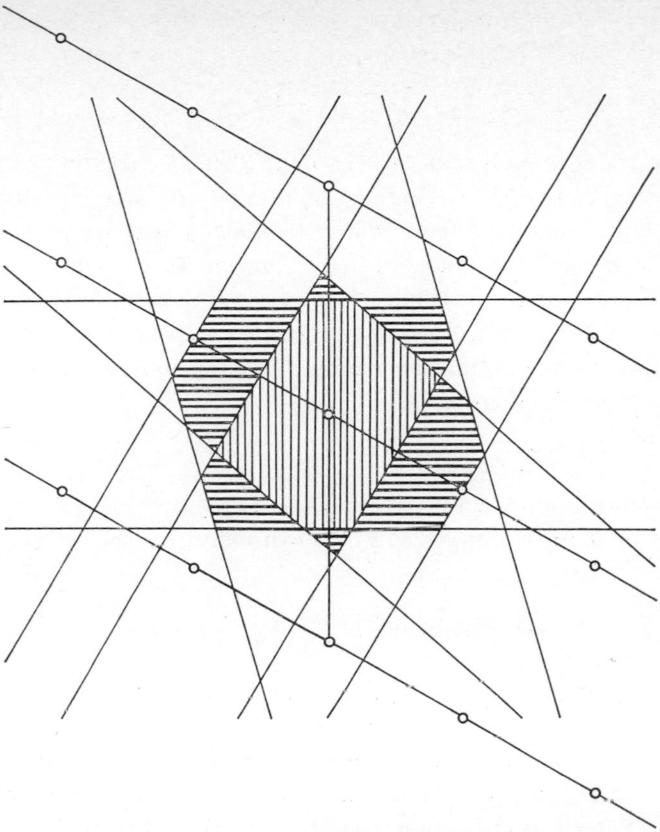

Fig. 18.4. Brillouin zones of the lattice shown in fig. 18.3.

three dimensions as well, with the only difference that equation (18.10) is replaced by

$$\langle p \rangle = \frac{m}{\hbar} \operatorname{grad}_{\boldsymbol{k}} E_{\boldsymbol{k}}. \tag{18.15}$$

8. Apart from the general properties of the wave functions and energy bands discussed in the one-dimensional case, which are governed by the *translational symmetry* of the crystal lattice, in the case of the three-dimensional lattice there are some additional features determined by its *rotational symmetry*. In particular, the rotational symmetry of the energy as a function of $\boldsymbol{k}$ is governed by that of the potential. To make this statement precise, and to justify it, let R be an operator representing a rotation of the spatial

coordinates. R is an orthogonal transformation. It can be represented by a matrix with elements R_{ij} such that

$$x_i' = \sum_j R_{ij} x_j . \tag{18.16}$$

Any such transformation can be regarded either as a rotation of the coordinate system in which a certain function is expressed, or as a rotation, in the opposite direction, of the "pattern" of the function with respect to the fixed coordinate system. The latter, or active, interpretation will be used here. The x_i' then represent new values of the coordinates which give the function the value $F(x, y, z)$. The axis system is regarded as fixed.

The application of the transformation R to a function F means a change in the form of the function so that

$$\mathsf{R}F(x, y, z) \equiv [\mathsf{R}F]\,(x, y, z,) \equiv F'(x, y, z) . \tag{18.17}$$

$[\mathsf{R}F]$ is enclosed in a bracket to emphasize the fact that it is a new function F', of the original variables x, y, z. The nature of the new function is given by the relationships

$$F'(x', y', z') \equiv [\mathsf{R}F]\,(\mathsf{R}x, \mathsf{R}y, \mathsf{R}z) = F(x, y, z) \tag{18.17a}$$

or

$$F'(x', y', z') = F(\mathsf{R}^{-1}x', \mathsf{R}^{-1}y', \mathsf{R}^{-1}z') . \tag{18.17b}$$

The meaning of these two versions of the transformation is the same, namely that if the coordinates x', y' and z' in $F'(x', y', z')$ are expressed in terms of x, y and z by the transformation (18.16), $F(x, y, z)$ will result.

The symmetry properties of a function $F(x, y, z)$ are described by its invariance under various transformations. A function $F(x, y, z)$ is invariant under a transformation R if

$$[\mathsf{R}F]\,(x, y, z) \equiv F'(x, y, z) = F(x, y, z) .$$

The kinetic energy operator being invariant under all translations, the symmetry properties of the lattice potential and rotations determine those of the total Hamiltonian. The transformations R_j under which the potential energy – and, consequently, the Hamiltonian – is invariant are called symmetry transformations, they form a group, called the point group of the lattice described by V. For a cubic lattice there are 48 orthogonal transformations under which the Hamiltonian is invariant. Not all of these transformations are true rotations. Half of them are "improper" rotations, or rotations plus inversions.

The importance of the group of the symmetry transformations is that its properties determine the degeneracy of the energy eigenvalues. This is because if an eigenfunction U satisfies the eigenvalue equation $HU = EU$, the application of a transformation to both sides does not disturb the equality, so that $\mathsf{R}_j HU = \mathsf{R}_j EU$. Then, if H is unchanged under the transformation, the only change is in the function U and

$$H(\mathsf{R}_j U) = E(\mathsf{R}_j U). \tag{18.18}$$

This shows that if U is an eigenfunction, $\mathsf{R}_j U$ is also an eigenfunction with the same energy. For a cubic lattice this leads to 48 solutions with the same energy, and for a lattice with lower symmetry to a smaller number. Not all of these functions will be different.

If the operator R_j is applied to the Bloch function $U_{\boldsymbol{k}}$, the result is

$$\mathsf{R}_j u_{\boldsymbol{k}}\, \mathrm{e}^{\mathrm{i}\boldsymbol{k}\cdot\boldsymbol{r}} = u_{\boldsymbol{k}}(r_1', r_2', r_3') \exp\{\mathrm{i}(k_1 r_1' + k_2 r_2' + k_3 r_3')\},$$

where $r_i' = \Sigma_m (\mathsf{R}_j)_{im} r_m$. To find the exact meaning of this operation, we must now express the transformed function in terms of the original coordinates,

$$\mathsf{R}_j u_{\boldsymbol{k}}\, \mathrm{e}^{\mathrm{i}\boldsymbol{k}\cdot\boldsymbol{r}} = u_{\boldsymbol{k}'}(r_1, r_2, r_3) \exp\{\mathrm{i}(k_1' r_1 + k_2' r_2 + k_3' r_3)\},$$

where

$$k_i' = \sum_m (\mathsf{R}_j)_{mi} k_m = \sum_m (\mathsf{R}_j^{-1})_{im} k_m .$$

In general the function $u_{\boldsymbol{k}'}$ is a different function from $u_{\boldsymbol{k}}$, but, since it also has the periodicity of the lattice, the resulting function is also a Bloch function, $\mathsf{R}_j U_{\boldsymbol{k}} = U_{\boldsymbol{k}'}$, i.e., the transformation may be regarded as a transformation from $\boldsymbol{k}$ to $\boldsymbol{k}' = \mathsf{R}_j^{-1}\boldsymbol{k}$. According to equation (18.18) $E_{\boldsymbol{k}} = E_{\boldsymbol{k}'}$.

Applying all the transformations belonging to the point group of the crystal to a particular Bloch function of wave vector $\boldsymbol{k}$ will generate what is called the star of $\boldsymbol{k}$: the set of all wavevectors which are transformed into each other by the members of the point group. The number of these vectors is at most equal to the number of symmetry transformations – if the given $\boldsymbol{k}$ vector is invariant under some symmetry transformations, it is less. From the above argument it follows that each energy eigenvalue $E_{\boldsymbol{k}}$ is at least as many times degenerate as there are $\boldsymbol{k}$ vectors in the star of $\boldsymbol{k}$.

PROBLEM 3. Write the matrix representing a rotation about the x-axis through an angle ϑ. Pay particular attention to the sign of ϑ. Then consider the function

$$F(x, y, z) = A(y^6 - 2y^4z^2 + y^2z^4 + x^2)$$

and determine if there are any angles of rotation about the x-axis under which $F(x, y, z)$ is invariant.

PROBLEM 4. Show that for any orthogonal transformation R, $\boldsymbol{k} \cdot (\mathsf{R}\boldsymbol{r}) = (\mathsf{R}^{-1}\boldsymbol{k}) \cdot \boldsymbol{r}$.

2. General properties of the energy spectrum

Having discussed the general properties of the wave functions of electrons in a periodic potential, we now turn to the discussion of the energy spectrum. In the previous section we considered wave functions appropriate to an infinite crystal, and we have shown that (in the reduced zone scheme) $E_{n,\boldsymbol{k}}$ for each band n is a continuous function of the wave vector $\boldsymbol{k}$. This implies that for each band n there is a continuous range of allowed energy values between the maximum and minimum values of $E_{n,\boldsymbol{k}}$. These energy bands may overlap or may be separated by forbidden bands, also called energy gaps.

Within each band the energy spectrum can be characterized by the (differential) *density of states* $N(E)$, which is defined by saying that the number of energy eigenstates in an infinitesimal energy range δE is $N(E)\delta E$. For an infinite crystal this number is clearly either infinity (within the bands) or zero (in the forbidden bands), and in a finite, macroscopic piece of solid it is proportional to the volume of the solid. Therefore, it is useful to introduce the density of states per unit volume. If we choose the volume per atom as the volume unit, this will be the same as the density of states per atom.

The density of states per unit volume is independent of the size and shape of the solid, i.e., independent of the boundary conditions, provided the size chosen is macroscopic, that is, it contains a large number of unit cells. The most convenient and most commonly used boundary condition is the so called periodic boundary condition, which is appropriate to a crystal of the shape of the unit cell, of edges N_1a_1, N_2a_2 and N_3a_3 long, where $N = N_1N_2N_3$ is the number of unit cells. Imposing periodic boundary conditions on the wavefunctions amounts to requiring that they be periodic in the direction of each lattice vector $\boldsymbol{a}_i$ with the period $N_i\boldsymbol{a}_i$. This is easily visualized in the one-dimensional case: it amounts to joining up the two ends of the chain which represents the one-dimensional "lattice" (if N is chosen large enough the curvature of the loop will be negligible). In three dimensions this cannot be done, but we can imagine that the whole space is filled with crystals identical to the one under consideration; the periodic boundary condition then means that we require in each of these crystals the wave-function to be identical.

Under the periodic boundary condition not every k value will be allowed. In the one-dimensional case, if the Bloch function (18.5) has to satisfy

$$U_k(x+Nd) = U_k(x),$$

$e^{ik(x+Na)} = e^{ikx}$ must hold, and hence kNa must be an integer multiple of 2π. The allowed values of k are thus $2\pi n/Nd$, where n is a positive or negative integer; we have exactly N allowed k values within the range $-\pi/d < k \leq +\pi/d$ defining the first Brillouin zone. The number of states in any range δk is $\delta k/(2\pi/Nd)$; the corresponding energy range being $\delta E = (\mathrm{d}E/\mathrm{d}k)\delta k$, the density of states is $N(E) = Nd/(2\pi\ \mathrm{d}E/\mathrm{d}k)$.

Along similar lines one can show that in three dimensions the allowed k vectors form a regular lattice within the first Brillouin zone; a general lattice point being given by $(2\pi/N_1)n_1\boldsymbol{b}_1+(2\pi/N_2)n_2\boldsymbol{b}_2+(2\pi/N_3)n_3\boldsymbol{b}_3$, with n_1, n_2 and n_3 integers. Again, we have $N = N_1N_2N_3$ allowed k vectors within each Brillouin zone, i.e., N states per band (not counting spin degeneracy). The density of states is given by

$$N(E) = \frac{Na_1a_2a_3}{(2\pi)^3}\int_E \mathrm{d}\boldsymbol{S}_k \frac{1}{|\nabla_k E_k|}, \tag{18.19}$$

where the surface integral in $\boldsymbol{k}$ space has to be taken over the surface defined by $E_{\boldsymbol{k}} = E$. As $Na_1a_2a_3$ is the volume of the crystal, we have verified that the density of states is proportional to the volume.

3. Nearly free electron and tight binding approximations

The determination of the Bloch functions appropriate to a particular crystal involves two problems: (i) the determination of the one-electron potential which most accurately represents the sum of the electron–ion interactions and the average of the electron–electron interactions, and (ii) the solution of the Schroedinger equation for this potential. As the distribution of electronic charge in the ground state of the crystal is determined by the Bloch functions themselves, the problem has to be solved self-consistently: the potential representing the electron–electron interactions has to be identical with the one due to the resulting electronic charge distribution. Actual band calculations are seldom carried to self consistency, but several computational methods have been developed, which give fairly reliable bands, i.e., $E_{n,\boldsymbol{k}}$ functions.

In this section we discuss two approximations which are not very widely used in actual computer calculations of band structures, but give a very clear insight into the nature of Bloch functions in two limiting cases, those of nearly free electrons and of tightly bound electrons.

3.1. *Nearly free electron approximation*

As one can expect from its name, this approximation is based on the assumption that electrons in a crystal behave very much like free electrons in a box of the size of the crystal. This assumption is supported by a number of observations, e.g., the high conductivity of metals, which corroborates the high mobility of electrons. The free electron model of metals, in which the effect of the crystal potential is totally neglected, was worked out by SOMMERFELD [1928]. This model serves as a zero-order solution in the nearly free electron approximation, where the Bloch functions are expressed as linear combinations of plane waves.

3.1.1. *One dimensional case.* Let the periodic potential be expressed as a Fourier series,

$$V = \sum_{r=-\infty}^{\infty} v_r \, \mathrm{e}^{2\pi \mathrm{i} rx/d} .$$

The form of this expression guarantees the periodicity, and otherwise it is arbitrary, except that since V is real, $v_{-r} = v_r^*$. Then let the periodic function u_k be expanded in a similar series

$$u_k = \sum_{s=-\infty}^{\infty} a_s^k \, \mathrm{e}^{2\pi \mathrm{i} sx/d} .$$

Substitution of these two series in equation (18.6) leads to

$$\sum_s a_s^k \left\{ - \frac{4\pi^2 s^2}{\mathrm{d}^2} - \frac{4\pi sk}{d} + \frac{2m}{\hbar^2} E_k - k^2 - \frac{2m}{\hbar^2} \sum_r v_r \mathrm{e}^{2\pi \mathrm{i} rx/d} \right\} \mathrm{e}^{2\pi \mathrm{i} sx/d} = 0 .$$

For such an expression to be zero for all values of x, it is necessary that the coefficient of each exponential be zero. Hence

$$a_s^k \left[E_k - \frac{\hbar^2}{2m} \left(k + \frac{2\pi s}{d} \right)^2 \right] - \sum_r a_r^k v_{s-r} = 0 . \tag{18.20}$$

A solution of the set of simultaneous equations (18.20) would lead to an exact solution of equation (18.6), but to solve (18.20) approximations must be made.

To obtain an approximate solution, it is first assumed that the coefficients v_r for $r = 0$ are small enough to be neglected in the evaluation of the determinant for the energy values. The determinant will then vanish if one of the diagonal elements vanishes. This leads to

$$E_{s,k} = v_0 + \frac{\hbar^2}{2m}\left(k + \frac{2\pi s}{d}\right)^2 . \tag{18.21}$$

With each value of the energy, one of the coefficients $a_s^k = 1$, and the others are zero, except in special cases.

In the special case in which $k = n\pi/d$, the condition (18.21) will make at least two of the diagonal terms equal to zero. The diagonal terms for $s = 0$ and $s = -n$ will be identical. Similarly when $k = -n\pi/d$, the terms for $s = 0$ and $s = n$ will be the same. In these and similar cases, even a small nondiagonal term cannot be neglected compared with the diagonal terms which are actually zero. The two nondiagonal terms at the intersections of the rows and columns containing the diagonal terms in question must be taken into account. Hence the energy values will be determined by the pair of equations

$$\left[\frac{2m}{\hbar^2}(E - v_0) - \left(\frac{n\pi}{d} - \frac{2\pi n}{d}\right)^2\right] a_{-n}^{(n\pi/d)} - \frac{2m}{\hbar^2} v_{-n} a_0^{(n\pi/d)} = 0 ;$$

$$-\frac{2m}{\hbar^2} v_n a_{-n}^{(n\pi/d)} + \left[\frac{2m}{\hbar^2}(E - v_0) - \left(\frac{n\pi}{d}\right)^2\right] a_0^{(n\pi/d)} = 0 .$$

Since the potential energy is real, $v_{-n} = v_n^*$, and if the potential is symmetrical about the origin, these coefficients will be real. The two values of the energy are then

$$E = v_0 + \frac{\hbar^2}{2m}\left(\frac{n\pi}{d}\right)^2 \pm v_n ,$$

and the corresponding functions are

$$\begin{aligned} u_{\pi n/d}^{(a)} &= 1 + e^{-2\pi i n x/d}, \qquad U^{(a)} = e^{\pi i n x/d} + e^{-\pi i n x/d} \\ u_{\pi n/d}^{(b)} &= 1 - e^{-2\pi i n x/d}, \qquad U^{(b)} = e^{\pi i n x/d} - e^{-\pi i n x/d} . \end{aligned} \tag{18.22}$$

The same two values of the energy and the same two wave functions also appear for $k = -n\pi/d$.

It is not to be expected that this departure of the energy value from that for free electrons takes place only at this one value of k. Other values of k in the neighborhood also give similar results. Let $k = (n\pi/d)+k_1$. Then

$$\left[(E-v_0) - \frac{\hbar^2}{2m}\left(-\frac{n\pi}{d} + k_1\right)^2\right] a_{-n}^{(k)} - v_{-n} a_0^{(k)} = 0\,;$$

$$-v_n a_{-n}^{(k)} + \left[(E-v_0) - \frac{\hbar^2}{2m}\left(\frac{n\pi}{d} + k_1\right)^2\right] a_0^{(k)} = 0\,;$$

$$E = v_0 + \frac{\hbar^2}{2m}\left(\frac{n^2\pi^2}{d^2} + k_1^2\right) \pm \left[v_n^2 + \left(\frac{\hbar^2}{2m}\right)^2 \frac{4\pi^2 n^2 k_1^2}{d^2}\right]^{\frac{1}{2}}.$$

This shows that the energy is a quadratic function of the distance from $(n\pi/d)$ and that the gap in the energy at this point is determined by the corresponding term in the potential energy. Evaluation of the coefficients shows that the wave functions experience a gradual transition from the free-electron function at some distance from $(n\pi/d)$ to the form given in (18.22) at the critical value.

Such an approximation gives to each value of k a value of the energy close to the value for free electrons, but with a discontinuity at the boundary of each Brillouin zone.

3.1.2. *Three-dimensional case.* The three-dimensional case can be outlined very briefly since it follows closely the pattern of the one-dimensional case. The potential energy is expanded in a Fourier series by means of the reciprocal lattice as

$$V = \sum_{n_1, n_2, n_3} v_{n_1 n_2 n_3} \exp\{2\pi \mathrm{i}(n_1\boldsymbol{b}_1 + n_2\boldsymbol{b}_2 + n_3\boldsymbol{b}_3)\cdot\boldsymbol{r}\},$$

and the periodic part of the wave function likewise:

$$u_{\boldsymbol{k}} = \sum_{s_1, s_2, s_3} a_{s_1, s_2, s_3}^{(\boldsymbol{k})} \exp\{2\pi \mathrm{i}(s_1\boldsymbol{b}_1 + s_2\boldsymbol{b}_2 + s_3\boldsymbol{b}_3)\cdot\boldsymbol{r}\}.$$

Let the vectors of the reciprocal lattice be expressed in terms of their cartesian components

$$s_1\boldsymbol{b}_1 + s_2\boldsymbol{b}_2 + s_3\boldsymbol{b}_3 = \boldsymbol{B}_{s_1, s_2, s_3} = B_{s_1, s_2, s_3;1}\boldsymbol{i} + B_{s_1+s_2, s_3;2}\boldsymbol{j} + B_{s_1, s_2, s_3;3}\boldsymbol{k}.$$

The differential equation then leads to

$$\left[E_{\boldsymbol{k}} - \frac{\hbar^2}{2m}\left(\boldsymbol{k} + 2\pi\boldsymbol{B}_{s_1, s_2, s_3}\right)^2\right] a_{s_1, s_2, s_3}^{(\boldsymbol{k})} - \sum_{n_1, n_2, n_3} v_{n_1, n_2, n_3} a_{(s_1-n_1), (s_2-n_2), (s_3-n_3)}^{(\boldsymbol{k})} = 0.$$

The solution of these equations can first be approximated, as before, by neglecting the terms off the diagonal. The energies in this case are

$$E_{s,k} = v_{0,0,0} + \frac{\hbar^2}{2m}(\boldsymbol{k}+2\pi \boldsymbol{B}_{s_1,s_2,s_3})^2 .$$

Along the boundaries of the Brillouin zones, several of these diagonal terms will vanish at the same time and must be treated together. To illustrate the procedure, consider the body-centered lattice, and let

$$\boldsymbol{k}' = \left(\frac{\pi}{d} - \kappa\right)(\boldsymbol{i}-\boldsymbol{k})+\boldsymbol{k}_\alpha ,$$

where $\boldsymbol{k}_\alpha$ is an arbitrary vector perpendicular to the vector $(\pi/d)(\boldsymbol{i}-\boldsymbol{k})$ of the reciprocal lattice. The vector $\boldsymbol{k}'$ is a distance κ from the surface of the first Brillouin zone. The vector

$$\boldsymbol{k}'' = \left(-\frac{\pi}{d} - \kappa\right)(\boldsymbol{i}-\boldsymbol{k})+\boldsymbol{k}_\alpha$$

ends near the opposite side of the zone. Since $B_{000} = 0$, the two equations to be considered are

$$\left\{E_k - v_{0,0,0} - \frac{\hbar^2}{2m}\left[2\left(\frac{\pi^2}{d^2} - \frac{2\pi}{d}\kappa+\kappa^2\right) + k_\alpha^2\right]\right\} a_{0,0,0} - v_{1,0,0}\, a_{-1,0,0} = 0;$$

$$-v_{-1,0,0}\, a_{0,0,0} + \left\{E_k - v_{0,0,0} - \frac{\hbar^2}{2m}\left[2\left(\frac{\pi^2}{d^2} + \frac{2\pi}{d}\kappa+\kappa^2\right) + k_\alpha^2\right]\right\} a_{-1,0,0} = 0 .$$

The energy is

$$E_k = v_{0,0,0} + \frac{\hbar^2}{2m}\left(k_1^2 + \frac{2\pi^2}{d^2} + 2\kappa^2\right) \pm \left[v_{1,0,0} + \left(\frac{\hbar^2}{2m}\right)^2 \frac{16\pi^2\kappa^2}{d^2}\right]^{\frac{1}{2}},$$

and the wave functions can also be written down.

In case the vector $\boldsymbol{k}$ is near the intersection of two or more planes defining the Brillouin zones, it may be necessary to consider three or more simultaneous equations to determine the energy values and the characteristic functions.

3.2. *Tight binding approximation*

The tight binding approximation is in many respects complementary to the nearly free electron approximation. In this case the feature of the Bloch functions one makes use of is that in the close neighbourhood of a given atom they are like the tightly bound wave functions appropriate to electrons in a free atom. This approximation is suitable for the description of a crystal in which the atoms are very far apart, so that their interactions can be treated as a small perturbation. It is also applicable to the low-lying levels in a normal crystal, because the wave functions for these levels are almost entirely confined to the regions near the nuclei. Considering the well localized atomic wave functions cannot be approximated by a linear combination of a few plane waves, these are the cases when the nearly free electron approximation fails.

3.2.1. *One-dimensional case.* To illustrate the tight binding approximation, consider a one-dimensional case, in which the potential energy can be represented by

$$V = \sum_s V_s(x) = \sum_s V_0(x - sd).$$

It is assumed that $V_0(x)$ has a minimum for $x = 0$, it vanishes as $x \to \pm\infty$, and is symmetrical about $x = 0$. For a given form of potential, V_0, the approximation will be better the larger the value of d. Let $U^0_{n,s}(x) = U^0_n(x - sd)$ be an energy eigenfunction for an electron in the single potential minimum $V_0(x - sd)$. Let the associated unperturbed energy be E^0_n.

In the zero-order approximation, the desired solution for the periodic potential is a linear combination of the unperturbed solutions. Hence let

$$U_{n,k} = \sum_s a^k_s U^0_{n,s}. \qquad (18.23)$$

The significance of the index k will appear later. For the present it may be considered as indicating merely a method of designating the various members of the set of zero-order solutions. Substituting (18.23) into the eigenvalue equation, and making use of the properties of the unperturbed functions, leads to

$$\sum_s a^k_s \left(E^0_n - E_{n,k} + \sum_{s' \neq s} V_{s'} \right) U^0_{n,s} = 0.$$

Multiplying this equation by $U^0_{n,r}$ and integrating over x leads to

$$a_{r-1}[(E^0_n - E)\alpha + B] + a_r[E^0_n - E + A] + a_{r+1}[(E^0_n - E)\alpha + B] = 0, \qquad (18.24)$$

where

$$\alpha = \int U_{n,s+1} U^0_{n,s}\,\mathrm{d}x;\; A = 2\int U^0_{n,s} V_{s-1} U^0_{n,s}\,\mathrm{d}x;\; B = \int U^0_{n,s+1} V_{s+1} U^0_{n,s}\,\mathrm{d}x.$$

The functions $U_{n,s}$ are real so α, A, B are real. The integrals over products of wave functions separated by more than the distance d are neglected. In this neglect lies part of the approximation of the method. This neglect is more justified in one dimension than in three, since in three dimensions the number of neighbors increases rapidly as the distance increases.

Equation (18.24) is a standard type of difference equation, and its general solution can be given. It can first be put in simpler form by dividing through by $(E_n^0 - E_{n,k})\alpha + B$ and letting

$$\frac{[E_n^0 - E_{n,k} + A]}{(E_n^0 - E_{n,k})\alpha + B} = -2 \cos kd. \tag{18.25}$$

This definition gives real values to the index k if the ratio lies between 2 and -2. Then the equation is

$$a_{r-1} - 2 \cos (kd)\, a_r + a_{r+1} = 0,$$

whose general solution is

$$a_r = A_1\, e^{ikrd} + A_2\, e^{-ikrd}. \tag{18.26}$$

As in the case of a differential equation of the second order, the general solution represents a linear combination of two fundamental solutions. The standard form can then be taken as

$$U_{n,k} = \sum_s e^{iksd}\, U_{n,s}^0. \tag{18.27}$$

For this solution, it follows from the definition of k in equation (18.25) that

$$E_{n,k} = E_n^0 + \frac{A + 2B \cos kd}{1 + 2\alpha \cos kd}. \tag{18.28}$$

If equation (18.25) does not lead to a real value of k, the solution (18.26) will contain real exponentials. A form such as (18.27) will then increase without limit, and the solution will not be admissible.

It is clear from the form of equation (18.28) that the energy eigenvalues lie in a band around the unperturbed E_n^0. Since $A < 0$, the center of the band is below E_n^0. In many cases α is small enough to be neglected in the denominator. It is possible to see in these results an illustration of the properties listed in Section 1.1.

PROBLEM 5. Verify the general properties of Bloch functions and calculate the effective mass and the density of states appropriate to the energy spectrum (18.28).

3.2.2. *Three-dimensional case*. The details of the tight binding approximation depend, in the three-dimensional case, on the type of crystal lattice treated and on the angular momentum assumed in the unperturbed wave functions. These questions do not arise in the one-dimensional case, for it can have only one lattice form and the unperturbed functions are either symmetrical or antisymmetrical about the center.

For illustration let us take the case of a body-centered cubic lattice and the case in which the unperturbed functions are *s* functions.

The lattice points can be specified by three indices, λ, μ, ν, which will be all integers or all integers plus $\frac{1}{2}$. These indices give the cartesian coordinates of the lattice points in units of the edge of the cube, d. Then let

$$V = \sum V_{\lambda,\mu,\nu} = \sum V_0(\boldsymbol{r} - [\lambda\hat{\boldsymbol{\imath}} + \mu\hat{\boldsymbol{\jmath}} + \nu\hat{\boldsymbol{k}}]d) \tag{18.29}$$

where $V_{\lambda,\mu,\nu}$ has a spherical symmetry about the lattice point (λ, μ, ν), has a minimum at the lattice point, and approaches zero at large distances. Then let $U^0_{n,\lambda,\mu,\nu}$ be the spherically symmetrical energy function for an s-state with the potential $V_{\lambda,\mu,\nu}$. It will be real. The index n signifies the quantum number necessary to specify the unperturbed function. It is also assumed that the energy of the unperturbed state, E^0_n, is far enough from the energies of other states that they can be neglected, and the zero-order approximation will be of some interest. Let

$$U_{n,\boldsymbol{k}} = \sum_{\lambda,\mu,\nu} a^{n,\boldsymbol{k}}_{\lambda,\mu,\nu}\, U^0_{n,\lambda,\mu,\nu}\,.$$

Substitution of this form into equation (18.6) combined with equation (18.29) leads to

$$a^{n,\boldsymbol{k}}_{\lambda,\mu,\nu}\,[E^0_n - E_{n,\boldsymbol{k}} + \textstyle\sum' V_{\lambda',\mu',\nu'}]\, U^0_{n,\lambda,\mu,\nu} = 0\,.$$

The sum over $V_{\lambda',\mu',\nu'}$ is over all values except $\lambda' = \lambda$, $\mu' = \mu$, $\nu' = \nu$. The approximation is then introduced by the specification that only integrals of products of functions adjacent along the body diagonal of the cube will be considered. These are the nearest neighbors. With this specification, multiplication by $U^0_{n,r,s,t}$ and integration leads to

$$\sum_{\varepsilon_1\,\varepsilon_2\,\varepsilon_3} a^{n,k}_{(r+\varepsilon_1),\,(s+\varepsilon_2),\,(t+\varepsilon_3)}[E^0_n - E_{n,\boldsymbol{k}})\alpha + B] + a^{n,\boldsymbol{k}}_{r,s,t}[E^0_n - E_{n,\boldsymbol{k}} + A] = 0\,.$$

The sum is over the possible values of the ε's. Each can take on the values $\pm\frac{1}{2}$;

$$\alpha = \int U^0_{(r+\frac{1}{2}),\,(s+\frac{1}{2}),\,(t+\frac{1}{2})}\, U^0_{r,s,t}\, \mathrm{d}V\,;$$

$$B = \int U^0_{(r+\frac{1}{2}),\,(s+\frac{1}{2}),\,(t+\frac{1}{2})} V_{(r+\frac{1}{2}),\,(s+\frac{1}{2}),\,(t+\frac{1}{2})} U^0_{r,\,s,\,t}\,\mathrm{d}V;$$

$$A = \sum_{\varepsilon} \int U^0_{r,\,s,\,t} V_{(r+\varepsilon_1),\,(s+\varepsilon_2),\,(t+\varepsilon_3)} U^0_{r,\,s,\,t}\,\mathrm{d}V.$$

By division the equation can be reduced to

$$\sum_{\varepsilon_1,\,\varepsilon_2,\,\varepsilon_3} a^{n,\,k}_{(\lambda+\varepsilon_1),\,(\mu+\varepsilon_2),\,(\nu+\varepsilon_3)} + \frac{E_n - E_{n,\,k} + A}{[(E^0_n - E_{n,\,k})\alpha + B]}\, a_{\lambda,\,\mu,\,\nu} = 0.$$

The solution of this equation is similar to that for one dimension. For arbitrary values of $E_{n,\boldsymbol{k}}$ there is no solution such that the square of the absolute value of the wave function is periodic with the period of the lattice. However, if

$$\frac{E^0_n - E_{n,\,k} + A}{[(E^0_n - E_{n,\,k})\alpha + B]} = -8\cos\frac{k_1 d}{2}\cos\frac{k_2 d}{2}\cos\frac{k_3 d}{2}, \tag{18.30}$$

the fundamental solutions of the wave equation are

$$U_{n,\,k} = \sum_{\lambda,\,\mu,\,\nu} \exp\{\mathrm{i}(k_1\lambda + k_2\mu + k_3\nu)d\}\, U_{n,\,\lambda,\,\mu,\,\nu}. \tag{18.31}$$

Equation (18.30) gives

$$E_{n,\,k} = E^0_n + \frac{A + 8B\cos(\frac{1}{2}k_1 d)\cos(\frac{1}{2}k_2 d)\cos(\frac{1}{2}k_3 d)}{1 + 8\alpha\cos(\frac{1}{2}k_1 d)\cos(\frac{1}{2}k_2 d)\cos(\frac{1}{2}k_3 d)}.$$

If equation (18.30) is not satisfied, its left-hand side can always be set equal to a product of hyperbolic cosines and the coefficients in (18.31) are then real exponentials. This clearly leads to a solution of the equation that is not suitable as an energy eigenfunction.

Three properties of this energy as a function of k are worth noting.

1. For very small values of $\boldsymbol{k}$, the cosines may be expanded to give

$$E_{n,\,k} = E^0_n + \frac{A + 8B}{1 + 8\alpha} + 4\frac{A\alpha - B}{(1 + 8\alpha)^2}\, d^2 k^2.$$

The integrals A and B are negative if the energy V_0 is taken to be zero at infinity. The surfaces of constant energy in the $\boldsymbol{k}$ space are spheres near $\boldsymbol{k} = 0$.

2. The energy as a function of $\boldsymbol{k}$ has cubic symmetry everywhere. In particular it is invariant to a change of sign of any components of $\boldsymbol{k}$, to an interchange of any two components of $\boldsymbol{k}$, or to a cyclic permutation of all three.

For a general $\boldsymbol{k}$ there are 48 different vectors leading to the same energy value.

3. The approximate solutions and energies are invariant to a change of $\boldsymbol{k}$ by any vector of the reciprocal lattice. To show this, it is necessary to evaluate the reciprocal vectors for a body-centered cubic lattice. Let the lattice vectors be

$$\boldsymbol{a}_1 = d\hat{\imath}, \quad \boldsymbol{a}_2 = d\hat{\jmath}, \quad \boldsymbol{a}_3 = \tfrac{1}{2}d(\hat{\imath}+\hat{\jmath}+\hat{k}),$$

$\boldsymbol{a}_1$ and $\boldsymbol{a}_2$ lie along the edges of the cube and along the coordinate axes. The third vector is half of the body diagonal of the cube. Any point in the lattice can be located by a vector composed of integral multiples of these three vectors $\boldsymbol{a}_1, \boldsymbol{a}_2, \boldsymbol{a}_3$.

The reciprocal lattice can be constructed by means of the relationships (18.13),

$$\boldsymbol{b}_1 = \frac{1}{d}(\hat{\imath}-\hat{k}), \quad \boldsymbol{b}_2 = \frac{1}{d}(\hat{\jmath}-\hat{k}), \quad \boldsymbol{b}_3 = \frac{2}{d}\hat{k}.$$

Let $\boldsymbol{k}$ be transformed by the addition of an arbitrary vector of the reciprocal lattice so that

$$\begin{aligned}\boldsymbol{k}' &= \boldsymbol{k}+2\pi(n_1\boldsymbol{b}_1+n_2\boldsymbol{b}_2+n_3\boldsymbol{b}_3)\\ &= \left(k_1 + \frac{2\pi}{d}n_1\right)\hat{\imath}+ \left(k_2 + \frac{2\pi}{d}n_2\right)\hat{\jmath}+ \left[k_3 + \frac{2\pi}{d}(-n_1-n_2+2n_3)\right]\hat{k}\end{aligned}$$

where n_1, n_2 and n_3 are arbitrary integers. Then from equation (18.31)

$$\begin{aligned}a^{n,\,\boldsymbol{k}'}_{\lambda,\,\mu,\,\nu} &= \exp\left[\mathrm{i}\left\{\left(k_1 + \frac{2\pi}{d}n_1\right)\lambda + \left(k_2 + \frac{2\pi}{d}n_2\right)\mu + \right.\right.\\ &\qquad \left.\left. + \left[k_3 + \frac{2\pi}{d}(2n_3-n_1-n_2)\nu\right]\right\}d\right]\\ &= a^{n,\,\boldsymbol{k}}_{\lambda,\,\mu,\,\nu}.\end{aligned}$$

The last equality follows because λ, μ, ν are either all integers or all integers plus $\frac{1}{2}$. Similarly the energy can be shown to be invariant to such transformations.

4. The Fermi energy; conductors, insulators and semiconductors

So far we have considered the states of a single electron moving in the periodic potential due to its interaction with the ions of the lattice and the other electrons. In practice, however, we are unable to observe individual electrons; most of what we know about solids comes from macroscopic measurements on crystals containing $\sim 10^{23}$ electrons. Clearly, we cannot hope to deduce from results of such measurements sufficient information to draw conclusions regarding the state of all the electrons. Therefore, we have to apply statistics, if we want to compare the theory developed in the preceding sections with experiment.

The number of electrons occupying a given Bloch state $U_{n,\mathbf{k}}$ is given by the Fermi-Dirac distribution function derived in equation (17.44):

$$n_{n,\mathbf{k}}(T) = \{\exp [(E_{n,\mathbf{k}} - \zeta)/kT] + 1\}^{-1}, \tag{18.32}$$

where ζ, the Fermi energy, is determined by the requirement that the total number of electrons be constant:

$$n \equiv 2 \sum_{n,\mathbf{k}} n_{n,\mathbf{k}}(T) = \text{constant}$$

(the factor 2 takes account of the spin degeneracy). At $T = 0$ the total number of electrons is easily calculated, because the distribution function (18.32) is a simple step-function:

$$n_{n,\mathbf{k}}(0) = \begin{cases} 1 & \text{if } E_{n,\mathbf{k}} < \zeta \\ 0 & \text{if } E_{n,\mathbf{k}} > \zeta . \end{cases} \tag{18.33}$$

This property of the Fermi-Dirac distribution gives a very simple meaning to the Fermi energy ζ: at $T = 0$ all states with energy lower than ζ are occupied, and all states having energies higher than ζ are unoccupied. This is in accordance with what was said in Chapter 10 about the states of a system of n independent Fermions: the ground state is described by the antisymmetrized product (Slater determinant, see equation (10.6)) of the n lowest single-electron energy eigenstates.

Let us consider the physical consequences of a distribution like (18.33) in view of the nature of the energy spectrum of Bloch electrons. We have seen in Section 2 that the spectrum consists of bands, each containing $2N$ states, where N is the number of unit cells in the crystal. Note that the number of electrons n is not necessarily equal to N; the ratio n/N gives the number of electrons per unit cell. If the bands overlap, the Fermi energy must fall within at least one band, no matter what the number of valence electrons per unit cell is. If, however, the bands are separated by band gaps, and there is an even number of valence electrons per unit cell, the lowest $2N$, $4N$, etc. states will fall into the lowest 1, 2, etc. bands, respectively, and, consequently, the Fermi energy will fall within an energy gap. Depending on which of these two cases is realized, the crystal will turn out to be a metallic conductor, or a dielectric insulator, respectively.

The decisive role of the Fermi energy in determining the electric properties of the solid can be seen by considering the possible low-energy excitations.

If the Fermi energy ζ falls within the band, the unoccupied states right above ζ are available for electrons occupying states right below ζ. A state of the crystal which differs from the ground state by an electronic state of energy $\zeta-\varepsilon$ being unoccupied instead of occupied, while another one of energy $\zeta+\varepsilon$ being occupied instead of unoccupied, has an energy $E_g-(\zeta-\varepsilon)+(\zeta+\varepsilon) = E_g+2\varepsilon$, where E_g is the energy of the ground state. Since ε can be arbitrarily small (if the crystal is sufficiently large, cf. Section 2), we conclude that such a crystal has a continuum of excited states beginning at zero excitation energy. This leads to the two typical *metallic* properties of good conductivity and high optical reflectivity: an external perturbation like an electric field arising either from the voltage applied at the two ends of the crystal or through light falling on the crystal, gives rise to excitations, which provide a response (electric current or reflected light) to the perturbation.

On the other hand, if the Fermi energy ζ falls in a forbidden band of width ΔE, the smallest excitation energy at the cost of which an occupied state below ζ can be emptied into an unoccupied one above ζ is ΔE. Therefore, the crystal is unable to respond to perturbations that cannot provide this minimum excitation energy, a static voltage does not lead to a current, and there is no interaction with photons having energy $h\nu < \Delta E$: the crystal is an *insulator*, transparent for light of frequency $\nu < \Delta E/h$.

If $T > 0$, the above argument is not rigorously valid: the sharp division between occupied and unoccupied states is gone, and even in the absence of any external perturbation there are unoccupied states below the energy gap and occupied ones above it. It can be seen from the form of the Fermi-Dirac distribution function (18.32) that the probability of such thermal excitations is $\exp(-\Delta E/kT)$. To the extent that there are occupied states in the continuum of unoccupied states above the energy gap, the material is then a conductor. For the typical insulators the value of ΔE is several electron volts, $\Delta E/kT \approx 10^4$, and therefore at normal temperatures this thermally activated conductivity is negligible. However, there is another class of materials, for which $\Delta E/kT \approx 1$ even below room temperature, and which therefore have a weak conductivity. These are the *semiconductors*, characterized by the temperature dependence of their conductivity, which is proportional to $\exp(-\Delta E/kT)$.

PROBLEM 6. Show that if $T \neq 0$ the sharp discontinuity of the Fermi-Dirac distribution function at the Fermi energy disappears, and that the energy range over which the transition from $n_{n,\boldsymbol{k}}(T) = 1$ to $n_{n,\boldsymbol{k}}(T) = 0$ takes place is $\sim kT$.

MATHEMATICAL APPENDIX

A.1. The Hermite polynomials

The differential equation satisfied by the Hermite polynomials is

$$\frac{d^2 H_n}{dx^2} - 2x \frac{dH_n}{dx} + 2n\, H_n = 0. \tag{A.1}$$

The coefficients of this equation have no singularities for finite values of x, so the solution can be expanded in a power series about any point and will be valid everywhere. Consider the expansion about the origin and let

$$H_n(x) = \sum_{s=0}^{\infty} b_s x^s. \tag{A.2}$$

Substituting in the differential equation and equating to zero the coefficient of each power gives

$$(s+2)(s+1)b_{s+2} = (2s-2n)b_s. \tag{A.3}$$

From the recursion formula it can be seen that two independent series can be obtained by starting with $s = 0$ and with $s = 1$, and that each series will contain only even or only odd powers of x. Furthermore $b_{n+2} = 0$ so that the solution will be a polynomial in case the terms used have the same parity as n. Only this case is of interest for the problem of the harmonic oscillator.

A few of these polynomials are as follows:

$$H_0(x) = 1 \qquad H_1(x) = 2x$$
$$H_2(x) = 4x^2-2 \qquad H_3(x) = 8x^3-12x$$
$$H_4(x) = 16x^4-48x^2+12 \qquad H_5(x) = 32x^5-160x^3+120x$$
$$H_6(x) = 64x^6-480x^4+720x^2-120$$
$$H_7(x) = 128x^7-1344x^5+3360x^3-1680x.$$

A useful form for the polynomials is

$$H_n(x) = (-1)^n \, e^{x^2} \frac{d^n(e^{-x^2})}{dx^n}. \tag{A.4}$$

The polynomials written out above come directly from this form, which accounts for the rather large numerical coefficients. The form (A.4) can be shown by substitution to satisfy the differential equation (A.1).

A useful recursion formula which can be derived from equation (A.4) is

$$H_{n+1}(x) - 2xH_n(x) + 2nH_{n-1}(x) = 0. \tag{A.5}$$

Also

$$\frac{dH_n}{dx} = 2xH_n - H_{n+1} = 2nH_{n-1}. \tag{A.6}$$

The Hermite polynomials are not orthogonal as they stand. As do all polynomials they become infinite for infinite values of the argument. However they do satisfy the relationship

$$\int_{-\infty}^{\infty} e^{-x^2} H_n(x) H_m(x) \, dx = 2^n \, n! \, \pi^{\frac{1}{2}} \delta(n, m). \tag{A.7}$$

A.2. The associated Legendre polynomials

For real values of the argument and for integral values of l and m ($l \geq m$), the associated Legendre polynomials may be defined by

$$P_l^m(x) = \frac{(-1)^m}{2^l l!} (1-x^2)^{\frac{1}{2}m} \frac{d^{l+m}}{dx^{l+m}} (x^2 - 1)^l. \tag{A.8}$$

These functions satisfy the differential equation

$$(1-x^2) \frac{d^2 P_l^m}{dx^2} - 2x \frac{dP_l^m}{dx} + \left[l(l+1) - \frac{m^2}{(1-x^2)} \right] P_l^m = 0 \tag{A.9}$$

and are defined for both $\pm m$. They also satisfy the integral relationship

$$\int_{-1}^{1} P_l^m(x) \, P_{l'}^m(x) \, dx = \frac{2(l+m)!}{(2l+1)(l-m)!} \delta(l, l'). \tag{A.10}$$

A few of these functions are as follows:

$P_0^0 = 1$	$P_2^2 = 3\,(1-x^2)$
$P_1^1 = -(1-x^2)^{\frac{1}{2}}$	$P_2^1 = -3(1-x^2)^{\frac{1}{2}}\,x$
$P_1^0 = x$	$P_2^0 = \frac{1}{2}\,(3x^2-1)$
$P_1^{-1} = \frac{1}{2}\,(1-x^2)^{\frac{1}{2}}$	$P_2^{-1} = \frac{1}{2}\,(1-x^2)^{\frac{1}{2}}\,x$
	$P_2^{-2} = \frac{1}{8}\,(1-x^2)$
$P_3^3 = -15\,(1-x^2)^{\frac{3}{2}}$	$P_4^4 = 105\,(1-x^2)^2$
	$P_4^3 = -105\,(1-x^2)^{\frac{3}{2}}\,x$
$P_3^2 = 15\,(1-x^2)\,x$	$P_4^2 = \frac{15}{2}\,(1-x^2)\,(7x^2-1)$
$P_3^1 = -\frac{3}{2}\,(1-x^2)^{\frac{1}{2}}\,(5x^2-1)$	$P_4^1 = -\frac{5}{2}\,(1-x^2)^{\frac{1}{2}}\,(7x^3-3x)$
$P_3^0 = \frac{1}{2}\,(5x^2-3)\,x$	$P_4^0 = \frac{1}{8}\,(35x^4-30x^2+3)$
$P_3^{-1} = \frac{1}{8}\,(1-x^2)^{\frac{1}{2}}\,(5x^2-1)$	$P_4^{-1} = \frac{1}{8}\,(1-x^2)^{\frac{1}{2}}\,(7x^2-3)\,x$
$P_3^{-2} = \frac{1}{8}\,(1-x^2)\,x$	$P_4^{-2} = \frac{1}{48}\,(1-x^2)\,(7x^2-1)$
$P_3^{-3} = \frac{1}{48}\,(1-x^2)^{\frac{3}{2}}$	$P_4^{-3} = \frac{1}{48}\,(1-x^2)^{\frac{3}{2}}\,x$
	$P_4^{-4} = \frac{1}{384}\,(1-x^2)^2.$

The numerical coefficients of the above examples, and the signs, follow from the form (A.8); they are not required by the differential equation.

The integral relationship (A.10) provides a means of normalizing these functions between 0 and 1 which is useful in series of orthogonal functions. One may define

$$\Pi_l^m(x) = \left[\frac{(2l+1)\,(l-m)!}{2(l+m)!}\right]^{\frac{1}{2}} P_l^m. \tag{A.11}$$

Then

$$\int_{-1}^{1} \Pi_l^m\, \Pi_{l'}^m\, \mathrm{d}x = \delta(l, l').$$

A.3. Surface spherical harmonics

The associated Legendre polynomials can be used to form a complete set of surface spherical harmonics. These normalized and orthogonal functions may be defined as

$$Y(\vartheta, \varphi) = \Pi_l^m\,(\cos\vartheta)\,\mathrm{e}^{im\varphi}/(2\pi)^{\frac{1}{2}}. \tag{A.12}$$

Using the above definitions these functions may be formed with both positive and negative values of m in a satisfyingly symmetric way. They satisfy the differential equation, which can be obtained from equation (A.9),

$$\frac{1}{\sin\vartheta}\frac{\partial}{\partial\vartheta}\left(\sin\vartheta\frac{\partial Y_l^m}{\partial\vartheta}\right)+\frac{1}{\sin^2\vartheta}\frac{\partial^2 Y_l^m}{\partial\varphi^2}=l(l+1)Y_l^m. \tag{A.13}$$

A few of these normalized functions are as follows:

$\Pi_0^0 = (\tfrac{1}{2})^{\frac{1}{2}}$

$\Pi_1^1 = -(\tfrac{3}{4})^{\frac{1}{2}}\sin\vartheta$

$\Pi_1^0 = (\tfrac{3}{2})^{\frac{1}{2}}\cos\vartheta$

$\Pi_1^{-1} = (\tfrac{3}{4})^{\frac{1}{2}}\sin\vartheta$

$\Pi_3^{-3} = -(\tfrac{35}{32})^{\frac{1}{2}}\sin^3\vartheta$

$\Pi_3^2 = (\tfrac{105}{16})^{\frac{1}{2}}\sin^2\vartheta\cos\vartheta$

$\Pi_3^1 = -(\tfrac{21}{32})^{\frac{1}{2}}\sin\vartheta(5\cos^2\vartheta-1)$

$\Pi_3^0 = (\tfrac{7}{8})^{\frac{1}{2}}(5\cos^3\vartheta-3\cos\vartheta)$

$\Pi_3^{-1} = (\tfrac{21}{32})^{\frac{1}{2}}\sin\vartheta(5\cos^2\vartheta-1)$

$\Pi_3^{-2} = (\tfrac{105}{16})^{\frac{1}{2}}\sin^2\vartheta\cos\vartheta$

$\Pi_3^{-3} = (\tfrac{35}{32})^{\frac{1}{2}}\sin^3\vartheta$

$\Pi_2^2 = (\tfrac{15}{16})^{\frac{1}{2}}\sin^2\vartheta$

$\Pi_2^1 = -(\tfrac{15}{4})^{\frac{1}{2}}\sin\vartheta\cos\vartheta$

$\Pi_2^0 = (\tfrac{5}{8})^{\frac{1}{2}}(3\cos^2\vartheta-1)$

$\Pi_2^{-1} = (\tfrac{15}{4})^{\frac{1}{2}}\sin\vartheta\cos\vartheta$

$\Pi_2^{-2} = (\tfrac{15}{16})^{\frac{1}{2}}\sin^2\vartheta$

$\Pi_4^4 = (\tfrac{315}{256})^{\frac{1}{2}}\sin^4\vartheta$

$\Pi_4^3 = -(\tfrac{315}{32})^{\frac{1}{2}}\sin^3\vartheta\cos\vartheta$

$\Pi_4^2 = (\tfrac{45}{64})^{\frac{1}{2}}\sin^2\vartheta\cos\vartheta(7\cos^2\vartheta-1)$

$\Pi_4^1 = -(\tfrac{45}{32})^{\frac{1}{2}}\sin\vartheta(7\cos^3\vartheta-3\cos\vartheta)$

$\Pi_4^0 = (\tfrac{9}{128})^{\frac{1}{2}}(35\cos^4\vartheta-30\cos^2\vartheta+3)$

$\Pi_4^{-1} = (\tfrac{45}{32})^{\frac{1}{2}}\sin\vartheta(7\cos^3\vartheta-3\cos\vartheta)$

$\Pi_4^{-2} = (\tfrac{45}{64})^{\frac{1}{2}}\sin^2\vartheta(7\cos^2\vartheta-1)$

$\Pi_4^{-3} = (\tfrac{315}{32})^{\frac{1}{2}}\sin^3\vartheta\cos\vartheta$

$\Pi_4^{-4} = (\tfrac{315}{256})^{\frac{1}{2}}\sin^4\vartheta.$

There are numerous relationships among these functions which are often useful. Among them are

$$\cos\vartheta\,\Pi_l^m = \left[\frac{(l+m+1)(l-m+1)}{(2l+1)(2l+3)}\right]^{\frac{1}{2}}\Pi_{l+1}^m + \left[\frac{(l+m)(l-m)}{(2l-1)(2l+1)}\right]^{\frac{1}{2}}\Pi_{l-1}^m \tag{A.14}$$

$$\cos\vartheta\,\Pi_l^m = \left[\frac{(2l+1)(l-m+1)}{(2l+3)(l+m+1)}\right]^{\frac{1}{2}}\Pi_{l+1}^m - \left[\frac{l-m}{l+m+1}\right]^{\frac{1}{2}}\sin\vartheta\,\Pi_l^{m+1} \tag{A.15}$$

$$\sin\vartheta\,\Pi_l^m = -\left[\frac{(l+m+1)(l+m+2)}{(2l+3)(2l+1)}\right]^{\frac{1}{2}}\Pi_{l+1}^{m+1} + \left[\frac{(l-m-1)(l-m)}{(2l-1)(2l+1)}\right]^{\frac{1}{2}}\Pi_{l-1}^{m+1} \tag{A.16}$$

$$\sin\vartheta\,\Pi_l^m = \left[\frac{(l-m+1)(l-m+2)}{(2l+1)(2l+3)}\right]^{\frac{1}{2}} \Pi_{l+1}^{m-1} - \left[\frac{(l+m)(l+m-1)}{(2l+1)(2l-1)}\right]^{\frac{1}{2}} \Pi_{l-1}^{m-1} \tag{A.17}$$

$$\sin\vartheta\,\frac{\mathrm{d}\,\Pi_l^m}{\mathrm{d}\,\vartheta} = l\cos\vartheta\,\Pi_l^m - \left(\frac{2l+1}{2l-1}\right)^{\frac{1}{2}} (l^2-m^2)^{\frac{1}{2}}\,\Pi_{l-1}^m \tag{A.18}$$

$$\sin\vartheta\,\frac{\mathrm{d}\,\Pi_l^m}{\mathrm{d}\,\vartheta} = -(l+1)\cos\vartheta\,\Pi_l^m + \left(\frac{2l+1}{2l+3}\right)^{\frac{1}{2}} [(l+m+1)(l-m+1)]^{\frac{1}{2}}\,\Pi_{l+1}^m\,. \tag{A.19}$$

In addition to equation (A.11) the functions $Y_l^m(\vartheta,\varphi)$ also satisfy the following relationships:

$$\begin{aligned}(l_x+\mathrm{i}l_y)Y_l^m &= \frac{\hbar}{\mathrm{i}}\,\mathrm{e}^{\mathrm{i}\varphi}\left(\mathrm{i}\,\frac{\partial}{\partial\vartheta} - \cot\vartheta\,\frac{\partial}{\partial\varphi}\right) Y_l^m \\ &= \hbar[(l-m)(l+m+1)]^{\frac{1}{2}}\;Y_l^{m+1}\,, \end{aligned} \tag{A.20}$$

$$\begin{aligned}(l_x-\mathrm{i}l_y)Y_l^m &= -\frac{\hbar}{\mathrm{i}}\,\mathrm{e}^{-\mathrm{i}\varphi}\left(\mathrm{i}\,\frac{\partial}{\partial\vartheta} + \cot\vartheta\,\frac{\partial}{\partial\varphi}\right) Y_l^m \\ &= \hbar[(l+m)(l-m+1)]^{\frac{1}{2}}\;Y_l^{m-1}\,. \end{aligned} \tag{A.21}$$

Another important relationship between spherical harmonics is the addition theorem. If $(\vartheta,\,\varphi)$ and $(\vartheta',\,\varphi')$ are the coordinates of two directions and θ is the angle between them

$$\cos\theta = \cos\vartheta\cos\vartheta' + \sin\vartheta\sin\vartheta'\cos(\varphi-\varphi') \tag{A.22}$$

and

$$\{\tfrac{1}{2}(2l+1)\}^{\frac{1}{2}}\,P_l(\cos\theta) = \sum_{m=-l}^{l} \Pi_l^m(\cos\vartheta')\,\Pi_l^m(\cos\vartheta)\,\mathrm{e}^{\mathrm{i}m(\varphi-\varphi')}. \tag{A.23}$$

A.4. The Laguerre polynomials

The Laguerre polynomials may be defined as

$$L_k(x) = \mathrm{e}^x\,\frac{\mathrm{d}^k}{\mathrm{d}x^k}\,(x^k\,\mathrm{e}^{-x})\,. \tag{A.24}$$

This satisfies the differential equation

$$x\,\frac{\mathrm{d}^2L_k}{\mathrm{d}x^2} + (1-x)\,\frac{\mathrm{d}L_k}{\mathrm{d}x} + kL_k = 0\,. \tag{A.25}$$

Further differentiation of this equation leads to

$$x \frac{d^2 L_k^s}{dx^2} + (s+1-x) \frac{dL_k^s}{dx} + (k-s)L_k^s = 0 \tag{A.26}$$

where

$$L_k^s(x) = \frac{d^s}{dx^s} L_k(x) \tag{A.27}$$

and is called the associated Laguerre polynomial. Equation (A.26) has the form of equation (4.17).

A few of these polynomials are as follows:

$L_0 = 1$

$L_1 = 1-x$ $\quad L_1^1 = -1$

$L_2 = 2-4x+x^2$ $\quad L_2^1 = -4+2x$ $\quad L_2^2 = 2$

$L_3 = 6-18x+9x^2-x^3$ $\quad L_3^1 = -18+18x-3x^2$ $\quad L_3^2 = 18-6x$ $\quad L_3^3 = -6$

$L_4 = 24-96x+72x^2-16x^3+x^4$ $\quad L_4^1 = -96+144x-48x^2+4x^3$

$L_4^2 = 144-96x+12x^2$ $\quad L_4^3 = -96+24x$ $\quad L_4^4 = 24.$

The associated Laguerre polynomials satisfy the orthogonality relationship

$$\int_0^\infty L_k^s(x) L_{k'}^s(x)\, e^{-x} x^{s+1}\, dx = \frac{(2k-s+1)\,(k!)^s}{(k-s)!}\, \delta(k, k'). \tag{A.28}$$

A.5. Bessel functions

The Bessel functions satisfy the differential equation

$$x^2 \frac{d^2 y}{dx^2} + x \frac{dy}{dx} + (x^2 - m^2)y = 0. \tag{A.29}$$

Solutions of this equation, valid for all values of x, can be expressed in power series about the origin. Hence the usual definition of a Bessel function is

$$J_m(x) = \frac{1}{\Gamma(m+1)} \left(\frac{x}{2}\right)^m \left\{1 - \frac{1}{m+1}\left(\frac{x}{2}\right)^2 + \frac{1}{(m+1)\,(m+2)}\frac{1}{2!}\left(\frac{x}{2}\right)^4 + \ldots\right. \tag{A.30}$$

When m is an integer the power series for $(-m)$ cannot be used and other methods must be employed to find the second solution. It will not be finite at $x = 0$. When m is not an integer, a second solution may be taken to be

$$Y_m(x) = \frac{\cos(m\pi)J_m(x) - J_{-m}(x)}{\sin m\pi}. \qquad (A.31)$$

When m is half integral, the Bessel functions take on a particularly useful form in terms of trigonometric functions,

$$J_{\frac{1}{2}}(x) = \left(\frac{2}{\pi x}\right)^{\frac{1}{2}} \sin x \qquad J_{-\frac{1}{2}} = \left(\frac{2}{\pi x}\right)^{\frac{1}{2}} \cos x$$

$$J_{\frac{3}{2}}(x) = \left(\frac{2}{\pi x}\right)^{\frac{1}{2}} \frac{\sin x}{x} - \cos x \qquad J_{-\frac{3}{2}} = \left(\frac{2}{\pi x}\right)^{\frac{1}{2}} - \frac{\cos x}{x} - \sin x$$

$$J_{\frac{5}{2}}(x) = \left(\frac{2}{\pi x}\right)^{\frac{1}{2}} \left(\frac{3}{x^2} - 1\right) \sin x - \frac{3}{x} \cos x$$

$$J_{-\frac{5}{2}} = \left(\frac{2}{\pi x}\right)^{\frac{1}{2}} \frac{3}{x} \sin x + \left(\frac{3}{x^2} - 1\right) \cos x.$$

As might be inferred from these forms, the half integral order Bessel functions approach a trigonometric function divided by $x^{\frac{1}{2}}$ as $x \to \infty$. This behavior is general and

$$J_n(x) \to \left(\frac{2}{\pi x}\right)^{\frac{1}{2}} [\cos\{x - \tfrac{1}{2}(n+\tfrac{1}{2})\pi\} - \frac{4n^2-1}{8x} \sin\{x - \tfrac{1}{2}(n+\tfrac{1}{2})\pi\}] \qquad (A.32)$$

as $x \to \infty$.

Among the numerous relationships between various Bessel functions are the following

$$J_m(x) = \frac{x}{2m}\{J_{m-1}(x) + J_{m+1}(x)\} \qquad (A.33)$$

$$\frac{dJ_m}{dx} = \tfrac{1}{2}\{J_{m-1}(x) - J_{m+1}(x)\} \qquad (A.34)$$

$$\frac{d}{dx}[x^l J_m(x)] = \frac{x^l}{2m}\{(l+m)J_{m-1}(x) + (l-m)J_{m+1}(x)\} \qquad (A.35)$$

$$\frac{d}{dx}[x^{-l} J_m(x)] = \frac{-x^{-l}}{2m}\{(l-m)J_{m-1}(x) + (l+m)J_{m+1}(x)\}. \qquad (A.36)$$

Since the Bessel functions approach zero only as $1/x^{\frac{1}{2}}$ as $x \to \infty$, integral relationships are useful especially when the integration is carried out between roots of the functions, not when carried to infinity. One such relationship is

$$\int_a^b J_m(kx)\, J_m(lx)\, x\, \mathrm{d}x = \frac{1}{k^2-l^2}\{lx\, J_m(kx)\, J_m'(lx) - kx\, J_m'(kx)\, J_m(lx)\}\,\Big|_a^b. \tag{A.37}$$

If the limits of integration ($x = a,\ x = b$) are such that the right-hand side vanishes,

$$\int_a^b J_m(kx)\, J_m(lx)\, x\, \mathrm{d}x = 0, \quad l \neq k. \tag{A.38}$$

If $k = l$

$$\int [J_m(kx)]^2 x\, \mathrm{d}x = \left\{\tfrac{1}{2}\left(x^2 - \frac{m^2}{k^2}\right)[J_m(kx)]^2 + \tfrac{1}{2}x^2[J_m'(kx)]^2\right\}. \tag{A.39}$$

A.6. Spherical Bessel functions

The Bessel functions described above frequently appear in problems of cylindrical symmetry. To treat problems of spherical symmetry Morse introduced the spherical Bessel functions defined by

$$j_n(x) = \left(\frac{\pi}{2x}\right)^{\frac{1}{2}} J_{n+\frac{1}{2}}(x). \tag{A.40}$$

These satisfy the differential equation

$$\frac{\mathrm{d}^2 j_n(x)}{\mathrm{d}x^2} + \frac{2}{x}\frac{\mathrm{d}\, j_n(x)}{\mathrm{d}x} + \left[1 - \frac{n(n+1)}{x^2}\right] j_n(x) = 0. \tag{A.41}$$

A few of these functions with $n \geq 0$ are

$$j_0(x) = \frac{1}{x}\sin x \qquad j_1(x) = \frac{1}{x}\left(\frac{\sin x}{x} - \cos x\right)$$

$$j_2(x) = \frac{1}{x}\left\{\left(\frac{3}{x^2} - 1\right)\sin x - \frac{3}{x}\cos x\right\}$$

$$j_3(x) = \frac{1}{x}\left\{\left(\frac{15}{x^3} - \frac{6}{x}\right)\sin x - \left(\frac{15}{x^2} - 1\right)\cos x\right\}.$$

There are various relationships which follow from the corresponding relationships between the standard Bessel functions. Among them are

$$j_m(x) = \frac{x}{2m+1}\{j_{m-1}(x)+j_{m+1}(x)\} \tag{A.42}$$

$$\frac{\mathrm{d}j_m(x)}{\mathrm{d}x} = \frac{1}{2m+1}\{mj_{m-1}(x)-(m+1)j_{m+1}(x)\} \tag{A.43}$$

$$\frac{\mathrm{d}}{\mathrm{d}x}[x^l j_m(x)] = \frac{x^l}{2m+1}\{(l+m)j_{m-1}(x)+(l-m-1)j_{m+1}(x)\} \tag{A.44}$$

$$\frac{\mathrm{d}}{\mathrm{d}x}[x^{-l} j_m(x)] = -\frac{x^{-l}}{2m+1}\{(l-m)j_{m-1}(x)+(l+m+1)j_{m+1}(x)\} \tag{A.45}$$

$$\int j_m(kx)j_m(lx)x^2\,\mathrm{d}x = \frac{x^{\frac{3}{2}}}{k^2-l^2}\{j_m(kx)\frac{\mathrm{d}}{\mathrm{d}x}[x^{\frac{1}{2}}j_m(x)] - j_m(lx)\frac{\mathrm{d}}{\mathrm{d}x}[x^{\frac{1}{2}}j_m(kx)]\} \tag{A.46}$$

$$\int [j_m(kx)]^2x^2\,\mathrm{d}x = \left(\tfrac{1}{2}x^3 - \frac{(m+\frac{1}{2})^2x}{2k^2}\right)[j_m(kx)]^2 + \frac{x^2}{2k^2}\left\{\frac{\mathrm{d}}{\mathrm{d}x}[x^{\frac{1}{2}}j_m(kx)]\right\}^2. \tag{A.47}$$

A.7. Evaluation of integrals containing the Fermi function

The Fermi function may be written as

$$f(u, u_0) = \frac{1}{\exp(u-u_0)+1} \tag{A.48}$$

where $u_0 > 0$. Figure A.1 shows $f(u, u_0)$ as a function of u, but the effective shape of the function really depends on the range of u in which one is interested. If the region in which u is less than or not much greater than u_0 is of importance, the constant part of the function and its rapid drop at $u = u_0$ are the main features. On the other hand, if values of $u \gg u_0$ are of importance, the function rapidly approaches $\exp(u_0-u)$ and may be treated as such. We shall treat these two cases separately in evaluating the integral

$$J(u_0, \Phi) = \int_0^\infty \frac{\mathrm{d}\Phi/\mathrm{d}u}{\exp(u-u_0)+1}\mathrm{d}u, \tag{A.49}$$

where $\Phi(0) = 0$.

Case I (u_0 small). In this case the integral between $u = 0$, and $u = u_0$ is

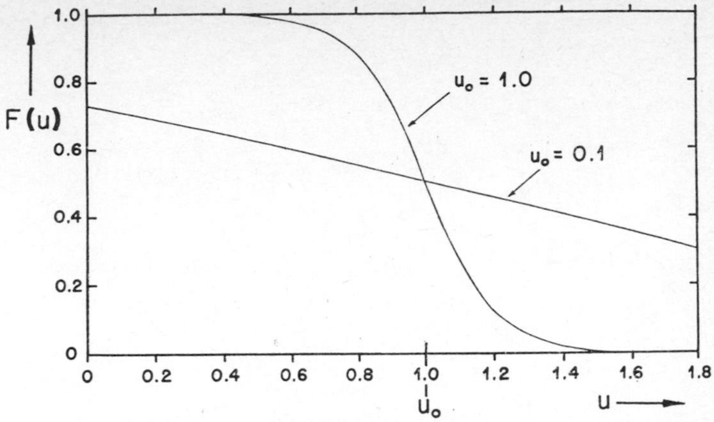

Fig. A.1. The Fermi function plotted for $u_0 = 1$ and $u_0 = 0.1$. As $u_0 \to 0$ the function approaches a simple exponential.

small compared with the remainder. The Fermi function can then be expanded to give

$$J(u_0, \Phi) = \int_0^\infty \frac{(\mathrm{d}\Phi/\mathrm{d}u)\exp(u_0-u)\,\mathrm{d}u}{1+\exp(u_0-u)} = \sum_{n=1}^{\infty}(-1)^{n+1}\int_0^\infty \frac{\mathrm{d}\Phi}{\mathrm{d}u}\,\mathrm{e}^{n(u_0-u)}\,\mathrm{d}u\,.$$

If the function $\Phi(u)$ can be expanded about the origin

$$\frac{\mathrm{d}\Phi}{\mathrm{d}u} = \sum_{s=1}^{\infty} s\,a_s u^{s-1}$$

and

$$J(u_0, \Phi) = \sum_{n=1}^{\infty}(-1)^{n+1}\mathrm{e}^{nu_0}\sum_{s=1}^{\infty} s\,a_s\int_0^\infty u^{s-1}\,\mathrm{e}^{-nu}\,\mathrm{d}u$$

$$= \sum_{n=1}^{\infty}\sum_{s=1}^{\infty}(-1)^{n+1}\mathrm{e}^{nu_0}\,a_s\frac{\Gamma(s)}{n^s}. \qquad (\mathrm{A.50})$$

This double sum will converge for suitable values of the a_s.

Case II (u_0 large). This is the case which is of principal importance in studying the behavior of electrons in solids. The process of integration takes advantage of the fact that the Fermi function is almost a constant for $u < u_0$ and then drops rapidly to zero when $u > u_0$. It is also assumed that $\Phi(u)$ is analytic and can be expanded about the point $u = u_0$,

$$J(u_0, \Phi) = \int_0^{u_0}\frac{\mathrm{d}\Phi}{\mathrm{d}u}\,\mathrm{d}u + \int_0^{u_0}\left[\frac{1}{\exp(u-u_0)+1} - 1\right]\frac{\mathrm{d}\Phi}{\mathrm{d}u}\,\mathrm{d}u \; +$$

$$+\int_{u_0}^{\infty} \frac{1}{\exp(u-u_0)+1} \frac{\mathrm{d}\Phi}{\mathrm{d}u}\,\mathrm{d}u = \Phi(u_0)+J_1+J_2. \qquad (A.51)$$

To evaluate J_1 let $u = u_0(1-t)$. Then

$$J_1 = -\int_0^{u_0} \frac{1}{1+\exp(u_0-u)} \frac{\mathrm{d}\Phi}{\mathrm{d}u}\,\mathrm{d}u = -u_0 \int_0^1 \frac{1}{1+\mathrm{e}^{u_0 t}} \frac{\mathrm{d}\Phi(u_0-u_0 t)}{\mathrm{d}u}\,\mathrm{d}t. \qquad (A.52)$$

To evaluate J_2 let $u = u_0(1+t)$. Then

$$J_2 = u_0 \int_0^{\infty} \frac{1}{1+\mathrm{e}^{u_0 t}} \frac{\mathrm{d}\Phi(u_0+u_0 t)}{\mathrm{d}u}\,\mathrm{d}t \qquad (A.53)$$

and

$$\begin{aligned} J_1+J_2 &= u_0 \int_0^{\infty} \frac{1}{1+\mathrm{e}^{u_0 t}} \left[\frac{\mathrm{d}\Phi(u_0+u_0 t)}{\mathrm{d}u} - \frac{\mathrm{d}\Phi(u_0-u_0 t)}{\mathrm{d}u}\right] \mathrm{d}t \\ &= 2u_0 \int_0^{\infty} \frac{\mathrm{d}t}{1+\mathrm{e}^{u_0 t}}\, \Phi''(u_0)u_0 t + \Phi^{(\mathrm{iv})}(u_0)\frac{{u_0}^3 t^3}{3!} \\ &\quad + \Phi^{(\mathrm{vi})}(u_0)\frac{{u_0}^5 t^5}{5!} + \ldots\Big\} \\ &= 2\sum_{r=1}^{\infty} \frac{{u_0}^{2r}}{(2r-1)!} \Phi^{(2r)}(u_0) \int_0^{\infty} \frac{t^{2r-1}\,\mathrm{d}t}{1+\mathrm{e}^{u_0 t}}. \end{aligned} \qquad (A.54)$$

The integral J_2 is extended to ∞ since the contribution of values of $t > 1$ is negligible. These integrals can be expanded as in case I to give

$$\int_0^{\infty} \frac{t^{2r-1}\,\mathrm{d}t}{1+\mathrm{e}^{u_0 t}} = \sum_{n=1}^{\infty} (-1)^{n-1} \int_0^{\infty} t^{2r-1}\,\mathrm{e}^{-nu_0 t}\,\mathrm{d}t = \sum_{n=1}^{\infty} \frac{(2r-1)!(-1)^{n-1}}{(nu_0)^{2r}}.$$

Then

$$\begin{aligned} J(u_0, \Phi) &= \Phi(u_0)+2\sum_{r=1}^{\infty} \Phi^{(2r)}(u_0) \sum_{n=1}^{\infty} \frac{(-1)^{n-1}}{n^{2r}} \\ &= \Phi(u_0)+2\sum_{r=1}^{\infty} e_{2r}\Phi^{(2r)}(u_0) \end{aligned} \qquad (A.55)$$

where

$$e_{2r} = \sum_{n=1}^{\infty} \frac{(-1)^{n-1}}{n^{2r}}, \quad e_2 = \frac{\pi^2}{12}, \quad e_{2r} = \frac{\pi^{2r}(2^{2r-1}-1)}{(2r)!} B_r.$$

B_r are the Bernoulli numbers. This series converges only when the derivatives diminish sufficiently rapidly.

A.8. The Clebsch-Gordon coefficients

The Clebsch-Gordon coefficients are used to form an eigenket of J and J_z as combinations of eigenkets of j and j_z of individual components. Let the coefficient be designated by

$$C^{J,M}_{j_1, j_2, m_1, m_2} = \langle j_1, m_1, j_2, m_2 \mid J, M \rangle . \tag{A.56}$$

J and M indicate that the linear combination is an eigenket of J^2 with the eigenvalue $J(J+1)\hbar^2$ and of J_z with the eigenvalue $M\hbar$. The component kets are eigenkets of j_1 and j_2 and j_{1z} and j_{2z} respectively. The combination is then given by equation (9.64),

$$\mid J, M \rangle = \sum_{m_1, m_2} C^{J,M}_{j_1, j_2, m_1, m_2} \mid j_1, m_1, j_2, m_2 \rangle . \tag{A.57}$$

The following tables give a few of the values of these coefficients.

	$j, M-\frac{1}{2}; \frac{1}{2}, \frac{1}{2} J, M$	$j, M+\frac{1}{2}; \frac{1}{2}, -\frac{1}{2} J, M$
$J = j+\frac{1}{2}$	$\dfrac{j+\frac{1}{2}+m^{\frac{1}{2}}}{2j+1}$	$\dfrac{j+\frac{1}{2}-M^{\frac{1}{2}}}{2j+1}$
$J = j-\frac{1}{2}$	$\dfrac{j+\frac{1}{2}-M^{\frac{1}{2}}}{2j+1}$	$\dfrac{j+\frac{1}{2}+M^{\frac{1}{2}}}{2j+1}$

	$j, M-1; 1, 1 J, M$	$j, M; 1, 0\ J, M$	$j, M+1; 1, -1 J, M$
$J = j+1$	$\dfrac{(j+1+M)(j+M)^{\frac{1}{2}}}{2(j+1)(2j+1)}$	$\dfrac{(j+1+M)(j+1-M)^{\frac{1}{2}}}{(j+1)(2j+1)}$	$\dfrac{(j-M)(j+1-M)^{\frac{1}{2}}}{2(j+1)(2j+1)}$
$J = j$	$-\dfrac{(j+M)(j+1-M)^{\frac{1}{2}}}{2j(j+1)}$	$\dfrac{M}{[j(j+1)]^{\frac{1}{2}}}$	$\dfrac{(j-M)(j+1+M)^{\frac{1}{2}}}{2j(j+1)}$
$J = j-1$	$\dfrac{(j-M)(j+1-M)^{\frac{1}{2}}}{2j(2j+1)}$	$\dfrac{(j-M)(j+M)^{\frac{1}{2}}}{j(2j+1)}$	$\dfrac{(j+M)(j+1+M)^{\frac{1}{2}}}{2j(2j+1)}$

REFERENCES

BOHR, N., 1935, Phys. Rev. (2) **48**, 696–762.
CHURCHILL, R. V., 1944, *Modern Operational Mathematics in Engineering* (New York).
COMPTON, A. H., 1923, Phys. Rev. (2) **21**, 483–502.
COMPTON, A. H. and A. W. SIMON, 1925, Phys. Rev. (2) **26**, 289–299.
DAVISSON, C. and L. H. GERMER, 1927, Phys. Rev. (2) **30**, 705–740.
EHRENFEST, P., 1927, Z. Physik **45**, 455–457.
EINSTEIN, A., 1905, Ann. d. Physik (4) **17**, 132–148.
ESTERMANN, I. and O. STERN, 1930, Z. Physik **61**, 95–125.
FRANK, J. and G. HERTZ, 1914, Verhandl. Deut. Phys. Ges. **16**, 457–461.
GEIGER, H. and E. MARSDEN, 1913, Phil. Mag. (6) **25**, 604–623.
GERLACH, W. and O. STERN, 1922, Z. Physik **9**, 349–352.
HEISENBERG, W., 1927, Z. Physik **43**, 172–198.
MATHEWS, J. and R. L. WALKER, 1964, *Methods of Mathematical Physics* (W. A. Benjamin, New York) pp. 26–37.
MILLIKAN, R. A., 1916, Phys. Rev. (2) **7**, 355–388.
MILLIKAN, R. A., 1917, *The Electron* (Univ. of Chicago Press).
PAULING, L. and E. B. WILSON, 1935, *Introduction to Quantum Mechanics* (McGraw-Hill).
PLANCK, M., 1901, Ann. d. Physik (4) **4**, 553–563.
ROBERTSON, H. P., 1930, Phys. Rev. (2) **35**, 667.
RUTHERFORD, E., 1911, Phil. Mag. (6) **21**, 669–688.
SCHROEDINGER, E., 1930, Sitz. ber. Preuss. Akad. Berlin, pp. 300–303.
THOMPSON, G. P., 1928, Nature **122**, 279–282.
THOMPSON, J. J., 1897, Phil. Mag. (5) **44**, 293–316.
WILSON, C. T. R., 1912, Proc. Roy. Soc. **A87**, 277–292.

SUBJECT INDEX

A
absorption, dipole, 208
– of energy, 202
– – radiation, 205
–, quadrupole, 208, 214
adjoint operator, 119
alpha particles, scattering of, 3
angular momentum of electromagnetic field, 189
angular momentum operator, 53, 72, 87, 140
– – –, matrix representation of, 143
– –, vector addition of, 159
annihilation operator, 137, 179, 181
anomalous Zeeman effect, 229
antisymmetric state, 163
atomic model, 4
– nucleus, 4
azimuthal quantum number, 86

B
Balmer formula, 90, 222
band, 335
basic vector, 117
Bessel function, 363
– –, spherical, 365
black-body radiation, 7
– – –, energy distribution in, 10
– – –, Planck's law of, 12
Bloch function, 334
Bohr magneton, 147
Born approximations, 274
– –, first, 290
– Oppenheimer theorem, 331
– series, 274
Bose-Einstein distribution, 329
– – statistics, 329
– gas, ideal, 329
bra, eigen, 122
– vector, 117
Brillouin zone, 335, 339

C
canonical density operator, 319
– distribution, 312, 319
– transformations, 24
– –, infinitesimal, 27
causality, 294
center of mass coordinates, 83, 264
central field, 85
– –, angular momentum in, 87
– –, electron in, 155
– –, parity in, 88
– potential, 290, 304
characteristic function, 63
– value, 64
charge, 1
charged particles in radiation field, 203
– –, scattering of, 3
charge, elementary, 2
classical approximation, 75
– mechanics, applicability of, 30, 36
Clebsch-Gordan coefficients, 159, 369
cloud chamber, 4
collimation, 31
combination principle, Ritz, 17
commutation relations and angular momentum operators, 140
– rule, 125
– – and selection rule, 260
commutator, 57
commuting operators, 121, 123
Compton effect, 13, 216
configuration, pp, 243, 246
–, sl, 251, 255
–, sp, 243, 246

–, two electron, 253
conservation law, 57
coordinate expectation value, 48
– representation, 46, 279
core electron, 331
Coulomb field, 89
coupling, *jj*, 237, 255
–, *LS*, 237, 248
–, Russel–Saunders, 237, 248
creation operator, 137, 179, 181
cross-section, scattering, 289, 300
crystal momentum, 336

D
degeneracy, 70
degenerate eigenstate, 70
density matrix, 317
– of states, 344
– operator, 315
– –, canonical, 319
destruction operator, 137, 179, 181
diffraction, 6
– of particle beams, 15
dipole absorption, 208
– radiation, 208, 260
– transition, 208
Dirac delta function, 118
– notation, 115
dispersion relation, 299
distribution, canonical, 312, 319
– function, Bose-Einstein, 330
– –, Fermi-Dirac, 329, 355, 366
dual space, 116
duality, 15

E
echo, resonance, 295
effective mass, 265, 334
Ehrenfest's theorem, 75
eigenfunction, 63
–, energy, 67
– expansion, 104
–, orthogonality of, 65
eigenstate, 63, 65, 122
–, degenerate, 70
eigenvalue, 64, 122
– equation, 64
eigenvector, 122
–, simultaneous, 123
Einstein's photoelectric law, 12
electromagnetic field, angular momentum of, 189
– –, energy of, 9, 188
– –, Hamiltonian function of, 188
– –, Hamiltonian operator of, 60
– –, momentum of, 190
– –, quantization of, 187
electron charge, 2
–, equivalent, 243
– in central field, 155
– – rotating magnetic field, 149
–, orbital functions of three, 169
–, scattering by a free, 126
– spin, 146
– – and spectrum, 227
–, spin functions of three, 171
–, system of three, 167
– wave length, 16
electrostatic interaction, 238, 244
emission of energy, 201
– – radiation, 205, 213
energy, conservation of, 200
– eigenfunction, 67
– gap, 344
–, negative, 89
–, positive, 91
– shell, off the, 278, 286
– spectrum in nearly free electron approximation, 349
– – – tight binding approximation, 353
ensemble, Gibbs, 309
even state, 259
exchange integral, 167
exclusion principle, 164, 243, 326
expectation value, 47

F
Fermi-Dirac distribution, 329, 355, 366
– – statistics, 326
– energy, 329, 355
– gas, ideal, 327
Feynman diagram, 274
fields and particles, interactions between, 204, 206
fine structure constant, 228
Floquet's theorem, 334
forced transition, 202
Franck-Hertz experiment, 18
free energy, 312, 323
– particle, motion of, 76

G
gamma ray microscope, 32
gap, energy, 344

gauge invariance, 60
– transformation, 61
generating function of a canonical transformation, 25, 28
– – – unitary transformation, 133
generator of a transformation, 28
Gibbs ensemble, 309

H
Hamiltonian function, 23
– – of electromagnetic field, 188
– operator, 55
– – and electromagnetic field, 60
Hamilton-Jacobi equation, 29
– equations, 23
– principle, 22
harmonic oscillator, one demensional, 68, 80, 136
– –, two dimensional, 70
– –, system of, 324
Heisenberg equation, 135
– operator, 153
– representation, 135
– uncertainty relation, 35, 36
Hermitian operator, 56, 120
Hermite polynomial, 69, 358

I
ideal Bose gas, 329
– Fermi gas, 326
identical particles, scattering of, 306
– –, system of, 162, 326
– –, system of two, 162
identity operator, 128
independent particle approximation, 163
indetermination principle, 35, 50, 184
insulator, 356
intensity, spectral line, 227, 234
interaction between particles and fields, 204, 206
–, electrostatic, 238, 244
–, spin-orbit, 237, 248, 255
interchange operator, 162
interference, 6
intermediate state, 113, 277

J
jj coupling, 237, 255

K
ket, eigen, 122
–, standard, 127
– vector, 116

L
laboratory coordinates, 265
Lagrangian and electromagnetic field, 58
– equations, 22
– function, 22
Laguerre polynomials, 362
Landé *g*-factor, 231
– interval rule, 253
Legendre polynomials, associated, 359
Levinson's theorem, 300
lifetime of atomic state, 214
line intensity, spectral, 227, 234
– series, 223
Liouville theorem, 311
Lippmann-Schwinger equation, 270
local potential, 290
Lorentz equations, Maxwell-, 5, 187
Lyman series, 222

M
magnetic quantum number, 86
mass, effective, 334
– tensor, 338
matrix, density, 317
–, R, 292
– representation of angular momentum operator, 143
– – an operator, 120
–, S, 292
Maxwell-Lorentz equations, 5, 187
measurement, 31, 73
– of position, 32
metal, 356
mixed state, 313
molecular field theory, 332
momentum, crystal, 336
– expectation value, 48
– in wave mechanics, 38
– of electromagnetic field, 190
– operator, 52, 143
– representation, 46, 131, 279
– shell, on the, 286
– states, scattering of, 284
multiplicity, 243, 262

N
nearly free electron approximation, 346
normalization, vector, 118
normal operator, 122
nucleus, atomic, 4

O

odd state, 259
off the energy shell, 278, 286
operator, 52
–, adjoint, 119
–, angular momentum, 140
–, annihilation, 137
–, canonical density, 319
–, coordinate, 52
–, creation, 137
–, density, 315
–, down stepping, 142
–, Hamiltonian, 55
–, Heisenberg, 153
–, Hermitian, 56, 120
–, interchange, 163
–, linear, 119
–, matrix representation of, 120
–, momentum, 52
–, normal, 122
–, parity, 259
–, R, 292
–, S, 292
–, self adjoint, 120
–, spin, 146
–, time development, 133, 150
–, unitary, 132
–, up-stepping, 142
optical theorem, 286, 290
orbital functions of three electrons, 169
– motion and spin motion, 165
– spin functions, combined, 172
orthogonal vector, 118

P

patity, 69, 72, 259
– in central field, 88
– operator, 259
partial wave decomposition, 295
– – phase shift, 298
particle beam, collimation of, 31
– –, diffraction of, 15
particles and fields, interaction between, 204, 206
Paschen-Back effect, 232
Paschen series, 222
Pauli exclusion principle, 164, 243, 326
periodic potential, wave function in, 332
perturbation theory, first order, 104
– –, time dependent, 108
– –, time independent, 100
– –, zero order, 103
phase shift, 293, 300
– –, partial wave, 298
–, space, 23
– –, classical, 309
phonon, 179, 181
–, localized, 182
photoelectric effect, 12
– law, Einstein's, 12
photon, 13, 191
– energy, 13
–, localized, 191
– momentum, 13
Planck's constant, 11
– –, expansion in terms of, 78
– law of black-body radiation, 12
point group, 342
Poisson bracket, 26
potential, central, 290, 304
–, constant – with discontinuities, 93
Poynting vector, 5
propagator, 267
– at large times, 285
pure state, 124, 313

Q

quadrupole absorption, 208, 214
– interaction, 214
– radiation, 208, 214
quantum number, 86
– –, azimuthal, 221
– –, magnetic, 221
– –, total, 221
quantum rules, 18

R

radiation, 205
–, atomic, 207
–, dipole, 208, 260
– field, 204
– –, charged particles in, 203
–, quadrupole, 208, 214
– temperature, 7
reciprocal lattice, 338
relative coordinates, 84, 264
– motion, 84
representation, coordinate, 279
–, momentum, 131, 279
– of a state, 126
–, Schroedinger, 129
resonance, 295
R operator, 292
Russel-Saunders coupling, 237, 248

S

scattering amplitude, 289, 292
– –, partial wave decomposition of, 297
– cross-section, 288, 300
– of charged particles, 3
– of identical particles, 306
– – momentum states, 284
– – radiation, 205, 217
– potential, central, 304
– probability, 289
– process, 263, 269
– state, unitarity of, 286
–, wave function treatment of, 301
Schroedinger equation, 54, 67, 134
– representation, 129
selection rule, 226, 233, 260
self adjoint operator, 120
semiconductor, 356
series, sharp line, 223
–, principal line, 223
singlet, 243
Slater determinant, 164
S operator, 292
spectral line, 17
– – intensity, 227, 234
– term, 220
spectrum, line, 17
–, one electron, 220
–, two electron, 236
spherical Bessel function, 365
– coordinates, 302
– harmonics, surface, 360
– polar coordinates, Schroedinger equation in, 85
spin and spectrum, 227
–, electron, 146
– functions of three electrons, 171
–, interaction with central field, 155
–, interaction with external field, 147
–, interaction with rotating magnetic field, 149
– motion and orbital motion, 165
– operator, 146
spin-orbital functions, combined, 172
– orbit interaction, 237, 248, 255
spontaneous transition, 202
standard ket, 127
state, even, 259
– in classical mechanics, 21
– in quantum mechanics, 41, 44
–, intermediate, 113, 277
–, mixed, 313
–, odd, 259
– of wave vector, 343
–, pure, 124, 313
–, representation of a, 126
–, s, p, d, f, 223
– vector, 116
– – and wave function, 126
–, virtual, 113
stationary state, 67
Stern-Gerlach experiment, 18
stimulated transition, 202
string, energy of vibrating, 177
–, Hamiltonian of vibrating, 177
–, particle and vibrating, 193
–, quantum mechanics of vibrating, 179
–, vibrating, 176
sum over states, 320
– – – of Bose gas, 330
– – – – Fermi gas, 329
symmetric state, 163
symmetry transformation, 342

T

target particle, 267
temperature, radiation, 7
term, spectral, 220
tight binding approximation, 349
time delay in scattering, 294
– development operator, 133, 150, 267
– – –, Laplace transform of, 269
trace, 315
transformation, unitary, 132
transition, dipole, 208
–, forbidden, 113, 214, 226
–, forced, 202
– probability, 112, 203, 205
– –, dipole, 210
–, spontaneous, 202
–, stimulated, 202
triplet, 243
two body problem, 83

U

uncertainty relation, 35, 36
unitarity of scattering states, 286
unitary operator, 132
– transformation, 132
– –, generating function of, 133

V

valence electron, 331
variation method, 106

vector and wave function, state, 126
–, basic, 117
–, bra, 117
– field, Fourier analyses of, 8
–, ket, 116
– model, 242
– potential, electromagnetic, 7
–, state, 116
– –, equation of motion of, 133
vibrating string, 176
– – and particle, 193
– –, energy of, 177
– –, Hamiltonian of, 177
– –, quantum mechanics of, 179
virtual state, 113

W
wave function, 41
– – and state vector, 126
– –, equation of motion for, 54
– –, interpretation of, 44
– –, normalization of, 45
– number, 333
– packet, 40
– vector, 339
– –, star of, 343
Wigner coefficient, 159
W. K. B. method, 97

Z
Zeeman effect, 224
– –, anomalous, 224, 229
zero point energy, 183
zone scheme, extended, 335
– –, reduced, 335